Inductance Calculations
Working Formulas and Tables

Frederick W. Grover

DOVER PUBLICATIONS, INC.
Mineola, New York

Copyright

Copyright © 1946, 1973 by Frederick W. Grover
All rights reserved.

Bibliographical Note

This Dover edition, first published in 2009, is an unabridged republication of the 2004 Dover edition of the work first published in 1946 by D. Van Nostrand Co., New York.

Library of Congress Cataloging-in-Publication Data

Grover, Frederick W. (Frederick Warren), 1876–1973.
    Inductance calculations : working formulas and tables / Frederick W. Grover.
      p. cm.
    Originally published: New York : D. Van Nostrand, 1946.
    Includes bibliographical references.
    ISBN-13: 978-0-486-47440-3
    ISBN-10: 0-486-47440-2
    1. Inductance. I. Title.

QC638.G78 2009
621.37'42—dc22

2009020775

Manufactured in the United States by RR Donnelley
47440205   2015
www.doverpublications.com

PREFACE

The design of inductors to have a given inductance or the calculation of the inductance of existing circuits are problems of importance in electrical engineering and especially in the field of communication.

Collections of formulas for the calculation of inductance and mutual inductance for different types of coils and other inductors are to be found in various electrical engineering handbooks and notably in the publications of the National Bureau of Standards.

It has, however, been the observation of the author of the present work, who has participated in the preparation of the Bureau of Standards collections, that certain difficulties are experienced in the use of this material. The engineer who has occasion to calculate an inductance is likely to be overwhelmed by the very wealth of the formulas offered him, and especially is this true in the more common types of inductor. Furthermore, certain formulas require the use of elliptic integrals or allied functions, others zonal harmonic functions or hyperbolic functions. Other formulas appear in the form of infinite series and it is necessary to choose from among those offered that formula whose degree of convergence will best suit the problem in question. Undoubtedly these complexities discourage the computer in many cases and lead to the substitution of empirical formulas or rough approximations for the accurate formulas.

The present work has been prepared with the idea of providing for each special type of inductor a single simple formula that will involve only the parameters that naturally enter together with numerical factors that may be interpolated from tables computed for the purpose. It has been found possible to accomplish this end in all the more important cases, and, even in the more complex arrangements of conductors, to outline a straightforward procedure. For the accomplishment of this end extensive tables have had to be calculated. Fortunately, certain of the tables are useful in more than a single case, but even so the tables represent a vast amount of computation. The tabular intervals are chosen so that where possible linear interpolation or at worst the inclusion of second order differences suffice. An accuracy of a part in a thousand is aimed at in general, but for the most part the tables lead to a better precision.

Illustrative examples are included with each case and where possible the numerical values found have been checked by other known formulas or methods. Procedure for the design of the more usual types of inductor has been included.

It is believed that all the more important forms of inductor and circuit elements have been covered, but in any new case it is usual to build a formula or method from the basic formulas by the general methods that are explained in the introductory chapters.

F. W. G.

Union College,
Schenectady, N. Y.
October 1945.

CONTENTS

	PAGE
Introduction	xiii

CHAPTER 1

General Principles 1

CHAPTER 2

Methods of Calculating Inductances 6
1. Basic Formulas, 6; 2. Formulas for Actual Circuits and Coils, 9; (a) Integration of Basic Formulas over the Cross Section of the Winding, 9; (b) Taylor's Series Expansions, 10; (c) Rayleigh Quadrature Formula, 11; (d) Lyle Method of Equivalent Filaments, 12; (e) Sectioning Principle, 13; (f) Geometric Mean Distance Method, 14; (g) Correction for Insulating Space, 15.

CHAPTER 3

Geometric Mean Distances 17
Equal Rectangles with Corresponding Sides Parallel, 18; Table 1. G.m.d. of Equal Parallel Rectangles, Longer Sides in Same Straight Line, 19; Table 2. G.m.d. of Equal Parallel Rectangles, Longer Sides Perpendicular to Line Joining Their Centers, 20, 21; G.m.d. of Line of Length a from Itself, 21; Circular Area of Radius a from Itself, 21; Ellipse with Semiaxes a and b, 21; Rectangle, sides B and C, 21; Table 3. Values of Constants for G.m.d. of a Rectangle, 22; G.m.d. of an Annulus from Itself, 22; Circular Area of Radius a, 22; Table 4. G.m.d. of an Annulus, 23; G.m.d. of Point or Area from an Annulus, 23; G.m.d. of One Circular Area from Another, 24; General Relation among Geometric Mean Distances, 24.

CHAPTER 4

Construction of and Method of Using the Collection of Working Formulas 26

PART I. CIRCUITS WHOSE ELEMENTS ARE STRAIGHT FILAMENTS

CHAPTER 5

Parallel Elements of Equal Length 31
 Mutual Inductance of Two Equal Parallel Straight Filaments, 31; Table 5. Values of Q for Use in Formula (2), 32, 33; Example 1, 34; Mutual Inductance of Two Equal Parallel Conductors, 34; Example 2, 35; Self-inductance of a Straight Conductor, 35; Example 3, 36; Inductance of Multiple Conductors, 37; Examples 4 and 5, 37, 38; Inductance of Return Circuit of Parallel Conductors, 39; Examples 6 and 7, 40, 41; Return Circuit of Two Tubular Conductors One Inside the Other, 41; Example 8, 42; Polycore Cable, 42; Example 9, 43; Inductance of Shunts, 43.

CHAPTER 6

Mutual Inductance of Unequal Parallel Filaments 45
 General Case, 45; Special Cases, 46; Filaments with Their Ends in Common Perpendicular, 46; Examples 10 and 11, 46, 47; Mutual Inductance of Parallel Conductors of Unequal Length However Placed, 47.

CHAPTER 7

Mutual Inductance of Filaments Inclined at an Angle to Each Other . . 48
 Equal Filaments Meeting at a Point, 48; Table 6. Values of Factor S in Formula (39) for the Mutual Inductance of Equal Inclined Filaments, 49; Example 12, 50; Unequal Filaments Meeting at a Point, 50; Table 7. Unequal Filaments Meeting at a Point. Values of S_1 to Be Used in Formula (45), 51; Example 13, 51; Unequal Filaments in the Same Plane, Not Meeting, 52; Examples 14 and 15, 53, 54; Mutual Inductance of Two Straight Filaments Placed in Any Desired Positions, 55; Example 16, 57.

CHAPTER 8

Circuits Composed of Combinations of Straight Wires 59
 General Formula for the Inductance of a Triangle of Round Wire, 59; Rectangle of Round Wire, 60; Regular Polygons of Round Wire, 60; General Formula for Calculation of the Inductance of Any Plane Figure, 60; Table 8. Values of α for Certain Plane Figures, 61; Table 9. Data for the Calculation of Inductance of Polygons of Round Wire, 62; Examples 17, 18, and 19, 63, 64; Inductance of Circuits Enclosing Plane Curves, 65; Example 20, 65.

CONTENTS vii

CHAPTER 9

PAGE

Mutual Inductance of Equal, Parallel, Coaxial Polygons of Wire . . . 66
Table 10. Ratios for Calculating the Mutual Inductance of Coaxial Equal Polygons, 67; Example 21, 69.

CHAPTER 10

Inductance of Single-layer Coils on Rectangular Winding Forms . . . 70
Formulas for Different Cases, 70; Table 11. Coefficients, Short Rectangular Solenoid, 72; Table 12. Values of Coefficients, Rectangular Solenoids, 72; Examples 22 and 23, 73, 74.

PART II. COILS AND OTHER CIRCUITS COMPOSED OF CIRCULAR ELEMENTS

CHAPTER 11

Mutual Inductance of Coaxial Circular Filaments 77
Formulas for Filaments of Unequal Radii, 77; Examples 24 and 25, 78; Table 13. Values of Factor f in Formula (77), 79; Table 14. Auxiliary Table for Circles Very Close Together, 81; Table 15. Auxiliary Table for Circles Very Far Apart, 82; Special Case. Circles of Equal Radii, 82; Table 16. Values of f for Equal Circles Near Together, 83; Table 17. Values of f for Equal Circles Far Apart, 84; Example 26, 85; Table 18. Auxiliary Table for Equal Circles Very Near Together, 85; Table 19. Auxiliary Table for Equal Circles Very Far Apart, 86.

CHAPTER 12

Mutual Inductance of Coaxial Circular Coils 88
Mutual Inductance of Coaxial Circles of Wire, 88; Example 27, 89; Mutual Inductance of Coaxial Circular Coils of Rectangular Cross Section, 89; Special Case. Equal Coils of Square Cross Section, 90; Example 28, 91; Mutual Inductance of Brooks Coils, 91; Table 20. Coupling Coefficients of Brooks Coils, 92; Design of Equal Coaxial Coils of Square Cross Section, 92; Example 29, 93; Design of Mutual Inductance Composed of Two Equal Brooks Coils, 93.

CHAPTER 13

Self-inductance of Circular Coils of Rectangular Cross Section 94
Nomenclature, 94; Inductance of Circular Coils of Square Cross Section, 95; Table 21. Values of Constant P_0' in Formula (91) for Coils of Square Cross Section, 96; Example 30, 97; Brooks Coils, 97; Cor-

CONTENTS

rection for Insulating Space, 98; Example 31, 99; Design of Brooks Coil to Obtain a Desired Inductance with a Chosen Size of Wire, 99; Example 32, 100; Design of a Brooks Coil to Obtain a Chosen Inductance and Time Constant, 101; Example 33, 105; Inductance of Circular Coil with Rectangular Cross Section of Any Desired Proportions, 105; Table 22. Values of k for Thin, Long Coils, Formula (99), 106; Table 23. Values of k for Short, Thick Coils, Formula (99), 107; Table 24. Values of F for Disc Coils, Formula (100), 108; Table 25. Values of F for Thin, Long Coils, Formula (100), 109; Interpolation in Tables 22, 23 and 24, 25 (Double Interpolation), 110; Examples 34 and 35, 111; Table 26. Values of P for Disc Coils, Formulas (100) and (100a), 113.

CHAPTER 14

Mutual Inductance of Solenoid and a Coaxial Circular Filament . . . 114

Basic Case. Circle in the End Plane of the Solenoid, 114; Table 27. Values of Q_0 for Mutual Inductance Solenoid and Circle, Formula (103), 115; Table 28. Values of R_0 for the Mutual Inductance of Solenoid and Coaxial Circle, Formula (104), 116; General Case. Circle Not in the End Plane, 117; Example 36, 117; Campbell Form of Mutual Inductance Standard, 119; Example 37, 120.

CHAPTER 15

Mutual Inductance of Coaxial Single-layer Coils 122

Nomenclature, 122; General Formula, 123; Example 38, 127; Table 29. Values of B_n as Function of α and ρ^2, Formula (108), 124; Table 30. Values of B_n. Auxiliary Table for Large α and ρ^2, 126; Example 38, 127; Concentric Coaxial Coils, 128; Examples 39 and 40, 128, 133; Table 31. Values of Polynomial $\lambda_6(\gamma^2)$ as Function of γ^2, 129; Table 32. Values of Polynomial λ_4 (γ^2) as Function of γ^2, 130; Table 33. Values of Polynomial $\lambda_6(\gamma^2)$ as Function of γ^2, 131; Table 34. Values of Polynomial $\lambda_8(\gamma^2)$ as Function of γ^2, 132; Example 40, 133; Principle of Interchange of Lengths, 133; Example 41, 134; Loosely Coupled Coils, 134; Examples 42 and 43, 135, 136; Coaxial Coils of Equal Radii, 137; Table 35. Values of B_n for Coils of Equal Radii ($\alpha = 1$), 138; Example 44, 139; Concentric Coils of Equal Length, 139; Example 45, 140.

CHAPTER 16

Single-layer Coils on Cylindrical Winding Forms 142

Basic Current Sheet Formulas, 142; Inductance of Ring Conductor, 143; Table 36. Values of K for Short Single-layer Coils, Formula

(118), 144; Table 37. Values for K for Long Single-layer Coils, Formula (118), 146; Correction for Insulating Space, 149; Example 46, 149; Table 38. Correction Term G in Formulas (120) and (135), 148; Correction for Insulating Space, 149; Example 46, 49; Table 39. Correction Term H in Formulas (120) and (135), 150; General Design of Single-layer Coils on Cylindrical Forms, 151; Table 40. Design Data. Single-layer Coils. $r = \frac{d}{l}$, 152; Problem A. Given Diameter, Length, and Winding Density; To Calculate the Inductance, 153; Table 41. Design Data for Short Single-layer Coils, $R = \frac{l}{d} \gtreqless 1$, 154, 155; Examples 47 and 48, 154, 156; Problem B. Given Inductance, Length, and Winding Density; To Calculate the Diameter, 156; Example 49, 157; Problem C. Given Inductance, Diameter, and Winding Density; To Calculate the Length, 157; Example 50, 158; Problem D. Given Inductance, Diameter, and Length; To Calculate the Winding Density, 158; Example 51, 159; Problem E. Given Inductance and the Shape Ratio; To Calculate Length and Diameter, 159; Example 52, 160; Inductance as a Function of the Number of Turns, 160; Example 53, 160; Most Economical Coil Shape, 161; Examples 54 and 55, 162.

CHAPTER 17

Special Types of Single-layer Coil 163
Helices of Conductor of Large Cross Section, 163; Helices of Round Wire, 163; Example 56, 163; Helices of Rectangular Strip, 164; Example 57, 166; Flat Spirals of Strip, 167; Example 58, 168; Toroidal Coils, 169; Closely Wound Single-layer Coil on a Torus, 169; Toroidal Coils of Rectangular Turns, 170; Example 59, 170; Single-layer Polygonal Coils, 170; Table 42. Data for Calculations of Polygonal Single-layer Coils, 172, 173; Example 60, 174; Series Formulas for Short Polygonal Coils, 175; Example 61, 175; Flat Spirals with Polygonal Turns, 176.

CHAPTER 18

Mutual Inductance of Circular Elements with Parallel Axes 177
Mutual Inductance of Circular Filaments of Equal Radii and with Parallel Axes, 177; Examples 62 and 63, 178; Table 43. Values of F for Equal Circles with Parallel Axes, Formula (159), 179; Table 44. Angular Position for Zero Mutual Inductance of Parallel Equal Circles, 180; Mutual Inductance of Coplanar Circular Filaments of Equal Radii, 180; Table 45. Constants for Equal Coplanar Circular Elements,

x   CONTENTS

Formulas (160) and (161), 181; Example 64, 182; Mutual Inductance of Circular Filaments Having Parallel Axes and Unequal Radii, 182; Case 1. Distant Circles, 182; Examples 65 and 66, 183, 184; Case 2. Circles Close Together, 184; Graphical Solution for Circular Filaments with Parallel Axes, 187; Examples 67 and 68, 186, 188; Mutual Inductance of Eccentric Circular Coils, 191.

CHAPTER 19

Mutual Inductance of Circular Filaments Whose Axes Are Inclined to One Another. 193
Circular Filaments Whose Axes Intersect at the Center of One of the Coils, 193; Examples 69 and 70, 194; Best Proportions for a Variometer, 195; Table 46. Values of Constant R for Inclined Circles, Formula (168), 196–200; Calculation in the Most General Case, 201; Examples 71 and 72, 201, 203; Mutual Inductance of Inclined Circular Filaments Whose Axes Intersect but Not at the Center of Either, 204; Example 73, 204; General Method of Treatment, 205; Most General Case. Inclined Circular Filaments Placed in Any Desired Position, 206; Mutual Inductance of Circular Coils of Small Cross Section with Inclined Axes, 207.

CHAPTER 20

Mutual Inductance of Solenoids with Inclined Axes, and Solenoids and Circular Coils with Inclined Axes. 209
Inclined Solenoids with Center of One on the End Face of the Other, 209; Concentric Solenoids with Inclined Axes, 210; Unsymmetrical Cases, 210; Solenoid and Circular Filament with Inclined Axes, 211; Examples 74 and 75, 212, 213.

CHAPTER 21

Circuit Elements of Larger Cross Sections with Parallel Axes 215
Solenoid and Circular Filament, 215; Table 47. Values of K_n in Formulas (183) and (185), 216; Examples 76 and 77, 217; Solenoids with Parallel Axes, 219; Examples 78, 79 and 80, 221, 222, 223; Solenoids with Parallel Axes Having Zero Mutual Inductance, 224; Example 81, 224; Solenoid and Coil of Rectangular Cross Section with Parallel Axes, 224; Table 48. Corrections for Coil Thickness. Coils with Parallel Axes, Formulas (188), (190), and (192), 226; Example 82, 227; Two Coils of Rectangular Cross Sections with Parallel Axes, 228; Examples 83 and 84, 229, 231; Mutual Inductance of Disc Coils with Parallel Axes, 234; Example 85, 235.

CHAPTER 22

Auxiliary Tables of Functions which Appear Frequently in Inductance
Formulas 236
  Auxiliary Table 1. Natural Logarithms of Numbers, 236; Auxiliary
  Table 2. For Converting Common Logarithms into Natural Logarithms, 237; Auxiliary Table 3. Values of Zonal Harmonic Functions,
  238, 241; Series for Zonal Harmonics, 242; Differential Coefficients,
  242; Auxiliary Table 4. Values of Differential Coefficients of Zonal
  Harmonics, 244, 247.

CHAPTER 23

Formulas for the Calculation of the Magnetic Force between Coils . . 248
  Force between Two Coaxial Circular Filaments, 248; Maximum
  Value of the Force, 249; Table 49. Values of P. Force between
  Coaxial Circular Filaments, 250; Table 50. Values of q_1 (or q) for
  Values of k'^2 (or k^2), 251; Table 51. Spacing Ratio and Force for Maximum Position Coaxial Circular Filaments, 252; Examples 86, 87 and 88,
  253; Force between Two Coaxial Coils of Rectangular Cross Section,
  253; Example 89, 254; Direction of the Force, 254; Force between
  Solenoid and a Coaxial Circular Filament, 255; Center of Circle at
  Center of End Face of Coil, 255; Filament Outside the Solenoid,
  255; Filament Inside the Solenoid, 256; Force between a Single-layer Coil and a Coaxial Coil of Rectangular Cross Section, 256;
  Example 90, 257; Force between Two Coaxial Single-layer Coils, 258;
  Examples 91 and 92, 259.

CHAPTER 24

High Frequency Formulas 261
  General Considerations, 261; Straight Cylindrical Conductor, 264;
  Table 52. High Frequency Resistance and Inductance of Straight
  Wires, 266; Table 53. Values of x_0 for Copper Wire 1 mm. Diameter,
  Frequencies 1 to 100 kc, 267; Table 54. Maximum Diameter of Conductors in Cm. for Resistance Ratio 1.01, 268; Example 93, 269;
  Table 55. Limiting Fractional Change of Inductance with Frequency,
  270; Isolated Tubular Conductor, 271; Table 56. Values of Factor
  F_0 for Approximate Calculations Based on (215) and (227), 272;
  Example 94, 273; Go-and-return Circuit of Round Wire, 274; Example 95, 276; Table 57. Correction Factor for Proximity Effect Parallel Round Wires, 276; Go-and-return Circuit of Parallel Tubular Con-

ductors, 277; Table 58. Correction Factor for Proximity Effect in Parallel Tubular Conductors, 277; Coaxial Cable, 278; High Frequency Resistance of Coaxial Cable, 278; Example 96, 279; High Frequency Inductance of Coaxial Cable, 280; Example 97, 281.

References 283

INTRODUCTION

Formulas for the calculation of self-inductance and mutual inductance are of practical importance in electrical applications. The calculation of single-layer coils, coils with rectangular cross section, current-limiting reactors, transmission lines, antennas, inductance standards may be named among the cases where a knowledge of the constants of existing circuits are required or where the problem is to design circuits to give a stated inductance.

In the absence of magnetic materials, mutual and self-inductances are parameters that are independent of the value of the current and depend only on the geometry of the system. The literature of the subject provides an abundance of formulas covering the more important cases occurring in practice but, for the most part, formulas adapted to routine calculation are not available.

For certain simple, ideal cases, exact solutions for the inductance have been found, but the expressions are complicated. For example, the expressions for circuits composed of straight filaments involve inverse hyperbolic and inverse trigonometric functions. For coaxial circular filaments, helices, and cylindrical current sheets, elliptic integrals are the normal functions. Tables of these functions are, of course, available but, in many practical computations, where these inductance formulas have to be used, the individual terms in the calculation nearly cancel, so that a high degree of accuracy must be attained in the separate terms, if the resulting calculated value of the inductance is to have even a moderate precision.

This difficulty and that of working with the more complex functions may be avoided to a great extent by the use of series developments of the exact formulas, but the computer is then met with the necessity of choosing a series that shall converge with sufficient rapidity for the problem in question. In some cases also, he is embarrassed by the wealth of series formulas available. For instance, some scores of series formulas, having different degrees of convergence, are known for the two cases of the mutual inductance of two coaxial circular filaments and the self-inductance of single-layer coils.

Furthermore, actual circuits are made up of wires, not filaments of negligible cross section, and are wound in layers or channels of rectangular cross section with insulating material between the wires. To cover practical cases, it is therefore necessary to combine solutions, which hold for the ideal cases, by methods of summation or integration in order to allow for these facts. Because of these complexities, the engineer is often deterred from making any inductance calculations at all, or is driven to the use of empirical formulas of rough accuracy and uncertain range of applicability, whose only recommendation lies in their comparative simplicity.

The present work, which is the result of years of research in this field, has for its purpose the simplification of routine calculations of mutual and self-inductances. For each case considered, so far as is possible, a single simple working formula is provided, in which appear, in addition to the given dimensions, numerical constants that may be interpolated from tables in which the shape ratios are the arguments. (Curves can, of course, be drawn from the tabular data, but the interpolation from the tables is simpler and more accurate than that obtainable from the curves.) An accuracy of a part in a thousand is in general obtainable and, in certain important cases, the results are more accurate than this. Errors in measurement of the dimensions of existing apparatus will usually be the limiting factors. Solutions of illustrative examples accompany each section of the work.

The formulas, except where otherwise stated, are for low frequencies. This does not, however, detract from their usefulness, since the effect of skin effect on inductance is small, while to take into account the effect of the capacitance of a coil on the apparent inductance, the low frequency inductance value, together with the self-capacitance of the coil, is what is required.

It is hoped that the references cited throughout the work cover sufficiently the sources of the material; much of it has not previously been published. However, no attempt has been made to present a complete bibliography of the subject. The bibliography given in Hak's *Eisenlose Drosselspulen* published by K. F. Koehler, Leipzig (1938), which is probably the most inclusive yet published, includes more than five hundred references.

Chapter 1

GENERAL PRINCIPLES

The electromotive force induced in a circuit A when the current in a circuit B is changed is proportional to the rate of change of the linkages of the flux set up by the current in B with the turns of the circuit A. If the circuits are linked through a core of iron or other magnetic material, nearly all of the flux ϕ, produced by the current, will link with the N turns of circuit A and the induced electromotive force is quite closely $-N\dfrac{d\phi}{dt}$. With magnetic materials, however, it is necessary to know the permeability of the material, which is a function of the magnetizing current and has to be determined by measurement for the current in question. Furthermore, although the knowledge of the permeability permits the reluctance of a complete magnetic circuit of iron to be estimated, the case of a straight magnetic core with the flux lines completed through the air is still further complicated by the difficulty of estimating the reluctance of the air path. It is, therefore, impracticable to do more than to make the roughest of calculations of the flux and therefore of the mutual inductance of circuits coupled by cores of magnetic materials. The treatment of standard apparatus employing complete magnetic circuits of iron, or circuits in which only a short air gap is included, is based on measurements of exciting current and leakage reactance.

With circuits free from iron, the case is different. The magnetic induction at any point due to current in a circuit B is directly proportional to the current i and, although the linkages of flux with the elements of a circuit A will vary, in general, from point to point, the total linkage with the circuit A is capable of being expressed as a constant M times the current. Thus, the electromotive force induced in A may be written as $-M\dfrac{di}{dt}$. The constant M is known as the coefficient of mutual induction or the *mutual inductance*. If the induced electromotive force is expressed in volts and the current in amperes, then M is expressed in henrys. A mutual inductance of one henry gives rise to an induced electromotive force of one volt, when the inducing current is changing at the rate of one ampere per second. For many simple circuits of only a few turns of wire, a more convenient unit of mutual inductance is the millihenry (mh), which is one thousandth of a henry, or the microhenry (μh), which is the millionth part of a henry. The latter is especially appropriate for expressing the mutual inductance of straight conductors or small coils of few turns.

The adjective "mutual" emphasizes the fact that if the electromotive force induced in circuit A by a current changing at the rate of one ampere per second in circuit B is equal to e, the same emf e is induced in circuit B when a current is made to change at the rate of one ampere per second in circuit A.

The mutual inductance may also be considered as the number of flux linkages with the circuit A due to unit current in circuit B. In the simple case where B has N_1 turns and circuit A, N_2 turns, the windings being concentrated, it is evident that the magnetic induction at any point due to unit current in B is proportional to N_1 and, therefore, the linkages with each turn of A are proportional to N_1. The total flux linkages with A, due to unit current in B are, consequently, proportional to $N_1 N_2$. If, on the other hand, unit current is set up in coil A, the linkages with each turn of B are proportional to N_2, but there are N_1 turns in B, so that the total number of linkages with B is also proportional to $N_1 N_2$. In general, the magnetic induction is a function of the dimensions of the inducing circuit and the number of linkages with this is a function of the dimensions of the linking circuit.

When the rôles of the two circuits are interchanged, the change in one of these factors is exactly compensated by the change in the other, and the mutual inductance is the same, whichever is the inducing circuit and whichever the circuit in which the electromotive force is induced.

The total electromotive force induced in a circuit at any moment is equal to the algebraic sum of the electromotive forces induced in the various elements of the circuit, opposing electromotive forces being regarded as of opposite signs. If we confine the consideration to frequencies such that the circuit dimensions are negligible with regard to the wave length, the magnetic induction at every point of the field is in phase with the current. In consequence, the induced electromotive forces are at all points in phase.

The total induced emf $-M \dfrac{di}{dt}$ may be considered as a summation of the elementary induced emfs around the circuit. This consideration defines what is meant by the mutual inductance of a circuit on a part of another circuit. The *partial* mutual inductance is the contribution made by the element to the *total* mutual of the circuit of which it forms a part.

Furthermore, the magnetic flux linked with a circuit element may be considered as the resultant of the fluxes contributed by the separate elements of the inducing circuit. Since, under the quasi-stationary condition assumed,

GENERAL PRINCIPLES

the currents in all the elements are in phase, so are the flux contributions of the separate elements. That is, the mutual inductance of an element of a circuit with the inducing circuit is the algebraic sum of the mutual inductances of the separate elements of the inducing circuit with the circuit element of the second circuit.

Assuming one circuit to be made up of elements A, B, C in series, and the other of elements a, b, c in series, the total mutual inductance is

$$M = M_{Aa} + M_{Ab} + M_{Ac} + M_{Ba} + M_{Bb} + M_{Bc} + M_{Ca} + M_{Cb} + M_{Cc},$$

and so on, for any number of sections of each circuit.

The concept of mutual inductance is not restricted to two *separate* circuits; every element of a single circuit has a mutual inductance on every other of its elements. A familiar case is presented by two coils in series carrying a current. Each coil induces an electromotive force into the other when the current is changing. In addition, the change of current induces in each coil alone an electromotive force due to the changing flux linkages of its own turns with its self-produced magnetic field. The coil is said to have *self-inductance* and the induced electromotive is commonly written $-L\dfrac{di}{dt}$, the coefficient L being designated as the *self-inductance*, or more commonly, the *inductance* of the coil. Self-inductance is merely a special case of mutual inductance. Each turn of the coil links with the magnetic field produced, not only by its own current, but by the current in the other turns also. We may, in fact, consider the self-inductance of a coil as equal to the summation of the mutual inductances of all the pairs of filaments of which the coil may be regarded as composed. Naturally, self-inductance is measured in the same unit, the henry, as is mutual inductance.

The summation principle applied to two coils A and B in series carrying the same current gives for the induced emf in the whole circuit

$$e = -L_A\frac{di}{dt} - L_B\frac{di}{dt} \mp M_{AB}\frac{di}{dt} \mp M_{BA}\frac{di}{dt}.$$

But $M_{BA} = M_{AB}$, so that

$$e = -(L_A + L_B \pm 2M_{AB})\frac{di}{dt}.$$

The inductance of the whole circuit is therefore

$$L = L_A + L_B \pm 2M_{AB}.$$

The double sign calls attention to the fact that the induced emf $-M_{AB}\dfrac{di}{dt}$ may either add to the self-induced emfs $-L_A\dfrac{di}{dt}$ and $-L_B\dfrac{di}{dt}$, or it may oppose them. By Lenz's law, the self-induced emfs are in such a direction

4 CALCULATION OF MUTUAL INDUCTANCE AND SELF-INDUCTANCE

as to oppose the change of current that gives rise to them, a fact taken into account by the negative sign. The impressed emf necessary to overcome these is, of course, $+L_A \frac{di}{dt} + L_B \frac{di}{dt}$. If the induced emf $-M \frac{di}{dt}$ adds to $-L_A \frac{di}{dt}$, the flux linkages of coil B with A add to the linkages of coil A with its own flux. The coils are said to be joined in "series aiding"; the coil A acts as though it had a self-inductance of $L_A + M_{AB}$ and the coil B as though it had an inductance $L_B + M_{AB}$. A simple interchange of the connections of one coil to the other leads to an opposing condition in which the magnetomotive force of one coil is opposed to that of the other. The apparent inductances of the coils are now $L_A - M_{AB}$ and $L_B - M_{AB}$, respectively.

When a current i_0 is established in a circuit or element of a circuit, the rise of current induces an electromotive force that opposes the rise of current. Thus, energy has to be expended by the source, in order to keep the current flowing against the induced emf. If we denote by i the current at any moment, the power expended in forcing this current against the induced emf $e = -L \frac{di}{dt}$ is $p = Li \frac{di}{dt}$. Thus the total energy supplied in raising the current to the final value i_0 is

$$W = \int_0^T Li \frac{di}{dt} dt = \int_0^{i_0} Li\, di = \tfrac{1}{2} L i_0^2,$$

in which T is the time interval for the establishment of the current. This energy is stored in the magnetic field and becomes available in the circuit when the current is broken. It may be shown that energy is stored in each volume element dV of the field to the amount, $\frac{H^2}{8\pi} dV$, where H is the magnetic field intensity at the point in question.

If while the current i_0 was being established in circuit 1 a current I_0 is maintained in a circuit 2 that has a mutual inductance M with circuit 1, then, during the rise of i an emf $-M \frac{di}{dt}$ is induced in circuit 2. To force the current I_0 against this emf, energy equal to $W_2 = \int_0^2 \left(M \frac{di}{dt} \right) I_0\, dt = M I_0 i_0$ is required. If the induced emf is in such a direction that it aids the flow of the current I_0, then energy is returned to the source of I_0 and M is to be considered as negative.

The energy of a system consisting of two circuits 1 and 2, in which currents I_1 and I_2, respectively, have been established, may be calculated by supposing the current I_1 in one circuit to be made first. Then the other current is supposed to rise from zero to I_2, while I_1 is held constant. First,

GENERAL PRINCIPLES

with circuit 2 open, the rise of the current in circuit 1 from zero to I_1 involves the storing of energy $\frac{1}{2}L_1I_1^2$ in the magnetic field. As the current in circuit 2 then rises from zero to I_2, energy $\frac{1}{2}L_2I_2^2$ is supplied by the source 2 and, at the same time, source 1 has to supply MI_1I_2 to maintain current I_1 unchanged. The total energy of the system is, therefore,

$$W = \tfrac{1}{2}L_1I_1^2 + MI_1I_2 + \tfrac{1}{2}L_2I_2^2.$$

If there be n circuits carrying currents $I_1, I_2, \cdots I_n$, having mutual inductances M_{12}, M_{13}, etc., the energy of the whole system is the sum of terms of the form $\frac{1}{2}L_sI_s^2$, one for each circuit, and a term $M_{rs}I_rI_s$ for each pair of coupled circuits. The magnetic field intensity at each point is, of course, the resultant of the components due to the individual circuits.

Chapter 2

METHODS OF CALCULATING INDUCTANCES

1. Basic Formulas. Although the inductances and mutual inductances of circuit elements not associated with magnetic materials are independent of the value of the current and dependent only on the geometry of the system, it is only in the simplest cases that these constants can be calculated exactly. Fortunately, from these basic formulas for ideal cases, formulas applicable to the more important circuit elements met in practice may be built up by general synthetic methods. A brief survey of the methods employed in deriving the basic formulas will first be given and, following this, a treatment of methods of procedure for building up solutions of the problem for actual circuits.

(a) The most direct method for calculating inductances is based on the definition of flux linkages per ampere. To calculate the flux linkages, it is necessary to write the expression for the magnetic induction at any point of the field and then to integrate this expression over the space occupied by the flux that is linked with the element in question. By the Biot-Savart law, the magnetic field intensity dH, due to a current i flowing in a straight circuit element of length ds, is, at any point P of the field (see Fig. 1),

$$dH = \frac{i\,ds}{r^2} \sin \theta,$$

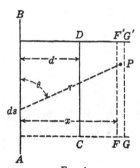

Fig. 1

in which r is the radius vector from ds to P and θ is the angle between the direction of ds and the radius vector. The magnetic flux is perpendicular to the plane through ds and the radius vector.

Suppose it is desired to calculate the mutual inductance of the straight parallel filaments AB and CD. The field intensity due to the current i in filament AB is found by integrating the expression for dH along the filament

METHODS OF CALCULATING INDUCTANCES 7

AB with respect to the position of ds. This integration is readily accomplished. Since the permeability is $\mu = 1$, the flux density B at P is numerically equal to the value of H. By passing planes KK' and LL' through the ends of filament CD perpendicular to AB, the mutual inductance will be found by integrating the expression for B between these planes from CD out to infinity. The flux lines are perpendicular to the plane of the paper. Through the plane $FF'GG'$ of width dx, the total flux is found by integrating the expression for the flux density, assuming x constant. The total flux linkages are then found by a second integration with respect to x between the limits $x = d$ and $x = \infty$. Placing $i = 1$ in this expression, there results the mutual inductance of the parallel filaments. If CD and AB have the same length l, the expression found is equation (1) below.

This procedure can be extended to find the self-inductance of a straight wire of radius of cross section ρ replacing the filament AB. The linkages of flux with this are found by integrating the expression for the flux density for the space between planes through the ends of AB perpendicular to it. The expression resulting is the same as in the preceding case with ρ in place of d. This represents the flux linkages external to the wire. To this must be added the inner linkages of the wire with the flux in the cross section of the wire, in order to find the self-inductance of the straight wire of length l and radius of cross section ρ.

To find the internal flux linkages, the flux through an elementary tube of length l, radius x, and thickness dx is $l\,dx$ times the flux density. The latter is twice the current inside the tube divided by the radius of the tube. If the whole current in the wire is i, that inside the tube is $\dfrac{x^2}{\rho^2}i$, so that the flux $\dfrac{2ixl}{\rho^2}dx$ passes through the tube and this links with $\dfrac{x^2}{\rho^2}$ of the current. Integrating the weighted expression $\left(\dfrac{2ixl}{\rho^2}\right)\left(\dfrac{x^2}{\rho^2}\right)dx$ with respect to x between zero and ρ, there are found the total flux linkages with current. Equating this to $L_i i$, the internal self-inductance L_i is found, which, added to the external inductance, gives the self-inductance of the straight wire.

(b) The internal inductance of the straight wire may be found also by noting that inside the conductor the flux density B is $\dfrac{2ix}{\rho^2}$. The volume element is $2\pi xl\,dx$. The energy in this volume is $\dfrac{B^2}{8\pi}(2\pi xl\,dx) = \dfrac{i^2 x^3 l}{\rho^4}dx$. Integrating this with respect to x between zero and ρ, the energy inside the conductor is found, and equating this to $\tfrac{1}{2}L_i i^2$, the same value $L_i = \dfrac{l}{2}$ is found as before.

8   CALCULATION OF MUTUAL INDUCTANCE AND SELF-INDUCTANCE

In general, this latter method does not work out as simply as the use of the Biot-Savart law, or the Neumann method, which follows.

(c) The Neumann formula for the mutual inductance of two circuit elements is given by

$$M = \iint \frac{\cos \epsilon}{r} \cdot ds\, ds', \tag{a}$$

in which ϵ is the angle of inclination between the two circuit line elements ds and ds', r is the radius vector between them, and the integration is to be taken over the contours of the two circuit elements.

This is the most general expression for finding the mutual inductance. It leads quite simply to a *formal* expression for the mutual inductance even though for most cases it is not possible to perform the integrations. However, in such cases also it is possible to obtain a numerical value for a specified case by mechanical integration, using Simpson's rule or the like. Naturally, however, this calculation may be tedious.

The mutual inductance of two parallel filaments may of course be found by the Neumann formula, but not as simply as by use of the Biot-Savart law. For inclined filaments, however, the Neumann formula has the advantage, and the formula for the mutual inductance of two straight filaments placed in any desired position has also been obtained by its use. This latter is a closed formula consisting of terms involving inverse hyperbolic and inverse trigonometric functions.

Other very important ideal cases solved by its use are the mutual inductance of two coaxial circular filaments,[1] the self-inductance of a helix,[2] the mutual inductance of a helix (or cylindrical current sheet) and a coaxial circle,[3] and the mutual inductance of two coaxial cylindrical current sheets.[4] For these cases elliptic integrals are the normal functions. For the coaxial circles and the solenoid, complete elliptic integrals of the first and second kinds appear; for the other cases elliptic integrals of all three kinds.

Calculations by means of these basic formulas depend upon tables of functions which are not difficult to use. Cases are often encountered, however, where the positive terms in the equations are nearly canceled by the negative, and care must be taken that the individual terms are calculated with a degree of accuracy sufficient to lead to the desired accuracy in the result. In other cases it is difficult to interpolate accurately the value of the elliptic integral. For many purposes, therefore, series expansions of the basic formulas are to be preferred to the basic expressions themselves. By their means, calculations can be made without recourse to special tables, and it is only necessary to select from those available a series formula whose convergence is satisfactory for the case in question.

A further advantage of the series expansions lies in their suitability for purposes of integration. It must be remembered, however, that no single

---

[1] References are numbered consecutively throughout the book and are found in a section beginning on page 283 at the end of the book.

series expansion can replace the basic formula. Some series converge well for small values of the variable, others for large values of the variable, while for intermediate cases no series may be entirely satisfactory. A further difficulty lies in the confusion caused by the large number of such expansions that have been found. For example, for the calculation of the mutual inductance of coaxial circles there are available, in addition to the basic elliptic integral formula, other elliptic integral formulas to other moduli, series in powers of the modulus, series in powers of the complementary modulus, series involving the shape ratio parameters, arithmetical-geometric mean series and q series developments. (For further details, see reference 5.)

(d) The list of basic cases already detailed includes those fundamentally most important. However, by integration of series developments of these cases, term by term, series expressions for other important cases are possible. For example, series expansions for the mutual inductance of cylindrical current sheets may be derived by the integration of a series expression for the mutual inductance of coaxial circles. Also a formula for the mutual inductance of circles with parallel axes may be derived from the coaxial case as a series involving zonal harmonics,[6] and a series for the mutual inductance of solenoids with parallel axes may be obtained from this by integration over the lengths of the two solenoids.[7] In all these cases it is necessary to select for integration a series in which the terms involve the variable of integration (a length, for example) directly.

2. **Formulas for Actual Circuits and Coils.** The basic formulas apply to the ideal cases of straight, circular, or helical filaments or to cylindrical current sheets of negligible thickness. Actual circuits are composed of conductors of appreciable cross section, of single-layer windings, of windings in channels of rectangular cross section. Not only is the current distributed over a finite cross section of conducting material, but the wires are separated by insulating spaces. Formulas for the inductance of actual circuit elements have to be derived by correcting the ideal basic formulas or by building up solutions by combinations of the formulas that hold for the ideal cases. Several fundamental methods are available for accomplishing these ends.

(a) *Integration of Basic Formulas over the Cross Section of the Winding.* Such direct integration is, in general, too difficult, but in a few instances has been accomplished, notably for the inductance of a circular coil of rectangular cross section, where the basic formula employed is that for the mutual inductance of coaxial circular filaments. Likewise, by integrating formulas for the mutual inductance of coaxial cylindrical currents the case of thick coaxial solenoids [8] has been treated. Also, from the basic formula for circles with parallel axes has been derived a formula for the mutual inductance of thick coils with parallel axes.[9] It should be noted that these formulas suppose the current to be uniformly distributed over the cross section, that is, no account is taken of the insulating space between the wires. Methods for applying this correction are given below, but fortunately the correction is of

10 CALCULATION OF MUTUAL INDUCTANCE AND SELF-INDUCTANCE

small importance, unless the wire has a large cross section and the insulation is thick.

(b) *Taylor's Series Expansions.* The mutual inductance of two windings having appreciable cross sectional dimensions may be referred to that of the central filaments of the coils by the use of a Taylor's series expansion around the positions of these filaments. Although the method, in principle, applies to cross sections of any shapes, the formula below applies only for rectangular cross sections. If N_1 and N_2 represent the numbers of turns on the two coils and M_0 is the mutual inductance of the central filaments OO' and PP', Fig. 2, then to a first approximation the mutual inductance of the coils is

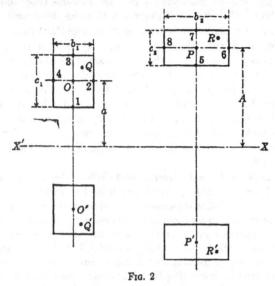

Fig. 2

$N_1 N_2 M_0$. Referring the coordinates of any filament QQ' to the center O of the section as origin, the mutual inductance of filament QQ' on the central filament PP' of the other coil may be expressed in a Taylor's series referred to M_0, in which the differential coefficients are the values that hold at the central points of the sections.

If this series be averaged over the cross section by integrating over the section and dividing by the area of the cross section, there is found the value of the mutual inductance (divided by $N_1 N_2$) of the left-hand coil of Fig. 2 and the central filament PP' of the other coil. If this value be denoted by M_1, the process may be repeated by referring the position of any filament RR' of the second coil to that of PP', and expressing the mutual inductance of RR' and the left-hand coil in a Taylor's series referred to M_1. Averaging this over the cross section of the second coil there results a value M_2, and the mutual inductance of the two coils of rectangular cross section is $M = N_1 N_2 M_2$.

The integrations are readily performed, since the differential coefficients are constants, that is, functions of the position of the central points to which they refer. This integration has been carried out to include differential coefficients of sixth order,[10] but the expressions become long and cumbersome. The expression including differential coefficients of second order only is

$$M = N_1 N_2 \left[M_0 + \frac{1}{24} \left\{ (b_1{}^2 + b_2{}^2) \frac{d^2 M_0}{dx^2} + c_1{}^2 \frac{d^2 M_0}{da^2} + c_2{}^2 \frac{d^2 M_0}{dA^2} \right\} \right]. \quad \text{(b)}$$

The calculation is thus made to depend upon the basic formula for the mutual inductance of the central filaments, the cross sectional dimensions, and the differential coefficients. The latter are obtained by differentiating the general basic formula for two filaments and then substituting the radii and axial distance corresponding to the central filaments.

This method has been used to find very accurate formulas for equal coaxial coils of rectangular cross section.[10] Its employment is not, however, restricted to coaxial elements, nor is it necessary that circular filaments be postulated. The closer the circuits and the greater the cross sectional dimensions, the more important are the terms involving higher order differential coefficients, but the work of Rosa [10] has shown that unfavorable cases require usually the inclusion of no order higher than the sixth.

These formulas assume that the current is uniformly distributed over the cross sections of the coil. The correction is negligible for windings of wire of sizes usual in practice. The special case where one or both of the cross sections has one of its cross sectional dimensions zero (simple layer winding) is also treated by the formulas derived above.

(c) *Rayleigh Quadrature Formula.* Rayleigh [11] has given another form to the expression derived for coils of rectangular cross section by the Taylor's series expansion. In this, the mutual inductance of the coils is made to depend upon an average of the mutual inductances of certain chosen filaments. Referring to Fig. 2, Rayleigh's expression is

$$M = \tfrac{1}{6}(M_{P1} + M_{P2} + M_{P3} + M_{P4} + M_{O5} + M_{O6} + M_{O7} + M_{O8} - 2M_{OP}), \quad \text{(c)}$$

where M_{P1} is the mutual inductance of filaments $11'$ and PP', M_{O5} is the mutual inductance of filaments OO' and $55'$, etc. Thus the calculation of the mutual inductance of the coils is reduced to finding the average of the values for certain filaments, each one of which may be calculated by the basic formula. With suitable tables this is easily accomplished. This formula is known as the Rayleigh quadrature formula. Its accuracy is, of course, limited by the sufficiency of the terms involving differential coefficients of second order in

12 CALCULATION OF MUTUAL INDUCTANCE AND SELF-INDUCTANCE

the Taylor's series from which it is derived. The quadrature formula is useful in any case where rectangular (or straight line) cross sections are involved, and the basic filament formula is available.

(d) *Lyle Method of Equivalent Filaments.* The use of the Rayleigh formula virtually replaces the coils of rectangular cross section by certain specially chosen filaments. A still more striking example of this method of solution is afforded by the Lyle method [12] of equivalent filaments. This is very accurate for coaxial coils of dimensions such that fourth order and higher order differential coefficients in the Taylor's expansion are negligible. The dimensions of the equivalent filaments in any case is illustrated by Fig. 3, which shows

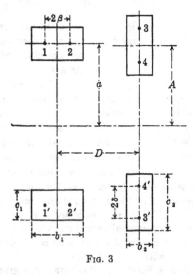

Fig. 3

two circular coils of rectangular cross section of mean radii a and A, axial dimensions b_1, b_2, radial dimensions c_1 and c_2 and spacing of median planes D.

Lyle's method replaces the left-hand coil $b_1 > c_1$ by two filaments 11' and 22' of radii $r_1 = a\left(1 + \dfrac{c_1^2}{24a^2}\right)$, spaced on each side of the median plane at a distance β, which is given by $\beta^2 = \dfrac{b_1^2 - c_1^2}{12}$.

The right-hand coil in Fig. 3 is replaced by two filaments 33' and 44' lying in the median plane having radii of $r_2 + \delta$ and $r_2 - \delta$, respectively, where $r_2 = A\left(1 + \dfrac{b_2^2}{24A^2}\right)$ and $\delta^2 = \dfrac{c_2^2 - b_2^2}{12}$.

Thus to calculate the mutual inductance of the coils of Fig. 3, the basic formula for the mutual inductance of coaxial circles is applied to find the

METHODS OF CALCULATING INDUCTANCES 13

mutual inductance of four pairs of filaments, each filament assumed to have half the turns of its coil. The mutual inductance of the coils is the sum of the four values. The following scheme shows the arrangement of the calculation:

Filaments	Product of Turns	Radii	Axial Spacing
11' and 33'	$\dfrac{N_1 N_2}{4}$	r_1 and $(r_2 + \delta)$	$D + \beta$
11' and 44'	$\dfrac{N_1 N_2}{4}$	r_1 and $(r_2 - \delta)$	$D + \beta$
22' and 33'	$\dfrac{N_1 N_2}{4}$	r_1 and $(r_2 + \delta)$	$D - \beta$
22' and 44'	$\dfrac{N_1 N_2}{4}$	r_1 and $(r_2 - \delta)$	$D - \beta$

A special case in the use of Lyle's method is that of a square cross section. For such a section β and δ are zero, so that the coil is replaced by a single filament of radius $r = a\left(1 + \dfrac{b^2}{24a^2}\right)$, where a and b are the mean radii of the coil and the side of the section, respectively.

(e) *Sectioning Principle.* Coils of large cross sectional dimensions, compared with their spacing and radii, require for accuracy the inclusion of higher order differential coefficients in the Taylor's series expansion, so that the Rayleigh formula and the Lyle method of equivalent filaments give only approximate values. The errors in practical cases are unimportant, especially since it is difficult to measure the dimensions and spacing of such coils with accuracy. However, in precision work, the errors of these two formulas may be reduced by sectioning the coils and applying the formulas of Rayleigh or Lyle to the individual sections. Thus, for example, in Fig. 3, each coil may be imagined to be divided into two sections giving coils P, Q and R, S. By the summation principle the total mutual inductance of the coils is then $M_{PR} + M_{PS} + M_{QR} + M_{QS}$. Each of the four terms may be calculated by replacing each section by equivalent filaments, or each pair of sections may be treated by the Rayleigh formula. Because of the decreased cross sections, the errors due to the neglect of higher order terms are much reduced. This process may be extended to a greater number of sections, but it is evident that thereby the number of pairs of filaments to be calculated increases rapidly. For example, if each coil in Fig. 3 is divided into two sections, each coil will be replaced by four filaments in the Lyle method, each carrying a quarter of the number of the turns of the coil. Sixteen pairs of filaments would have to be calculated.

By application of the summation principle, the mutual inductance of coils of large cross section in contact, or nearly in contact, may be found by basing the calculation on formulas for the *self*-inductance of coils.[13] For example, if it is desired to calculate the mutual inductance M_{AB} of the two coils A and B of Fig. 4, which have the same mean radius of winding and the same radial dimension c but different axial dimensions b_1 and b_2, the procedure is as follows: Imagine a coil C of rectangular cross section just filling in the space between A and B. Supposing a winding density of N turns per unit area of cross section, the mutual inductance of A and B under that assumption would be

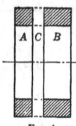

Fig. 4

$$2m_{AB} = (L_{ABC} + L_C) - (L_{AC} + L_{BC}), \qquad (d)$$

in which L_{ABC} is the self-inductance of the three coils in series, aiding, L_{AC} the self-inductance of A and C in series, aiding, etc. Each of the terms in the formula would be calculated from a formula for the self-inductance of a circular coil with rectangular cross section. The value m_{AB} gives the mutual inductance of the coils A and B under the assumption that they have NA_1 and NA_2 turns, respectively, $A_1 = b_1c$ and $A_2 = b_2c$ being the areas of the cross sections. If the actual numbers of turns of the coils are N_1 and N_2, then the actual mutual inductance of the coils is $M_{AB} = \left(\dfrac{N_1}{NA_1}\right)\left(\dfrac{N_2}{NA_2}\right) m_{AB}$. The quantities $\dfrac{N_1}{A_1}$ and $\dfrac{N_2}{A_2}$ are the actual winding densities on the coils, so that the ratio $\dfrac{M_{AB}}{m_{AB}}$ is the ratio of the product of the true winding densities to the square of the assumed density N. If N is taken equal to unity, M_{AB} is found by multiplying the value m_{AB} so calculated by the product of the actual winding densities.

The foregoing principle is applicable also to single layer coils and has been found useful in the solution of other problems where coils of large cross section are placed near together.[13]

(*f*) *Geometric Mean Distance Method.* The calculation of the mutual inductance of straight conductors of rectangular cross section may be treated by the Taylor's series expansion method or by the Rayleigh formula. A more useful method and one applicable to other shapes of cross section is the geometric mean distance method, treated below. This is essentially a method of replacing the conductors by equivalent filaments, so that the basic formulas for straight filaments or the series expressions for coaxial filaments close together can be used directly.

(g) *Correction for Insulating Space.* The preceding methods for calculating the inductances of windings of appreciable cross sections have assumed that the current is uniformly distributed over the cross section of the winding. This assumption leads to no sensible error in the calculation of the mutual inductance of ordinary windings, except for coils in contact. Self-inductance formulas need correction, especially where heavy conductors and thick spacing of insulation are employed. For calculating the correction the simplest and most accurate method proposed is that of Rosa.[14] The Rosa method may best be made clear by an example. Suppose we have a helical single-layer winding, Fig. 5, wound with round wire in a pitch p. The basic formula for the induct-

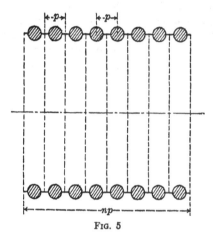

FIG. 5

ance of a single-layer coil or helix is that for a cylindrical current sheet. Suppose this to consist of a winding of metallic tape of negligible thickness with a number of turns such that each turn of round wire of the winding is at the center of a turn of tape. The turns of tape are supposed to be separated by insulating spaces of negligible thickness. If the winding has N turns, the length of the cylindrical current sheet is Np. This is, for the coil, the equivalent current sheet; evidently it will project beyond the winding slightly at both ends. The inductance of the coil is to a first approximation the same as that of the equivalent current sheet, which is calculated by the basic formula.

The inductance of the coil differs from that of the equivalent current sheet for two reasons:

1. The self-inductance of a turn of the wire is slightly different from that of a turn of the current sheet. The two turns have the same mean radius but different cross sectional shapes and areas. This small difference may be accurately calculated by use of geometric mean distance formulas, and the

difference multiplied by the number of turns gives the correction for this cause.

2. The mutual inductance of any two turns of the wire is different from that of the corresponding turns of tape. The mean radii and the mean spacing is the same for both pairs of turns, but the cross sectional shape differs in the two cases. The difference in these mutual inductances may be calculated by geometric mean distance formulas. This calculation has to be repeated for all the pairs of turns in the coil and the values added. The correction is a function of the number of turns. Evidently this method of correction may be used for wires of any shape of cross section for which geometric mean distance formulas are available. Rosa,[15] extending work of Maxwell,[15] gave tables for the correction for insulation for windings of round wire in single-layer coils and also for the case of a circular coil of round wires wound in a rectangular channel. The correction, in this last case, for the usual winding of thin wires with the customary enamel, silk, or cotton insulation is of no practical importance.

Chapter 3

GEOMETRIC MEAN DISTANCES

In calculating the mutual inductance of two conductors whose cross sectional dimensions are small compared with their distance apart, it suffices to assume that the mutual inductance is sensibly the same as the mutual inductance of the filaments along their axes and to use the appropriate basic formula for filaments to calculate the mutual inductance.

For conductors whose cross section is too large to justify this simplifying assumption it is necessary to average the mutual inductances of all the filaments of which the conductors may be supposed to consist. That is, the basic formula for the mutual inductance is to be integrated over the cross sections of the conductors.

For example, to calculate the mutual inductance of two equal straight conductors A and B of rectangular cross section (Fig. 6), the mutual inductance of two filaments perpendicular to the paper through P and Q may be calculated by formula (3), page 33. This formula has a principal term involving the logarithm of the distance between the filaments, that is, for this case $\log_e r$ of Fig. 6, and this term will vary with the positions of P and Q.

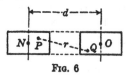

FIG. 6

The other terms of the formula are independent of the positions of the filaments. Integrating over cross section A to find the mutual inductance of conductor A on filament Q, and over cross section B to find the mutual inductance of A and B, is equivalent to averaging the basic formula (3) over the cross sections. The constant terms are of course unchanged; the average of $\log_e r$ will lead to a value $\log_e R$. Thus, the mutual inductance of the conductors is given by an expression that is of the same form as the basic filament formula. In other words, instead of being equal to the mutual inductance of the central filaments N, O (spacing d), the mutual inductance of the conductors of rectangular cross section is the same as that of two parallel filaments spaced at a distance R, which differs from d.

The distance R is called the *geometric mean distance* (g.m.d.)[16] of the two rectangles that represent the cross sections of the conductors. It differs, of course, from the *arithmetical* mean distance of the sections, which would be found by taking $\frac{1}{n}$ of the sum of the n distances between the n pairs of points of the two sections. To obtain the *geometric mean distance* we have to find $\frac{1}{n}$ of the sum of the n values of the logarithm of the distances between the n pairs of points.

Thus, the two conductors of rectangular cross section have been replaced by two filaments whose spacing is equal to the geometric mean distance of the two sections. Since no restriction has been laid on the shapes of the cross section, it is evident that the mutual inductance of any two conductors of any desired cross sections may be found in the same manner, provided only that the geometric mean distance of their sections may be calculated.

Since the self-inductance of a conductor is equal to the sum of the mutual inductances of all the pairs of filaments of which it is composed, it is evident also that the self-inductance of a straight conductor of any desired section is equal to the mutual inductance of two parallel straight filaments, spaced at the geometric mean distance of all the points of the section from each other. Thus is derived the idea of the *geometric mean distance of an area from itself.*

A knowledge of geometric mean distances is useful in other cases where straight circuit elements are involved. For circular coils also the effect of cross sectional dimensions may be closely evaluated by replacing the windings by coaxial circles, spaced at a distance equal to the geometric mean distance. This is more accurate, the greater the radius of curvature of the turns as compared with the cross sectional dimensions of the coil.

In the discussion of the integration of formula (3) for the case of the two parallel conductors of rectangular cross section, the term $\frac{d}{l}$ was ignored. In all cases where cross sectional dimensions are important, this is a relatively small term compared with the others. If taken into account it will be accurate enough to use the geometric mean distance in place of d, or even the distance between central filaments. Strictly, the value of the arithmetical mean distance of the points of the sections should be used for this term.

Formulas and tables are available for certain of the more frequently required geometric mean distances and the more important cases are given here.

Equal Rectangles with Corresponding Sides Parallel.[17] This is the most important case in practice. The general formula for $\log_e R$ is known,[18] but involves many terms and is ill suited for computations. Series formulas have been derived for simplifying the work and the results have been tabulated.

GEOMETRIC MEAN DISTANCES

TABLE 1. GEOMETRIC MEAN DISTANCES OF EQUAL PARALLEL RECTANGLES

Table gives values of $\log_e k$ in the equation $\log_e R = \log_e p + \log_e k$. (a) Longer sides of rectangles in same straight line.

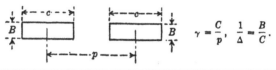

γ	$\frac{1}{\Delta}=0$	0.1	0.2	0.3	0.4	0.5	0.6	0.7	0.8	0.9	1.0
0	0	0	0	0	0	0	0	0	0	0	0
0.05	−0.0002	−0.0002	−0.0002	−0.0002	−0.0002	−0.0002	−0.0001	−0.0001	−0.0001	−0.0000	+0.0000
.10	.0008	.0008	.0008	.0008	.0007	.0006	.0005	.0004	.0003	.0002	.0000
.15	.0019	.0019	.0018	.0017	.0016	.0014	.0012	.0010	.0006	.0003	.0000
.20	.0034	.0033	.0032	.0030	.0028	.0025	.0021	.0017	.0012	.0006	.0000
0.25	−0.0053	−0.0052	−0.0051	−0.0048	−0.0044	−0.0039	−0.0034	−0.0027	−0.0019	−0.0010	+0.0000
.30	.0076	.0076	.0073	.0069	.0064	.0057	.0048	.0038	.0027	.0014	.0001
.35	.0105	.0104	.0100	.0095	.0087	.0078	.0066	.0052	.0036	.0018	.0002
.40	.0138	.0136	.0132	.0125	.0115	.0102	.0086	.0068	.0047	.0024	.0002
.45	.0176	.0174	.0169	.0159	.0146	.0130	.0110	.0086	.0059	.0029	.0003
0.50	−0.0220	−0.0217	−0.0210	−0.0198	−0.0182	−0.0161	−0.0136	−0.0106	−0.0073	−0.0036	+0.0005
.55	.0269	.0266	.0257	.0243	.0222	.0197	.0164	.0128	.0087	.0042	.0007
.60	.0325	.0321	.0310	.0292	.0267	.0235	.0196	.0152	.0103	.0048	.0010
.65	.0388	.0383	.0369	.0347	.0316	.0277	.0231	.0178	.0120	.0055	.0014
.70	.0458	.0452	.0435	.0408	.0370	.0324	.0269	.0207	.0137	.0062	.0019
0.75	−0.0536	−0.0529	−0.0509	−0.0476	−0.0431	−0.0375	−0.0310	−0.0237	−0.0156	−0.0070	+0.0023
.80	.0625	.0616	.0591	.0551	.0497	.0431	.0354	.0269	.0176	.0075	.0031
.85	.0725	.0714	.0683	.0634	.0569	.0491	.0401	.0302	.0195	.0081	.0037
.90	.0839	.0825	.0786	.0726	.0648	.0555	.0451	.0337	.0216	.0087	.0046
0.95	.0973	.0954	.0903	.0828	.0734	.0625	.0504	.0374	.0236	.0092	.0056
1.00	−0.1137	−0.1106	−0.1037	−0.0942	−0.0828	−0.0700	−0.0561	−0.0413	−0.0258	−0.0098	+0.0065

Table 1 gives $\log_e k$ in the formula $\log_e R = \log_e p + \log_e k$, in the arrangement shown in Fig. 7. The shape ratios used are $\gamma = \dfrac{C}{p}$ and $\dfrac{1}{\Delta} = \dfrac{B}{C}$, both ranging between zero and unity. It will be noted that $\log_e k$ is negative, that is, R is less than the distance p between centers, except for values of B nearly equal to C.

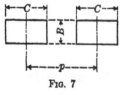

Fig. 7

Table 2 treats the case shown in Fig. 8. The parameters are here $\beta = \dfrac{B}{p}$ and $\Delta = \dfrac{C}{B}$, both of which range from zero to unity. Here $\log_e k$ is positive throughout, that is, R is greater than the distance between centers.

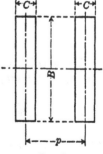

Fig. 8

CALCULATION OF MUTUAL INDUCTANCE AND SELF-INDUCTANCE

TABLE 2. GEOMETRIC MEAN DISTANCES OF EQUAL PARALLEL RECTANGLES

Values of $\log_e k$ in equation $\log_e R = \log_e p + \log_e k$. (b) Longer sides of rectangles perpendicular to line joining their centers.

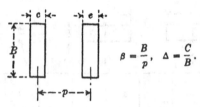

$$\beta = \frac{B}{p}, \quad \Delta = \frac{C}{B}.$$

β	Δ = 0	0.1	0.2	0.3	0.4	0.5	0.6	0.7	0.8	0.9	1.0
0	0	0	0	0	0	0	0	0	0	0	0
0.1	0.0008	0.0008	0.0008	0.0008	0.0007	0.0006	0.0005	0.0004	0.0003	0.0002	0.0000
.2	.0033	.0033	.0032	.0030	.0028	.0025	.0021	.0017	.0012	.0007	.0000
.3	.0074	.0073	.0071	.0067	.0062	.0056	.0048	.0038	.0027	.0015	.0001
.4	.0129	.0128	.0124	.0118	.0109	.0098	.0084	.0068	.0050	.0027	.0003
0.5	0.0199	0.0197	0.0191	0.0182	0.0169	0.0152	0.0131	0.0106	0.0077	0.0043	0.0005
.6	.0281	.0278	.0271	.0258	.0240	.0216	.0185	.0152	.0111	.0064	.0011
.7	.0374	.0371	.0361	.0344	.0320	.0290	.0251	.0206	.0155	.0090	.0019
.8	.0477	.0473	.0461	.0440	.0411	.0373	.0321	.0268	.0200	.0129	.0031
0.9	.0589	.0584	.0569	.0544	.0506	.0464	.0404	.0338	.0254	.0158	.0046
1.0	0.0708	0.0702	0.0685	0.0655	0.0614	0.0560	0.0492	0.0406	0.0313	0.0199	0.0065
0.9	.0847	.0841	.0821	.0787	.0738	.0675	.0596	.0501	.0382	0.0250	
.8	.1031	.1023	.0999	.0959	.0903	.0829	.0745	.0622	0.0485		
.7	.1277	.1268	.1240	.1192	.1125	.1037	.0925	0.0788			
.6	.1618	.1607	.1573	.1507	.1436	.1329	0.1194				
0.5	0.2107	0.2094	0.2053	0.1984	0.1886	0.1754					
.4	.2843	.2826	.2776	.2691	0.2567						
.3	.4024	.4003	.3942	0.3831							
.2	0.6132	0.6105	0.6021								
0.1	1.0787	1.1075									
1/β											

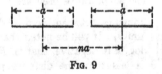

Fig. 9

Included in Table 1 is the case shown in Fig. 9 which may be calculated directly by the simple relation [19]

$$\log_e R = \log_e na - \left(\frac{1}{12n^2} + \frac{1}{60n^4} + \frac{1}{168n^6} + \frac{1}{360n^8} + \frac{1}{660n^{10}} + \cdots \right). \quad (e)$$

This is very convergent for all values of n except $n = 1$. For that case, Table 1 gives $\log_e R = \log_e a - 0.1137$.

TABLE 2. GEOMETRIC MEAN DISTANCES OF EQUAL PARALLEL RECTANGLES (*Concluded*)

For accurate interpolation in the case of broad rectangles, near together ($1/\beta$ small, and Δ small), write $\log_e R = \log_e B + \log_e k'$, and use values of $\log_e k'$ from the following table. (*Note:* All values of $\log_e k'$ in the table are negative.)

$1/\beta$	$\Delta=0$	0.05	0.10	0.15	0.20	0.25	0.30	0.35	0.40	0.45	$\Delta=0.50$
0	−1.5	—									
0.05	1.3542	1.3555	—								
0.10	1.2239	1.2248	1.2278	—							
0.15	1.1052	1.1060	1.1084	1.1125	—						
0.20	0.9962	0.9969	0.9989	1.0024	1.0073	—					
0.25	−0.8953	−0.8959	−0.8977	−0.9007	−0.9049	0.9105	—				
0.30	.8015	.8020	.8037	.8062	.8098	.8147	0.8208	—			
0.35	.7140	.7145	.7159	.7182	.7215	.7258	.7311	0.7375	—		
0.40	.6321	.6325	.6337	.6358	.6387	.6425	.6472	.6530	0.6596	—	
0.45	.5550	.5554	.5565	.5584	.5610	.5645	.5687	.5738	.5797	0.5865	—
0.50	−0.4825	−0.4828	−0.4838	−0.4855	−0.4879	−0.4910	−0.4948	−0.4994	−0.5046	−0.5109	0.5178

Geometric Mean Distance of a Line of Length a from Itself.

$$\log_e R = \log_e a - \tfrac{3}{2}, \tag{f}$$

or

$$R = a\epsilon^{-3/2} = 0.22313a. \tag{g}$$

Circular Area of Radius a from Itself.

$$\log_e R = \log_e a - \tfrac{1}{4}, \tag{h}$$

or

$$R = a\epsilon^{-1/4} = 0.7788a. \tag{i}$$

Ellipse with Semiaxes a and b.

$$\log_e R = \log_e \frac{a+b}{2} - \frac{1}{4}. \tag{j}$$

Rectangle, sides B and C. The exact expression is known but is not simple to calculate. However, whatever the values of B and C, the g.m.d. is nearly proportional to the perimeter and is approximately $R = 0.2235(B + C)$.

22   CALCULATION OF MUTUAL INDUCTANCE AND SELF-INDUCTANCE

Since, however, $\log_e R$ rather than R is required it is simple to make use of the relation

$$\log_e R = \log_e (B + C) - \tfrac{3}{2} + \log_e e, \tag{k}$$

or

$$R = K(B + C), \tag{l}$$

and to obtain the values of $\log_e e$ and K from Table 3 as a function of $\dfrac{B}{C}$ or $\dfrac{C}{B}$, whichever is less than unity.

TABLE 3. VALUES OF CONSTANTS FOR THE GEOMETRIC MEAN DISTANCE OF A RECTANGLE

Sides of the rectangle are B and C. The geometric mean distance R is given by $\log_e R = \log_e (B + C) - 1.5 + \log_e e$.

$$R = K(B + C), \quad \log_e K = -1.5 + \log_e e.$$

B/C or C/B	K	$\log_e e$	B/C or C/B	K	$\log_e e$
0	0.22313	0	0.50	0.22360	0.00211
0.025	.22333	0.00089	.55	.22358	.00203
.05	.22346	.00146	.60	.22357	.00197
.10	.22360	.00210	.65	.22356	.00192
.15	.22366	.00239	.70	.22355	.00187
.20	.22369	.00249	0.75	0.22354	0.00184
0.25	0.22369	0.00249	.80	.22353	.00181
.30	.22368	.00244	.85	.22353	.00179
.35	.22366	.00236	.90	.22353	.00178
.40	.22364	.00228	0.95	.223525	.00177
.45	.22362	.00219	1.00	0.223525	0.00177
0.50	0.22360	0.00211			

G.m.d. of an Annulus from Itself. The exact expression is known, but it is simpler to write

$$\log_e R = \log_e \rho_1 - \log_e \zeta, \tag{m}$$

in which ρ_1 is the outer radius, and to obtain $\log_e \zeta$ from Table 4 for the given ratio of the radii $\dfrac{\rho_2}{\rho_1}$.

Circular Area of Radius a. (a) G.m.d. of a point P outside the circle to the entire circular area is equal to the distance from P to the center of the circle.

(b) G.m.d. of point P to the entire circumference of the circle is also the distance from P to the center of the circle.

(c) G.m.d. from the center of the circle to the circumference is the radius a.

(d) G.m.d. of any point within the circle to the entire circumference is equal to the radius a.

GEOMETRIC MEAN DISTANCES

TABLE 4. GEOMETRIC MEAN DISTANCE OF AN ANNULUS

Outer and inner radii ρ_1 and ρ_2, respectively.

$\log_e R = \log_e \rho_1 - \log_e \zeta$.

ρ_2/ρ_1	$\log_e \zeta$	d_1	d_2	ρ_2/ρ_1	$\log_e \zeta$	d_1	d_2
0	0.2500			0.50	0.1603		
		−12				−147	
0.05	.2488		−24	.55	.1456		−5
		−36				−152	
.10	.2452		−21	.60	.1304		−4
		−57				−156	
.15	.2395		−18	.65	.1148		−3
		−75				−159	
.20	.2320		−16	.70	.0989		−3
		−92				−162	
0.25	0.2228		−14	0.75	0.0827		−2
		−105				−163	
.30	.2123		−12	.80	.0663		−1
		−116				−164	
.35	.2007		−10	.85	.0499		−1
		−127				−165	
.40	.1880		−8	.90	.0333		−1
		−135				−166	
.45	.1745		−7	0.95	0.0167		−1
		−142				−167	
0.50	0.1603		−6	1.00	0		

(e) G.m.d. of any point on the circumference from the whole circumference is also a.

(f) G.m.d. of a circular line of radius a from itself is also a.

G.m.d. of Point or Area from an Annulus. (a) For a point P_1 outside the annulus the g.m.d. of the point from the annulus equals the distance from P_1 to the center of the annulus.

(b) For an area S_1 and the annulus the g.m.d. is equal to the g.m.d. of the center of the annulus from the points of the area S_1.

(c) The g.m.d. of any point within the annulus from the annulus itself is given by

$$\log_e R = \frac{\rho_1^2 \log_e \rho_1 - \rho_2^2 \log_e \rho_2}{\rho_1^2 - \rho_2^2} - \frac{1}{2}. \quad (n)$$

(d) This same expression holds for the g.m.d. of the points of any area S_2 inside the annulus from the annulus.

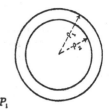

Fig. 10

24 CALCULATION OF MUTUAL INDUCTANCE AND SELF-INDUCTANCE

G.m.d. of One Circular Area from Another. The g.m.d. is the distance between their centers.

General Relation among Geometric Mean Distances. Assuming any number of figures having areas A, B, C, etc., whose geometric distances from another area S are R_A, R_B, R_C, etc., the geometric mean distance R of their sum from the area S is given by the general equation

$$\log_e R = \frac{A \log_e R_A + B \log_e R_B + C \log_e R_C + \cdots}{A + B + C + \cdots}. \quad (o)$$

This formula may be used to find the g.m.d. of other areas not directly treated above. For example, we may treat the case of the four equal squares in Fig. 11. If R_{12} equals g.m.d. of squares 1 and 2, R_4 the g.m.d. of square 4 from itself, and R_{s4} the g.m.d. of the whole area from square 4, then, assuming each small square to have an area A,

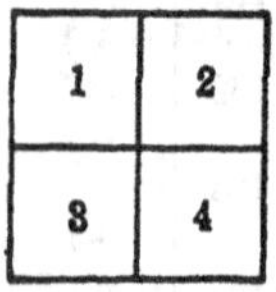

$$\log_e R_{s4} = \frac{A \log_e R_{14} + A \log_e R_{24} + A \log_e R_{34} + A \log_e R_4}{A + A + A + A}$$

$$= \frac{\log_e R_{14} + 2 \log_e R_{24} + \log_e R_4}{4}.$$

Fig. 11

But the g.m.d. of the whole area from square 4 is the same as for the whole area from any one of the squares. That is, the g.m.d. of the whole area from itself R_s is equal to R_{s4}. Therefore,

$$4 \log_e R_s = \log_e R_{14} + 2 \log_e R_{24} + \log_e R_4.$$

Furthermore, the same reasoning yields

$$2 \log_e r_{24} = \log_e R_4 + \log_e R_{24},$$

in which r_{24} is the g.m.d. of area 2 plus 4 from itself. Substituting in the equation above, we may find

$$\log_e R_{14} = 4 \log_e R_s - 4 \log_e r_{24} + \log_e R_4. \quad (p)$$

Thus the calculation for the obliquely situated squares is obtained in terms of the g.m.d. of a rectangle from itself. To find the g.m.d. of squares 1 and 3 in Fig. 12, the g.m.d. R_{s3} of the whole area from 3 is

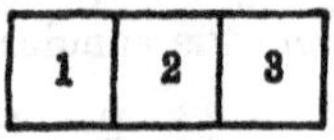

$$3 \log_e R_{s3} = \log_e R_{13} + \log_e R_{23} + \log_e R_3,$$

Fig. 12

and this is the same as the g.m.d. of the whole area on square 1. Likewise R_{s2} is given by

$$3 \log_e R_{s2} = \log_e R_{12} + \log_e R_2 + \log_e R_{23}.$$

The g.m.d. R_s of the whole area on itself is the weighted means of $\log_e R_{s3}$ and $\log_e R_{s2}$, so that

$$\log_e R_s = \frac{2 \log_e R_{s3} + \log_e R_{s2}}{3},$$

or
$$9 \log_e R_s = 2 \log_e R_{13} + 4 \log_e R_{12} + 3 \log_e R_1, \tag{q}$$

which gives finally

$$2 \log_e R_{13} = 9 \log_e R_s - 8 \log_e R_{12} + \log_e R_1.$$

Chapter 4

CONSTRUCTION OF AND METHOD OF USING THE COLLECTION OF WORKING FORMULAS

There follows a collection of working formulas for the calculation of self-inductance and mutual inductance so designed as to make possible routine calculations for the circuit elements and coils likely to be met in practice. These formulas are based for the most part on formulas existing in the literature but, as far as possible, the necessity of choosing from among several possible formulas or, in the case of series expressions, of selecting a series sufficiently convergent for the given problem, has been avoided. In certain cases new formulas have been built up from basic ideal cases by the methods treated in the introduction.

The collection falls naturally into two parts, viz.: circuits built up of straight conductors and circuits consisting of circular turns. Each section includes special cases, arranged so as to progress from the more common simple circuits to the more complex.

For each case, a single working formula is given in a form involving simple arithmetical operations on the given dimensions, the complication found in the majority of inductance formulas being avoided by the use of tables, in which the arguments are the given shape ratios. In the preparation of the tables, of course, a number of complicated formulas have had to be employed to cover the range of possible values of the shape ratios.

Except where the contrary is stated, dimensions are in *centimeters*, and inductances are given in *microhenrys* by the formulas. Shape ratios, of course, may be calculated from data given in any units, provided numerator and denominator are in the same unit.

The accuracy attainable is one part in 1000 in the worst cases and much better for the more important circuit elements. The use of five-place logarithms should suffice but, naturally, where the accuracy of the given data is only moderate, slide rule calculations will be preferred.

Natural logarithms appear in some of the formulas and to facilitate their use two tables are provided in the appendix. Auxiliary Table 1 gives directly the natural logarithms of integers up to 150. Where a number may be resolved into simple factors, the natural logarithms of the factors may be taken from this table and added to find the natural logarithm of the number. In the more general case, it is necessary to find the common logarithm and then to multiply this by 2.3026. This multiplication is readily accomplished by the aid of Auxiliary Table 2, which is a multiplication table up to 100 times this factor. By breaking up the common logarithm in groups of two figures and shifting the decimal point, the natural logarithm is easily found. This process is illustrated following Auxiliary Table 2 in the appendix.

For approximate calculations, simple interpolation in the tables will suffice. However, for more accurate work, the second order differences should be taken into account. The Newton general interpolation formula is

$$f(a + h) = f(a) + kd_1 + \frac{k(k-1)}{2!} \cdot d_2 + \frac{k(k-1)(k-2)}{3!} \cdot d_3 + \cdots, \quad (r)$$

in which a is the nearest entry in the table, δ is the tabular interval, and $k = \frac{h}{\delta}$ is the fraction of the interval for which interpolation is to be made, and d_1, d_2, and d_3 are the first, second, and third differences of the value in the table for the entry a. Including the second difference only, the formula becomes

$$f(a + h) = f(a) + k\left[d_1 - \frac{(1-k)}{2} \cdot d_2\right]. \quad (s)$$

As an example, suppose in Table 10 page 67, that the value of F for triangles is required for $\frac{d}{s} = 0.513$. From the tabular data we have

$\frac{d}{s}$	F	d_1	d_2
0.50	0.4792		
		−106	
.55	.4686		12
		−94	
0.60	0.4592		

The tabular interval is $\delta = 0.05$, $a = 0.50$, and $h = 0.013$, so that $k = \frac{0.013}{0.05}$ = 0.26, and $\frac{1-k}{2} = 0.37$. Therefore, $d_1 - \frac{(1-k)}{2} \cdot d_2 = -106 - 0.37(12)$ = −110, and $k\left[d_1 - \frac{(1-k)}{2} \cdot d_2\right] = -28.6$. Therefore, the required interpolated value is $0.4792 - 0.0029 = 0.4763$. In actual computation, interpolation is often carried out almost by inspection.

28 CALCULATION OF MUTUAL INDUCTANCE AND SELF-INDUCTANCE

Although a single working formula and tables cover completely the whole range of values of the variable, it is found that for extreme cases the tabular values are changing so rapidly as to make interpolation uncertain. It is, therefore, better in such cases to use an auxiliary formula for calculating directly the required quantity. For example, in calculating the mutual inductance of equal, parallel straight filaments, the routine formula (2), page 32, covers the whole range of values of $\frac{d}{l}$ or $\frac{l}{d}$ even when these quantities are very small, provided that the value of $\frac{d}{l}$ or $\frac{l}{d}$ in question is one of the values for which Table 5 was calculated. If, however, it is necessary to interpolate in the table for such cases, the fact that the quantity Q is changing in such a way that higher order differences are appreciable renders interpolation uncertain. To avoid this difficulty, the auxiliary formulas (3) and (4) are provided, and these are easy and accurate for calculating the quantity Q.

The tables where possible are inserted in the text in appropriate positions to facilitate their use. In certain instances, however, where the same table is used for more than one working formula, it is impossible to avoid separation of the working formula and the table in the text.

Examples of the solution of problems are introduced following each working formula, to illustrate the calculations.

Part I

CIRCUITS WHOSE ELEMENTS ARE STRAIGHT FILAMENTS

Chapter 5

PARALLEL ELEMENTS OF EQUAL LENGTH

The problem of the calculation of the inductance of circuits made up of straight elements of negligible cross section has been completely solved. For parallel elements the formulas are relatively simple. For inclined filaments the formulas are more complicated, but even so, the mutual inductances may be expressed exactly in terms of inverse trigonometric and inverse hyperbolic functions. The expressions are, however, often unsuitable for routine calculations, and much simpler formulas, derived by expanding the exact formulas in series, are to be preferred.

When, as is often the case, the cross sectional dimensions of the conductors of a circuit composed of straight elements are small compared with the distances between the elements, the mutual inductances of the elements are sensibly the same as those of their central filaments.

The self-inductances of the elements depend, however, on the shape and dimensions of the cross section, and the same is true in a lesser degree for the mutual inductance of thick conductors close together. An exact expression for the self-inductance of a straight wire of circular cross section is available, and certain other simple cases may be accurately treated by the use of geometric mean distance formulas.

In the formulas listed below, reference is made to the original sources, but in general the formulas presented have been adapted to the necessities of simplicity and convenience for routine calculation, and use is freely made of auxiliary tables.

Mutual Inductance of Two Equal Parallel Straight Filaments. This case is basic for the treatment of circuits made up of parallel elements. Assuming the lengths of the filaments to be l and their distance apart d (both expressed in centimeters), the exact formula for the mutual inductance is [20]

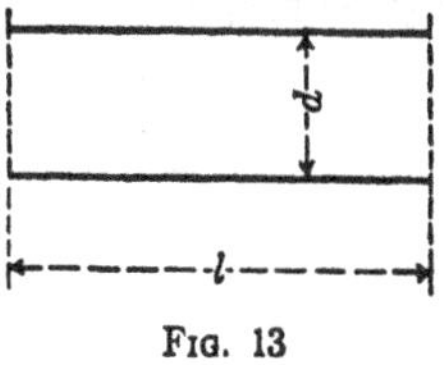

$$M = 0.002l \left[\log_e \left(\frac{l}{d} + \sqrt{1 + \frac{l^2}{d^2}} \right) - \sqrt{1 + \frac{d^2}{l^2}} + \frac{d}{l} \right] \quad (1)$$

Fig. 13

32 CALCULATION OF MUTUAL INDUCTANCE AND SELF-INDUCTANCE

In general, it will suffice to write

$$M = 0.002lQ. \quad \mu\text{h} \tag{2}$$

the value of Q being obtained from Table 5 for the given value of $\dfrac{d}{l}$ or $\dfrac{l}{d}$ (whichever is less than unity).

TABLE 5. VALUES OF Q FOR USE IN FORMULA (2)

d/l	Q	d_1	d/l	Q	d_1	d/l	Q	d_1
0.050	2.7382	−903	0.20	1.4926	−398	0.50	0.8256	−240
.055	2.6479	−822	.21	1.4528	−376	.52	.8016	−227
.060	2.5657	−752	.22	1.4152	−355	.54	.7789	−215
.065	2.4905	−693	.23	1.3797	−337	.56	.7574	−204
.070	2.4212	−642	.24	1.3460	−321	.58	.7370	−194
0.075	2.3570	−597	0.25	1.3139	−305	0.60	0.7176	−184
.080	2.2973	−558	.26	1.2834	−290	.62	.6992	−175
.085	2.2415	−524	.27	1.2544	−277	.64	.6817	−167
.090	2.1891	−493	.28	1.2267	−265	.66	.6650	−160
.095	2.1398	−466	.29	1.2002	−253	.68	.6490	−152
0.100	2.0932	−440	0.30	1.1749	−243	0.70	0.6338	−145
.105	2.0492	−418	.31	1.1506	−233	.72	.6193	−139
.110	2.0074	−397	.32	1.1273	−224	.74	.6054	−134
.115	1.9677	−379	.33	1.1049	−214	.76	.5920	−128
.120	1.9298	−361	.34	1.0835	−207	.78	.5792	−122
0.125	1.8937	−345	0.35	1.0627	−199	0.80	0.5670	−118
.130	1.8592	−330	.36	1.0429	−192	.82	.5552	−113
.135	1.8262	−318	.37	1.0238	−186	.84	.5439	−109
.140	1.7944	−305	.38	1.0052	−178	.86	.5330	−105
.145	1.7639	−293	.39	0.9874	−172	.88	.5225	−101
0.150	1.7346	−281	0.40	0.9702	−166	0.90	0.5124	−97
.155	1.7065	−271	.41	.9536	−161	.92	.5027	−93
.160	1.6794	−262	.42	.9375	−156	.94	.4934	−90
.165	1.6532	−253	.43	.9219	−151	.96	.4843	−87
.170	1.6279	−244	.44	.9068	−146	0.98	.4756	−84
0.175	1.6035	−236	0.45	0.8922	−141	1.00	0.4672	
.180	1.5799	−228	.46	.8781	−137			
.185	1.5571	−222	.47	.8644	−133			
.190	1.5349	−215	.48	.8511	−130			
.195	1.5134	−208	.49	.8381	−125			
0.200	1.4926		0.50	0.8256				

TABLE 5. VALUES OF Q FOR USE IN FORMULA (2) *(Concluded)*

l/d	Q	d_1	l/d	Q	d_1
1.00	0.4672		0.50	0.2451	
		−84			−94
0.98	.4588		.48	.2357	
		−83			−95
.96	.4505		.46	.2262	
		−84			−96
.94	.4421		.44	.2166	
		−85			−95
.92	.4336		.42	.2071	
		−85			−96
0.90	0.4251		0.40	0.1975	
		−85			−97
.88	.4166		.38	.1878	
		−86			−97
.86	.4080		.36	.1781	
		−87			−97
.84	.3993		.34	.1684	
		−87			−97
.82	.3906		.32	.1587	
		−87			−98
0.80	0.3819		0.30	0.1489	
		−88			−98
.78	.3731		.28	.1391	
		−88			−98
.76	.3643		.26	.1293	
		−89			−99
.74	.3554		.24	.1194	
		−90			−98
.72	.3464		.22	.1096	
		−90			−99
0.70	0.3374		0.20	0.0997	
		−90			−99
.68	.3284		.18	.0898	
		−91			−100
.66	.3193		.16	.0798	
		−91			−99
.64	.3102		.14	.0699	
		−92			−100
.62	.3011		.12	.0599	
		−93			−99
0.60	0.2918		0.10	0.0500	
		−92			−100
.58	.2826		.08	.0400	
		−93			−100
.56	.2733		.06	.0300	
		−93			−100
.54	.2640		.04	.0200	
		−94			−100
.52	.2546		0.02	0.0100	
		−95			−100
0.50	0.2451		0	0	

For filaments very close to one another, that is, $\dfrac{d}{l}$ very small, it is better to make use of the expansion of (1);

$$M = 0.002l\left[\log_e \frac{2l}{d} - 1 + \frac{d}{l} - \frac{1}{4}\frac{d^2}{l^2} + \cdots\right]. \quad (3)$$

The natural logarithm may be taken directly from Auxiliary Table 1 (page 236) or calculated from the common logarithm by use of the multiplication table, Auxiliary Table 2 (page 237).

34 CALCULATION OF MUTUAL INDUCTANCE AND SELF-INDUCTANCE

In the rare cases where the distance between the filaments is large compared with their lengths, the following series development is simple and accurate:

$$M = 0.002l \left(\frac{1}{2}\frac{l}{d}\right)\left[1 - \frac{1}{12}\frac{l^2}{d^2} + \frac{1}{40}\frac{l^4}{d^4} - \cdots\right]. \qquad (4)$$

Formulas (2), (3), and (4) cover all possible cases.

Example 1: The mutual inductance of two filaments each 10 feet long, spaced 6 inches apart, is required.

The length is $l = 10 \times 12 \times 2.54 = 304.8$ cm., $\frac{d}{l} = \frac{6}{120} = 0.05$, $\frac{2l}{d} = 40$. From Auxiliary Table 1, page 236, $\log_e 40 = 3.6889$.

Therefore, from (3), $M = 0.002(304.8)[3.6889 - 1 + 0.05 - 0.00006] = 1.6690\,\mu h$.

If the same filaments were placed 3 feet apart, $\frac{d}{l} = 0.3$, and from Table 5, $Q = 1.1749$, so that $M = 0.002(304.8)(1.1749) = 0.6999\,\mu h$. This value agrees also with that found for this case by formula (3).

In the extreme case that the filaments are separated by a distance of 40 feet, $\frac{l}{d} = \frac{1}{4}$, and Table 5 leads to the value $Q = 0.1244$.

Therefore,
$$M = 0.002(304.8)(0.1244) = 0.07583\,\mu h.$$

The series formula (4) gives

$$M = 0.6096(\tfrac{1}{8})[1 - (\tfrac{1}{12})(\tfrac{1}{16}) + (\tfrac{1}{40})(\tfrac{1}{256})]$$

$$= 0.0762[0.99489] = 0.07581\,\mu h.$$

Mutual Inductance of Two Equal Parallel Conductors. The mutual inductance of two conductors of circular cross section is sensibly the same as that of the filaments through the centers of their cross sections, even in the case of conductors nearly in contact. The formulas of the preceding section apply, therefore, without change.

For cross sections other than circular, the geometric mean distance R of the sections should be used in formula (2) in place of the distance d between the central filaments. Writing in (3) $\log_e R = \log_e d + \log_e k$, there results also, therefore, the general formula for conductors close together:

$$M = 0.002l \left[\log_e \frac{2l}{d} - \log_e k - 1 + \frac{d}{l} - \frac{1}{4}\frac{d^2}{l^2}\right]. \qquad (5)$$

For the practical case of conductors of equal rectangular cross sections with corresponding sides parallel, the value of $\log_e k$ may be taken from Tables 1 or 2, for equal parallel rectangles.

The mutual inductance of two *tubular* conductors is equal to that of the filaments along their axes.

PARALLEL ELEMENTS OF EQUAL LENGTH

Example 2: The mutual inductance of circular or tubular conductors 10 feet long, spaced 6 inches apart between centers, is as shown in Example 1, 1.6690 μh. If the cross sections, instead of being circular, were rectangles 1 inch by ¼ inch, with the long sides of the sections parallel, the value of $\log_e k$ obtained from Table 2, for $\beta = \frac{1}{8}$, $\Delta = \frac{1}{4}$, is 0.0022, so that formula (5) gives

$$M = 0.002(304.8)(2.7367) = 1.6682 \text{ μh},$$

which is not quite one part in 1000 less than for the circular conductors.

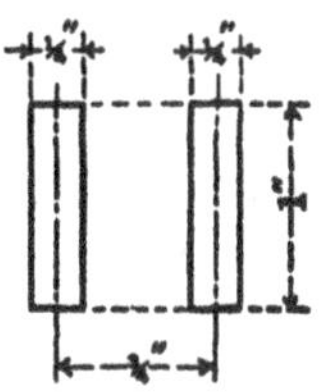

If the spacing between these rectangular conductors was reduced so that the centers of the cross sections were only ½ inch apart, we enter Table 2 with $\frac{1}{\beta} = \frac{1}{2}$ and $\Delta = \frac{1}{4}$, and $\log_e k$ is seen to be 0.2022. $\frac{d}{l} = \frac{1}{240}$, $\log_e \frac{2l}{d} = 6.1738$, and formula (5) gives

$$M = 0.002(304.8)[6.1738 - 0.2022 - 1 + 0.0042]$$
$$= 0.6096(4.9758) = 3.033 \text{ μh}.$$

Fig. 14

Even in this extreme case the mutual inductance of the rectangular conductors is only 4 per cent less than that of their central filaments.

Self-Inductance of a Straight Conductor. The self-inductance of a straight conductor of length l (in centimeters) is given, with an accuracy sufficient for all practical purposes, by the general formula [21]

$$L = 0.002l \left[\log_e \frac{2l}{r} - 1 + \frac{\delta_1}{l} \right], \quad (6)$$

in which r is the geometric mean distance and δ_1 the arithmetic mean distance of the points of the cross section. The last term is usually negligible.

For a round wire of radius ρ, this gives, neglecting the last term,[21]

$$L = 0.002l \left[\log_e \frac{2l}{\rho} - \frac{3}{4} \right]. \quad (7)$$

If, however, the material of the wire is magnetic, and has a permeability μ, the formula becomes

$$L = 0.002l \left[\log_e \frac{2l}{\rho} - 1 + \frac{\mu}{4} \right]. \quad (8)$$

The following formulas enable the inductance to be calculated for certain other simple cross sectional shapes. Nonmagnetic material is assumed.

Wire of rectangular cross section of sides B and C

$$L = 0.002l \left[\log_e \frac{2l}{B + C} + \frac{1}{2} - \log_e e \right]. \quad (9)$$

The last term is obtained from Table 3 for the given value of $\frac{B}{C}$.

36 CALCULATION OF MUTUAL INDUCTANCE AND SELF-INDUCTANCE

A conductor of elliptical cross section, the semiaxes of the ellipse being α and β, has an inductance

$$L = 0.002l \left[\log_e \frac{2l}{\alpha + \beta} - 0.05685 \right], \tag{10}$$

and for a straight piece of tubing, whose annular cross section has outer and inner radii ρ_1 and ρ_2, respectively,

$$L = 0.002l \left[\log_e \frac{2l}{\rho_1} + \log_e \zeta - 1 \right]. \tag{11}$$

Values of $\log_e \zeta$ are given in Table 4 as a function of the ratio $\frac{\rho_2}{\rho_1}$.

Example 3: A round wire of nonmagnetic material 10 feet (304.8 cm.) long and 2 mm. thick has an inductance by formula (7) of

$$L = 0.002(304.8) \left[\log_e \frac{609.6}{0.1} - \frac{3}{4} \right]$$

$$= 0.6096(7.965) = 4.855 \ \mu h.$$

A metal strip of the same length having a rectangular cross section 1 inch by $\frac{1}{4}$ inch will have a parameter value $\frac{2l}{B + C} = \frac{240}{1.25} = 192$, so that $\log_e \frac{2l}{B + C} = 5.2575$. The value of $\frac{C}{B}$ is $\frac{1}{4}$, so that Table 3 gives $\log_e e = 0.00249$. Using formula (9),

$$L = 0.6096(5.755) = 3.508 \ \mu h.$$

If this strip had the corners of its cross section rounded, so that an elliptical cross section of axes 1 inch and $\frac{1}{4}$ inch would result, then in formula (10), $\alpha = \frac{1}{2}$, $\beta = \frac{1}{8}$, $\frac{2l}{\alpha + \beta} = 384$, and

$$L = 0.6096(5.9506 - 0.0568) = 3.593 \ \mu h.$$

Likewise a tubular conductor 10 feet long with outer and inner diameters of 1 inch and 0.8 inch would have $\rho_1 = 0.5$ inch, $\frac{\rho_2}{\rho_1} = 0.8$, and Table 4 gives $\log_e \zeta = 0.0663$. The value of $\frac{2l}{\rho_1}$ is 480 and, from formula (11),

$$L = 0.6096(6.1738 + 0.0663 - 1) = 3.194 \ \mu h.$$

The minimum inductance possible, obtained by making the wall thickness vanishingly small, ($\log_e \zeta = 0$), is 3.153 μh. At the other extreme is the case where the wall thickness makes $\frac{\rho_2}{\rho_1} = 0$ (solid wire) and, for this, $\log_e \zeta = 0.25$, so that

$$L = 0.6096(5.4238) = 3.306 \ \mu h.$$

Note that any change of cross section that makes the average distance between the filaments smaller leads to a larger value of inductance.

Inductance of Multiple Conductors. The inductance of multiple conductors, that is, of groups of several members joined in parallel, may be found by fundamental circuit theory using the foregoing formulas for the self-inductance of a straight conductor and for the mutual inductance of parallel conductors. Only in symmetrical arrangements, where the contribution of each conductor to the total impedance of the combination is the same, will the resulting formula be simple. The following special cases [22] include equal round wires of radius ρ symmetrically placed.

Two equal parallel wires, separated by a distance d between centers, joined in parallel:

$$L = 0.002l \left[\log_e \frac{2l}{\sqrt{\rho d}} - \frac{7}{8} \right]. \tag{12}$$

Three equal wires, joined in parallel, situated at the corners of an equilateral triangle of side d:

$$L = 0.002l \left[\log_e \frac{2l}{(rd^2)^{1/3}} - 1 \right], \tag{13}$$

in which r is the geometric mean distance of the circular area of radius ρ that is, $\log_e r = \log_e \rho - \frac{1}{4}$.

For the general case of n equal round wires, spaced uniformly on a circle of radius a and connected in parallel, the inductance is found by substituting for R in the formula

$$L = 0.002l \left[\log_e \frac{2l}{R} - 1 \right] \tag{14}$$

the expression

$$R = (rna^{n-1})^{1/n}, \tag{15}$$

the value of r being the same as in (13).

Example 4: Consider a conductor 10 meters long made of wires of 2 mm. diameter joined in parallel. If two wires are used spaced $d = 0.5$ cm. apart, we have to use $2l = 2000$ cm., $\rho = 0.1$ cm. in (12), and $\frac{(2l)^2}{\rho d} = \frac{(2000)^2}{0.05} = 80 \times 10^6$, one-half of the logarithm of this is 3.95309, and using Auxiliary Table 2, the natural logarithm is 9.1023. Therefore, by (12),

$$L = 1000(0.002)(9.1023 - 0.875) = 16.454 \ \mu\text{h}.$$

A single wire of 2 mm. diameter gives, by formula (7),

$$L = 18.307 \ \mu\text{h}.$$

Three wires spaced 0.5 cm. apart will be treated by formula (13). From formula (h), page 21,

$$\log_e (rd^2)^{\frac{1}{3}} = \tfrac{1}{3}[\log_e \rho - \tfrac{1}{4} + 2\log_e d]$$

$$= -\tfrac{1}{3}\left[\log_e \tfrac{1}{\rho} + \tfrac{1}{4} + 2\log_e \tfrac{1}{d}\right]$$

$$= -\tfrac{1}{3}[\log_e 10 + \tfrac{1}{4} + 2\log_e 2] = -1.3130,$$

$$\log_e 2l = 7.6009.$$

Therefore, by (13),

$$L = 2[7.6009 + 1.3130 - 1] = 15.828 \ \mu h.$$

Summarizing, the inductance is

> one strand, 18.307 μh,
> two strands, 16.454,
> three strands, 15.828.

Example 5: If a conductor is composed of a number of strands twisted together, it is of interest to inquire how the stranding affects the inductance. The rigorous solution of the problem, especially when the number of strands is large, leads to results that are far from simple, because, in general, the impedance of any individual strand does not involve the same mutual inductances as all the other strands. Formula (15), however, enables a survey of the numerical relations that hold in the case of a symmetrical arrangement. Two cases are of interest: (a) when the strands used for different arrangements have all the same cross sectional radius ρ; (b) when the total cross sectional area is assumed to be the same, whatever the number of the strands. The formula (14) will be used for strands in contact, so that they are equally spaced on a circle of radius $a = \dfrac{\rho}{\sin \dfrac{\pi}{n}}$, and by comparing the results given by formula (14) with formula (7) for the inductance of a single wire, the equivalent radius ρ' of a single wire giving the same inductance will be found.

Case a. In the following table is shown the ratio of the radius a of the circle on which the strands are centered to the radius ρ of cross section of the strand. There is shown also the radius ρ' of a solid wire having the same inductance, and the radius ρ_1 of the solid wire having the same cross section as the stranded wire, both referred to ρ.

Number of Strands	$\dfrac{a}{\rho}$	$\dfrac{\rho'}{\rho}$	$\dfrac{\rho_1}{\rho}$	Calipered Radius ÷ ρ
3	1.159	1.875	1.732	
4	1.414	2.213	2.000	2.414
6	2.000	2.958	2.449	3.000
12	3.864	5.339	3.464	4.864

It is evident that in each case, the inductance of the stranded wire is less than that of a single strand. The inductance is not very different from that of a solid wire of about the same diameter as the calipered diameter of the stranded combination, but is decidedly less than that of the solid wire having the same cross sectional area.

PARALLEL ELEMENTS OF EQUAL LENGTH 39

Case b. The relations are clearer if the radii ρ_n of the strands are decreased as the number of strands n is increased, so as to keep the cross section of copper the same. That is, $n\pi\rho_n{}^2 = \pi\rho^2$. The following table has been calculated for different numbers of strands, arranged on a circle of such radius a that the strands are in contact.

Number of Strands	$\dfrac{\rho'}{\rho}$	$\dfrac{a}{\rho}$	Calipered Radius $+ \rho$
2	1.1333		
3	1.083	0.6667	1.2440
4	1.106	0.7071	1.2071
6	1.208	0.8164	1.2247
8	1.322	0.9243	1.2770
12	1.541	1.1154	1.4041

Here again is evident the effect in reducing the inductance of moving material away from the center of the cross section. The equivalent radius of the combination is not far different from the calipered dimension. However, in the practical case, there is no empty region at the center of the cross section, as is the case for these special conductors, and the value of ρ'/ρ for larger values of n would be nearer unity. A good approximation to the inductance of a solid stranded conductor will be to find the inductance of the solid wire having the same cross sectional area of copper. This furnishes the upper limit of the inductance of the stranded conductor.

Inductance of a Return Circuit of Parallel Conductors. An important case in practice is a loop formed of two parallel conductors whose length l is great compared with their distance d apart. The conductors carry equal, oppositely directed currents. Such a case is an ordinary transmission line.

In the simplest case of equal round wires of radius ρ, the inductance is given by [23]

$$L = 0.004l \left[\log_e \frac{d}{\rho} + \frac{1}{4} - \frac{d}{l} \right]. \qquad (16)$$

The last term is usually negligible. In long line theory the inductance per unit length (measured along the line) is required. This may be calculated from the preceding formula by expressing the desired unit of length, for example a foot or a mile, in centimeters and using this value for l.

If the wires are of magnetic material of permeability μ, the term $\frac{1}{4}$ in (16) is to be replaced by $\dfrac{\mu}{4}$.

In the more general case of wires of different cross sections, the self-inductances L_1 and L_2 of the conductors and the mutual inductance M_{12} are to be used in the expression $L = L_1 + L_2 - 2M_{12}$.

40 CALCULATION OF MUTUAL INDUCTANCE AND SELF-INDUCTANCE

If the geometric mean distances r_1 and r_2 of each of the two sections and the geometric mean distance R_{12} between the sections are known, the general formula becomes

$$L = 0.002l\,[2\log_e R_{12} - \log_e r_1 - \log_e r_2], \qquad (17)$$

or for equal cross sections

$$L = 0.004l \left[\log_e \frac{R_{12}}{r}\right]. \qquad (18)$$

For two equal *circular* cross sections this last formula goes over into (16). For two circular wires of unequal radii ρ_1 and ρ_2 it becomes

$$L = 0.002l \left[\log_e \frac{d^2}{\rho_1\rho_2} + \frac{1}{2}\right]. \qquad (19)$$

For the important case of equal parallel rectangular cross sections of dimensions B by C, C being the smaller dimension, spaced at a distance d between centers,

$$L = 0.004l \left[\log_e \frac{d}{B+C} + 1.5 + (\log_e k - \log_e e)\right], \qquad (20)$$

where $\log_e k$ and $\log_e e$ are given in Tables 1 or 2 and 3, respectively.

And since the geometric mean distances between two annuli is the distance between their centers, the inductance of a return circuit of two parallel tubular conductors spaced with a distance d between their axes is

$$L = 0.002l \left[\log_e \frac{d^2}{\rho_1\rho_3} + \log_e \zeta_1 + \log_e \zeta_3\right]. \qquad (21)$$

The outer radii of the tubes are assumed to be ρ_1 and ρ_3 and the corresponding inner radii ρ_2 and ρ_4. The values of $\log_e \zeta_1$ and $\log_e \zeta_3$ are taken from Table 4 for the given value of $\frac{\rho_2}{\rho_1}$ and $\frac{\rho_4}{\rho_3}$, respectively.

If the two tubes have the same dimensions, (21) becomes

$$L = 0.004l \left[\log_e \frac{d}{\rho_1} + \log_e \zeta\right]. \qquad (22)$$

Example 6: A line that has been much used for communications circuits consists of two wires 0.104 inch in diameter spaced 12 inches apart to form a return circuit. To calculate the inductance of a mile of such a circuit, we have to place for l in (16) $5280 \times 12 \times 2.54$ cm., $d = 12$, $\rho = \frac{1}{2}(0.104) = 0.052$ inch. $\text{Log}\frac{d}{\rho} = 2.36318$, $\log_e \frac{d}{\rho} = 5.4414$, and (16) gives

$$L = 0.004(5280 \times 12 \times 2.54)(5.6914) = 3.664 \text{ mh.}$$

This result agrees with the accepted value.[14]

PARALLEL ELEMENTS OF EQUAL LENGTH

Example 7: To find the inductance per 1000 feet of a return circuit composed of two strip conductors having a rectangular cross section $\frac{1}{4}$ inch by 1 inch. The strips are spaced $\frac{3}{4}$ inch between centers with their longer dimensions parallel (Fig. 14).

Here $B = 1$, $C = \frac{1}{4}$, $d = \frac{3}{4}$ inch, corresponding to p in Table 2. Therefore $\Delta = \frac{C}{B} = \frac{1}{4}$ and $\frac{1}{\beta} = \frac{d}{B} = \frac{3}{4}$. Table 3 gives $\log_e e = 0.0025$, and by interpolation from Table 2, $\log_e k = 0.1089$.

$$\log_e \frac{d}{B+C} = \log_e 0.6 = -0.51082.$$

Therefore, from (20),

$$L = 0.004(1000 \times 12 \times 2.54)[-0.51082 + 1.5 + 0.1089 - 0.0025]$$

$$= 133.6 \ \mu h.$$

The reactance at 60 cycles, found by multiplying this by 377, is 0.05039 ohms, or 0.0252 ohms per 1000 feet for each bar in the presence of the other.[25]

A circuit composed of round wires having the same cross section as the strip conductors would employ conductors having a radius of 0.282 inch. The inductance of the return circuit with a spacing $\frac{3}{4}$ inch comes out to be 149.7 μh. The round conductors are separated by only 0.186 inch at their nearest point, whereas the strip conductors have a clearance of 0.5 inch. Furthermore, the inductance of the strip is about 12 per cent smaller, and the difference in its favor would be still greater for equal clearances in the two cases. These facts, together with mechanical considerations, illustrate the superiority of strip conductors for heavy current work.

If, instead of equal solid round conductors, equal tubular conductors of the same outer diameter, 0.564 inch, and an internal radius of 0.1 inch were used, spaced 0.75 inch between their axes, the ratio $\frac{\rho_2}{\rho_1}$ is 0.3545, and for this value Table 4 gives $\log_e \xi = 0.1993$, so that the inductance by formula (22) comes out 143.5 μh. This is only 4 per cent less than for the solid round conductors, the clearance is the same, and the cross section is about 12 per cent smaller.

Return Circuit of Two Tubular Conductors One Inside the Other. The radii of the tubes arranged in order of decreasing magnitude are ρ_1, ρ_2, ρ_3, and ρ_4.

The inductance of the return circuit is[26]

$$L = 0.002l \left[\log_e \frac{\rho_1}{\rho_3} + \frac{2\left(\frac{\rho_2}{\rho_1}\right)^2}{1 - \left(\frac{\rho_2}{\rho_1}\right)^2} \cdot \log_e \frac{\rho_1}{\rho_2} - 1 + \log_e \zeta_1 + \log_e \zeta_3 \right]. \quad (23)$$

The values of $\log_e \zeta_1$ and $\log_e \zeta_3$ are obtained from Table 4 for the given values of $\frac{\rho_2}{\rho_1}$ and $\frac{\rho_4}{\rho_3}$.

It should be noted that the spacing between the axes of the tubes does not enter into the value of the inductance. That is, the inductance is the same whether the tubes are coaxial or eccentrically placed one within the other.

42 CALCULATION OF MUTUAL INDUCTANCE AND SELF-INDUCTANCE

It must be, however, expressly emphasized that this is not true for high frequency currents or in any case where the distribution of current in the cross section is not uniform.

If the inner conductor is solid, $\frac{\rho_4}{\rho_3}$ is zero, $\log_e \zeta_3 = \frac{1}{4}$, and

$$L = 0.002l \left[\log_e \frac{\rho_1}{\rho_3} + \frac{2\left(\frac{\rho_2}{\rho_1}\right)^2}{1 - \left(\frac{\rho_2}{\rho_1}\right)^2} \cdot \log_e \frac{\rho_1}{\rho_2} - \frac{3}{4} + \log_e \zeta_1 \right]. \quad (24)$$

Formula (24) applies to the important case of a coaxial cable carrying direct current. For the limiting case of very high frequency, where the current flows only on the outer surface of the inner conductor and on the inner surface of the outer conductor, formula (24) becomes

$$L = 0.002l \left[\log_e \frac{\rho_1}{\rho_3} \right]. \quad (25)$$

The general calculation for a coaxial cable carrying, as is usual, very high frequency currents, is treated on pages 280 ff.

Example 8: A coaxial line is constructed to have the same cross section as the line of strip conductor in Example 7. Its inner member is a solid round wire 0.564 inch in diameter. The outer conductor is a tube of outer diameter 2 inches and the inner diameter such as to give it a cross section of $\frac{1}{4}$ square inch. Calculate the inductance for 1000 feet of the line.

Here $\rho_3 = 0.282$ inch, $\rho_4 = 0$, $\rho_1 = 1$, and it is found that $\rho_2 = 0.9594$ inch, which gives the desired cross section. From Table 4, for this value of $\frac{\rho_2}{\rho_1}$, $\log_e \zeta = 0.0136$, $\frac{\rho_2^2}{\rho_1^2} = 0.9204$. Using formula (24),

$$L = 0.002(1000 \times 12 \times 2.54) \left[1.2655 + \frac{1.8408}{0.0796}(0.03914) - \frac{3}{4} + 0.0136 \right]$$
$$= 89.47 \ \mu h.$$

The carrying capacity, as far as cross section is concerned, is the same as for the strip conductor; the clearance between conductor is one-third greater; and the inductance is only about two-thirds as great. Of course the inner conductor is not so well ventilated as the strip, but the outer radiating surface of the tube is two and one-half times as great as the whole surface of the strip.

Russell [17] has derived formulas for the inductance of a number of complicated conductor systems of which two only will be included here.

Polycore Cable. The cable is composed of n round conductors, each of radius ρ, spaced uniformly on a circle of radius a, enclosed by a tubular sheath of outer and inner radii ρ_1 and ρ_2, respectively.

PARALLEL ELEMENTS OF EQUAL LENGTH

The current divides uniformly between the conductors and returns by the sheath. The length of the cable is l, and $\log_e \zeta$ is to be obtained from Table 4 for the ratio ρ_2/ρ_1. The inductance of the return circuit is

$$L = 0.002l \left[\log_e \frac{\rho_1}{a} + \frac{2\frac{\rho_2^2}{\rho_1^2}}{1 - \frac{\rho_2^2}{\rho_1^2}} \log_e \frac{\rho_1}{\rho_2} + \frac{1}{n} \log_e \frac{a}{n\rho} + \log_e \zeta + \frac{1}{4n} - 1 \right]. \quad (26)$$

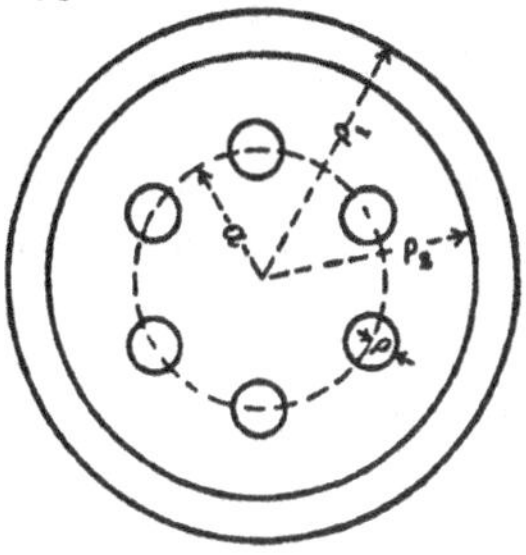

Fig. 15

A less common case is where the tubular conductor of radii ρ_1 and ρ_2 is surrounded by n round conductors of radii ρ, evenly spaced on a circle of radius a. In this case

$$L = 0.002l \left[\log_e \frac{a}{\rho_1} + \frac{1}{n} \log_e \frac{a}{n\rho} + \log_e \zeta + \frac{1}{4n} \right] \quad (27)$$

If the inner conductor is solid, $\log_e \zeta = \frac{1}{4}$.

Example 9: Assume that cable consists of 6 round conductors, each of radius of cross section of $\frac{1}{8}$ inch, spaced on a circle of radius $\frac{1}{2}$ inch. The coaxial sheath has inner and outer radii of $\frac{7}{8}$ inch and 1 inch, respectively. Calculate the inductance of the return circuit for a mile of cable.

Here $a = \frac{1}{2}$, $n = 6$, $\rho = \frac{1}{8}$, $\rho_1 = 1$, $\rho_2 = \frac{7}{8}$. For the value $\frac{\rho_2}{\rho_1} = \frac{7}{8}$, Table 4 gives $\log_e \zeta = 0.0416$. $\text{Log}_e \frac{\rho_1}{a} = 0.69315$, $\log_e \frac{\rho_1}{\rho_2} = 0.13353$, $\log_e \frac{a}{n\rho} = -0.40546$. Therefore, from formula (26),

$$L = 0.002(5280 \times 12 \times 2.54) \left[0.69315 + \frac{2(0.7656)}{0.2344} (0.13353) \right.$$
$$\left. - 0.06759 + 0.0416 + \frac{1}{24} - 1 \right]$$

$= 187.1\ \mu\text{h per mile,}$

or 35.43 μh for a 1000 foot length.

Inductance of Shunts. The use of shunts for measuring alternating currents requires designs of conductor arrangements that shall be as nearly

44 CALCULATION OF MUTUAL INDUCTANCE AND SELF-INDUCTANCE

noninductive as possible. Since the resistance of the shunt is usually only a small fraction of an ohm, an inductance of the order of only a few abhenrys suffices to give rise to an appreciable phase angle between the emf and the current.

Silsbee [28] has shown that the inductance of a shunt is equal to the mutual inductance between the potential circuit and the current circuit of the four-terminal conductor formed by the shunt, taking into account the fact that the two circuits have conductors in common. He has given a systematic procedure for expressing the inductance of the most complicated cases of shunts, composed of parallel elements, in terms of the geometric mean distances of the cross sections of the elements. The length and arrangement of the potential leads are important as well as the dimensions and spacings of the current carrying elements. The reader is referred to Silsbee's article for examples of shunts of different designs and arrangements. The inductance calculation follows lines already illustrated for the summation of elements composing circuits of rectangular or tubular elements.

Chapter 6

MUTUAL INDUCTANCE OF UNEQUAL PARALLEL FILAMENTS

General Case. Adopting the nomenclature of Fig. 16, the general formula for the mutual inductance of the filaments is [29]

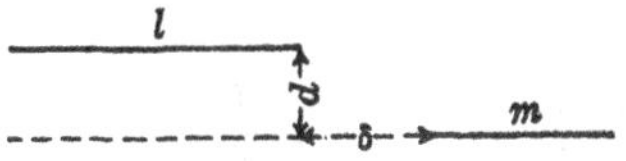

Fig. 16

$$M = 0.001 \left[\alpha \sinh^{-1}\frac{\alpha}{d} - \beta \sinh^{-1}\frac{\beta}{d} - \gamma \sinh^{-1}\frac{\gamma}{d} + \delta \sinh^{-1}\frac{\delta}{d} \right.$$
$$\left. - \sqrt{\alpha^2 + d^2} + \sqrt{\beta^2 + d^2} + \sqrt{\gamma^2 + d^2} - \sqrt{\delta^2 + d^2} \right], \quad (28)$$

where
$$\alpha = l + m + \delta, \quad \beta = l + \delta, \quad \gamma = m + \delta.$$

All lengths are in centimeters.

If the filaments overlap, the measured length δ is to be used with a negative sign, so that

$$\alpha = l + m - \delta, \quad \beta = l - \delta, \quad \gamma = m - \delta.$$

The form of formula (28) suggests that it merely expresses a result obtained from the more special case of equal parallel filaments by applying the laws of summation of mutual inductance, so that numerical computations may be based on the formulas and tables that apply to this special case.

Denoting by a subscript the lengths of two equal parallel filaments spaced at the distance d of the given problem, the general formula for the mutual inductance of the given filaments is found to be the result of combining the mutual inductances of four pairs of equal parallel filaments.

$$2M = (M_{l+m+\delta} + M_\delta) - (M_{l+\delta} + M_{m+\delta}), \quad (29)$$

and for overlapping filaments

$$2M = (M_{l+m-\delta} + M_\delta) - (M_{l-\delta} + M_{m-\delta}). \quad (30)$$

The individual terms are to be calculated by formulas (2), (3), or (4) using for l in these formulas the lengths given by the subscripts.

Special Cases. Unequal filaments placed as in Fig. 17:

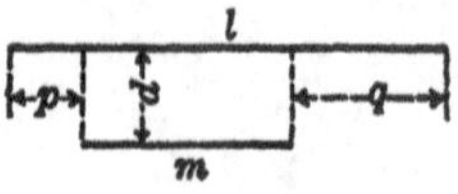

$$2M = (M_{m+p} + M_{m+q}) - (M_p + M_q), \quad (31)$$

and for the symmetrical case, $p = q$,

Fig. 17

$$M = M_{m+p} - M_p. \quad (32)$$

Filaments with Their Ends in the Common Perpendicular. Fig. 18(a) Filaments on the same side of the perpendicular:

$$2M = (M_l + M_m) - M_{l-m}. \quad (33)$$

Fig. 18(b) Filaments on opposite side of the perpendicular:

$$2M = M_{l+m} - (M_l + M_m). \quad (34)$$

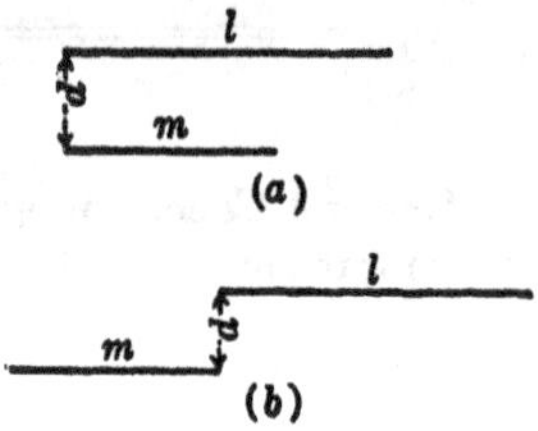

Fig. 18

Example 10: Two filaments of lengths 10 feet and 5 feet have a spacing between their axes of 6 inches and their ends are separated by 1 foot, measured between perpendiculars to the axes. The filaments extend in opposite directions (Fig. 16). Thus $l = 10$, $m = 5$, $d = \frac{1}{2}$, $\delta = 1$, $l + m + \delta = 16$, $l + \delta = 11$, $m + \delta = 6$, so that $\dfrac{d}{l+m+\delta} = \dfrac{1}{32}$, $\dfrac{d}{l+\delta} = \dfrac{1}{22}$, $\dfrac{d}{m+\delta} = \dfrac{1}{12}$, $\dfrac{d}{\delta} = \dfrac{1}{2}$. Of the four mutual inductances to be found, only $M_{m+\delta}$ and M_δ come within the range of Table 5. They give values of Q equal to 2.2597 and 0.8256, respectively.

For the other two we have to use formula (3). For $M_{l+m+\delta}$ we have to find $\log_e 64 - 1 + \frac{1}{32} - \frac{1}{4}(\frac{1}{32})^2 = 3.1899$, and for $M_{l+\delta}$, $\log_e 44 - 1 + \frac{1}{22} - \frac{1}{4}(\frac{1}{22})^2 = 2.8291$. Therefore, by (2) and (29),

$$2M = 0.002(12 \text{ in.} \times 2.54)[16(3.1899) + 0.8256 - 11(2.8291) - 6(2.2597)],$$

$$M = 0.03048[51.864 + 0.826 - 31.120 - 13.558]$$

$$= 0.2190 \ \mu\text{h}.$$

For the case of filaments whose axes are *in the same straight line*, the general method gives a simpler expression: [30]

$$M = 0.001[(l + m + \delta) \log_e (l + m + \delta) - (l + \delta) \log_e (l + \delta)$$

$$- (m + \delta) \log_e (m + \delta) + \delta \log_e \delta], \quad (35)$$

and if, further, their ends are in contact,

$$M = 0.001 \left[l \log_e \frac{l+m}{l} + m \log_e \frac{l+m}{m} \right]. \quad (36)$$

Formula (35) gives, nearly enough, the correct value of mutual inductance for filaments whose length is great compared with the distance between their axes, even when they are not exactly in the same straight line.

MUTUAL INDUCTANCE OF UNEQUAL PARALLEL FILAMENTS 47

Example 11: If, in the previous example, the distance d between the axes of the filaments was zero, the other dimensions being unchanged, the mutual inductance calculation is obtained from (35).

$$M = 0.001(12 \times 2.54)[16 \log_e 16 - 11 \log_e 11 - 6 \log_e 6]$$
$$= 0.03048[44.3614 - 26.3768 - 10.7506]$$
$$= 0.03048(7.2340) = 0.2205 \; \mu h,$$

or only 7 parts in 1000 larger than the value where the axes are 6 inches apart.

The mutual inductances calculated in the last two examples are very small because, on the average, the points of the two filaments are far apart. It is of interest to consider the increase in mutual inductance if the filaments are moved until the line joining their centers is a common perpendicular, the axial separation being kept as before (see Fig. 17).

Here $p = \dfrac{l-m}{2} = 2.5$, $m + p = 7.5$, $d = \frac{1}{2}$ ft., $\dfrac{d}{m+p} = \dfrac{1}{15}$, $\dfrac{d}{p} = \dfrac{1}{5}$. Table 5 gives $Q_{m+p} = 2.4668$, $Q_p = 1.4926$, and by formula (32) and (2),

$$M = 0.002(2.54 \times 12)[7.5(2.4668) - 2.5(1.4926)]$$
$$= 0.06096(18.501 - 3.731) = 0.9004 \; \mu h.$$

Mutual Inductance of Parallel Conductors of Unequal Length However Placed. For *round wires* the preceding formulas (28) to (36) for *filaments* apply without change, the positions of the filaments being taken to coincide with the axes of the wires. For other cross sectional shapes these same formulas are to be used with the difference that for the distance d there is to be placed the geometrical mean distance of the cross sections of the conductors, making use of Tables 1 or 2, for example, for rectangular cross sections. Except, however, when the conductors on the average are very close together, they may be replaced by their central filaments without appreciable error.

Chapter 7

MUTUAL INDUCTANCE OF FILAMENTS INCLINED AT AN ANGLE TO EACH OTHER

Equal Filaments Meeting at a Point. This is the most important case for practical purposes. Either the angle ϵ between their directions will be given, together with their common lengths l, or the length will be given and the measured distance R_1 between their ends (Fig. 19). The relation exists between ϵ and R_1 that

$$R_1^2 = 2l^2(1 - \cos \epsilon), \tag{37}$$

or

$$\cos \epsilon = 1 - \frac{R_1^2}{2l^2}$$

Fig. 19

The mutual inductance between the filaments is [31]

$$M = 0.004l \cos \epsilon \, \tanh^{-1} \frac{l}{l + R_1} \tag{38}$$

$$= 0.001 lS, \tag{39}$$

in which l is to be expressed in centimeters and the value of S is given as a function of $\cos \epsilon$ in Table 6.

For values of $\cos \epsilon$ approaching unity, that is, for filaments making an acute angle with each other, interpolation in Table 6 becomes uncertain, and it is better to make use of formula (38) directly. The value of the $\tanh^{-1}$ function may be obtained by interpolation in a table of hyperbolic functions, or it may be calculated directly from the relation

$$\tanh^{-1} \frac{l}{l + R_1} = \frac{1}{2} \log_e \frac{2l + R_1}{R_1}, \tag{40}$$

making use of Auxiliary Tables 1 or 2 for calculating natural logarithms.

From formula (38) and Table 6, it is evident that when the filaments subtend a right angle, their mutual inductance is zero, and that in passing from an acute angle to an obtuse angle M changes sign. The significance of the change of sign is that if the two filaments were joined in series, in one case

FILAMENTS INCLINED AT AN ANGLE TO EACH OTHER

TABLE 6. VALUES OF FACTOR S IN FORMULA (39) FOR THE MUTUAL INDUCTANCE OF EQUAL INCLINED FILAMENTS

cos ε	S	d_1	d_2	cos ε	S	d_1	d_2
0.95	3.7830			−0.	0		
		−7236				−867	
.90	3.0594		2274	−0.05	−0.0867		28
		−4462				−840	
.85	2.6132		1146	− .10	− .1707		25
		−3316				−815	
.80	2.2816		637	− .15	− .2523		22
		−2679				−793	
.75	2.0137		405	− .20	− .3316		21
		−2274				−772	
0.70	1.7863		283	−0.25	−0.4088		20
		−1991				−752	
.65	1.5872		211	− .30	− .4840		18
		−1780				−734	
.60	1.4092		162	− .35	− .5574		18
		−1618				−716	
.55	1.2474		130	− .40	− .6290		15
		−1488				−701	
.50	1.0986		106	− .45	− .6991		15
		−1382				−686	
0.45	0.9604		88	−0.50	−0.7677		15
		−1294				−671	
.40	0.8310		76	− .55	− .8348		13
		−1218				−658	
.35	0.7092		64	− .60	− .9006		13
		−1154				−645	
.30	0.5938		57	− .65	−0.9651		12
		−1097				−633	
.25	0.4841		49	− .70	−1.0284		11
		−1048				−622	
0.20	0.3793		45	−0.75	−1.0906		11
		−1003				−611	
.15	0.2789		40	− .80	−1.1517		10
		− 964				−601	
.10	0.1825		35	− .85	−1.2118		10
		− 929				−591	
0.05	0.0896		32	− .90	−1.2709		10
		− 896				−581	
0	0		30	−0.95	−1.3290		9
		− 867				−572	
			28	−1.00	−1.3862		

50 CALCULATION OF MUTUAL INDUCTANCE AND SELF-INDUCTANCE

the mutually induced emfs would act with the self-induced emfs and in the other would oppose them. As ϵ becomes smaller, they form a "return" circuit; the mutual inductance makes the total inductance of the combination less than the sum of their self-inductances, so that in the general formula applicable to a return circuit, $L = L_1 + L_2 - 2M_{12}$, the mutual inductance is to be taken as a positive quantity. However, if the angle ϵ is obtuse, the series combination has a greater total inductance than the sum of the self-inductances and M_{12} is to be considered as a negative quantity in the general equation, so as to give $L = L_1 + L_2 + 2M_{12}$. The algebraic signs in Table 6 are consistent with this usage.

The formulas for *filaments* may be used without sensible error for the inclined *conductors*, if each conductor is replaced by the filament along its axis.

Example 12: Two straight wires, each 500 feet long, meet at a point. The measured distance between their outer ends is 350 feet. Calculate their mutual inductance.

The value of $\dfrac{R_1}{l} = \dfrac{350}{500} = 0.7$, so that by (37) $\cos \epsilon = 1 - \tfrac{1}{2}(0.49) = 0.755$, which corresponds to $\epsilon = 40° 58'.5$.

Interpolating from Table 6, using first and second differences, $S = 2.0367$, or including third differences, 2.0385. Thus, from (39)

$$M = 0.001(500 \times 12 \times 2.54)(2.0385) = 31.07 \, \mu\text{h}.$$

Calculating S directly by (38) and (40), we find $\tfrac{1}{2} \log_e \dfrac{2.7}{0.7} = 0.67496$, and $S = 4(0.755)(0.67496) = 2.0384$, which gives an indication of the accuracy of Table 6.

Unequal Filaments Meeting at a Point. Assuming the lengths of the filaments to be l_1 and m_1 that they subtend an angle ϵ, and that the distance between their ends is R, then,[31]

$$M = 0.002 \cos \epsilon \left[l_1 \tanh^{-1} \frac{m_1}{l_1 + R} + m_1 \tanh^{-1} \frac{l_1}{m_1 + R} \right] \tag{41}$$

$$= 0.001 \, l_1 \cos \epsilon \left[\log_e \frac{1 + \dfrac{m_1}{l_1} + \dfrac{R}{l_1}}{1 - \dfrac{m_1}{l_1} + \dfrac{R}{l_1}} + \frac{m_1}{l_1} \log_e \frac{\dfrac{m_1}{l_1} + \dfrac{R}{l_1} + 1}{\dfrac{m_1}{l_1} + \dfrac{R}{l_1} - 1} \right]. \tag{42}$$

In practice, either the angle ϵ or the distance R will be given. The following relations enable one to be found from the other.

$$\cos \epsilon = \frac{l_1^2 + m_1^2 - R^2}{2 l_1 m_1}, \tag{43}$$

$$\frac{R^2}{l_1^2} = 1 + \frac{m_1^2}{l_1^2} - 2 \frac{m_1}{l_1} \cos \epsilon. \tag{44}$$

FILAMENTS INCLINED AT AN ANGLE TO EACH OTHER

For routine calculations it is convenient to write

$$M = 0.001 l_1 S_1, \qquad (45)$$

and to obtain S_1 from Table 7 for the given values of $\cos \epsilon$ and $\dfrac{m_1}{l_1}$, considering m_1 as the smaller length.

TABLE 7. UNEQUAL FILAMENTS MEETING AT A POINT.
VALUES OF S_1 TO BE USED IN FORMULA (45)

$\cos \epsilon$	$\dfrac{m_1}{l_1}=1$	0.9	0.8	0.7	0.6	0.5	0.4	0.3	0.2	0.1
0.95	3.7830	3.5786	3.3405	3.0683	2.7622	2.4221	2.0473	1.6348	1.1776	0.6598
.90	3.0594	2.8958	2.7095	2.4980	2.2597	1.9930	1.6957	1.3643	0.9918	.5630
.85	2.6132	2.4744	2.3178	2.1411	1.9422	1.7189	1.4690	1.1877	.8688	.4973
.80	2.2816	2.1609	2.0256	1.8735	1.7028	1.5108	1.2950	1.0512	.7727	.4452
0.75	2.0137	1.9071	1.7889	1.6562	1.5073	1.3399	1.1513	0.9374	0.6917	0.4008
.70	1.7863	1.6922	1.5876	1.4710	1.3402	1.1932	1.0272	.8386	.6209	.3615
.65	1.5872	1.5038	1.4113	1.3083	1.1931	1.0636	0.9172	.7504	.5572	.3258
.60	1.4092	1.3352	1.2534	1.1625	1.0609	0.9468	.8177	.6703	.4991	.2929
.55	1.2474	1.1820	1.1098	1.0297	0.9404	.8400	.7264	.5964	.4452	.2622
0.5	1.0986	1.0411	0.9776	0.9074	0.8291	0.7412	0.6417	0.5277	0.3947	0.2332
.4	0.8310	0.7876	.7398	.6870	.6283	.5625	.4880	.4024	.3020	.1794
.3	.5935	.5628	.5288	.4913	.4496	.4030	.3501	.2893	.2179	.1301
.2	.3793	.3595	.3378	.3140	.2876	.2580	.2244	.1856	.1404	.0842
0.1	0.1825	0.1730	0.1626	0.1512	0.1385	0.1244	0.1083	0.0898	0.0680	0.0410
0	0	0	0	0	0	0	0	0	0	0
−0.1	−0.1707	−0.1618	−0.1522	−0.1416	−0.1298	−0.1167	−0.1018	−0.0847	−0.0644	−0.0391
− .2	− .3316	− .3144	− .2956	− .2750	− .2523	− .2269	− .1982	− .1650	− .1257	− .0765
− .3	− .4840	− .4588	− .4314	− .4015	− .3684	− .3315	− .2898	− .2416	− .1844	− .1125
− .4	− .6290	− .5963	− .5608	− .5220	− .4791	− .4314	− .3772	− .3148	− .2406	− .1472
−0.5	−0.7677	−0.7278	−0.6845	−0.6372	−0.5850	−0.5268	−0.4611	−0.3852	−0.2948	−0.1808
− .6	−0.9006	− .8538	− .8031	− .7476	− .6865	− .6186	− .5416	− .4528	− .3470	− .2134
− .7	−1.0284	−0.9750	−0.9172	− .8540	− .7844	− .7070	− .6194	− .5182	− .3976	− .2450
− .8	−1.1517	−1.0919	−1.0272	−0.9563	− .8788	− .7922	− .6944	− .5814	− .4467	− .2758
−0.9	−1.2709	−1.2048	−1.1335	−1.0557	−0.9701	− .8748	− .7671	− .6428	− .4943	− .3058
−1.0	−1.3862	−1.3143	−1.2366	−1.1517	−1.0585	−0.9548	−0.8376	−0.7029	−0.5406	−0.3351

When the angle ϵ is small or, in general, when $\dfrac{m_1}{l_1}$ is small, accurate interpolation of S_1 is difficult and it is better to calculate M by formula (41) or, supposing tables of hyperbolic functions are not available, by (42).

Example 13: If two wires have lengths of 200 and 1200 cm. and subtend an angle whose cosine is 0.75, then $\dfrac{m_1}{l_1} = \dfrac{200}{1200} = \dfrac{1}{6}$, and interpolating in Table 7, using third differences,

$$S_1 = 0.4008 + \tfrac{2}{3}(0.2908) - \tfrac{1}{9}(-0.0452) + \tfrac{4}{81}(0.0136)$$

so that by (45),
$$= 0.4008 + 0.1939 + 0.0050 + 0.0007 = 0.6004,$$

$$M = 0.001(1200)(0.6004) = 0.7205 \ \mu\text{h}.$$

This interpolation is somewhat uncertain, on account of the size of the higher order differences. A more accurate value of S_1 may be found from formula (41) or (42). In this case, by (44), $\frac{R^2}{l_1^2} = 1 + \frac{1}{36} - \frac{1}{3} 0.75 = 0.77778$, or $\frac{R}{l_1} = 0.8819$, and $R = 1058.3$.

$$\tanh^{-1}\frac{m_1}{l_1 + R} = \tanh^{-1}\frac{200}{1200 + 1058.3} = 0.08880,$$

$$\tanh^{-1}\frac{l_1}{m_1 + R} = \tanh^{-1}\frac{1200}{200 + 1058.3} = 1.87085,$$

$$S_1 = 2\cos\epsilon\,[0.08880 + \tfrac{1}{6}(1.87085)] = 0.6008.$$

By formula (42),

$$\log\frac{1.1667 + 0.8819}{0.8333 + 0.8819} = 0.07712,$$

$$\log\frac{1.1667 + 0.8819}{0.8819 - .8333} = 0.27082$$

$$\overline{0.34794} = \text{sum},$$

which, multiplied by $\cos\epsilon$, equals 0.26096. Converted to a natural logarithm by Auxiliary Table 2, the result is $S_1 = 0.6009$. The result is slightly more accurate by the $\tanh^{-1}$ formula.

Unequal Filaments in the Same Plane, Not Meeting. The filaments have lengths l and m and when produced to their point of intersection the latter is distant from their nearer ends by the distances μ and ν, respectively. The four distances between the ends of the filaments are as shown in Fig. 20.

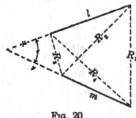

Fig. 20

The given data may include, in addition to the lengths l and m, the distances μ and ν and the angle ϵ, or, instead, the measured values of the four distances R_1, R_2, R_3, and R_4 may be specified. The equations connecting the two systems are

$$2\cos\epsilon = \frac{\alpha^2}{lm}, \text{ where } \alpha^2 = R_4{}^2 - R_3{}^2 + R_2{}^2 - R_1{}^2, \tag{46}$$

$$\mu = \frac{[2m^2(R_2{}^2 - R_3{}^2 - l^2) + \alpha^2(R_4{}^2 - R_3{}^2 - m^2)]l}{4l^2m^2 - \alpha^4}, \tag{47}$$

$$\nu = \frac{[2l^2(R_4{}^2 - R_3{}^2 - m^2) + \alpha^2(R_2{}^2 - R_3{}^2 - l^2)]m}{4l^2m^2 - \alpha^4}, \tag{48}$$

$$\left.\begin{aligned}
R_1{}^2 &= (\mu + l)^2 + (\nu + m)^2 - 2(\mu + l)(\nu + m)\cos\epsilon, \\
R_2{}^2 &= (\mu + l)^2 + \nu^2 - 2\nu(\mu + l)\cos\epsilon, \\
R_3{}^2 &= \mu^2 + \nu^2 - 2\mu\nu\cos\epsilon, \\
R_4{}^2 &= \mu^2 + (\nu + m)^2 - 2\mu(\nu + m)\cos\epsilon.
\end{aligned}\right\} \tag{49}$$

FILAMENTS INCLINED AT AN ANGLE TO EACH OTHER

The general formula for the mutual inductance is

$$\frac{M}{2\cos\epsilon} = 0.001\left[(\mu+l)\tanh^{-1}\frac{m}{R_1+R_2} + (\nu+m)\tanh^{-1}\frac{l}{R_1+R_4}\right.$$
$$\left. - \mu\tanh^{-1}\frac{m}{R_3+R_4} - \nu\tanh^{-1}\frac{l}{R_2+R_3}\right], \quad (50)$$

or lacking tables of hyperbolic functions, the $\tanh^{-1}$ functions may be calculated by the relation $\tanh^{-1}x = \frac{1}{2}\log_e\frac{1+x}{1-x}$.

Routine calculations may be based on Table 7. Since, by the summation principle,

$$M_{lm} = (M_{\mu+l,\,\nu+m} + M_{\mu\nu}) - (M_{\mu+l,\,\nu} + M_{\nu+m,\,\mu}), \quad (51)$$

the subscripts denoting the lengths of the filaments entering into each term, it will be seen that each of the four terms in (51) is the mutual inductance of two filaments meeting at a point and may be calculated by (45) using Table 7.

Example 14: Assume two wires of lengths $l = 1000$ feet, $m = 600$ feet, $\mu = 200$ feet, and $\nu = 400$, the angle of inclination being $\cos\epsilon = 0.7$.

For the solution by the summation method, we have for calculating the individual terms of (51) to use in (45) the following constants:

	l_1	m_1	m_1/l_1	S_1
$M_{\mu+l,\,\nu+m}$	1200	1000	$\frac{5}{6}$	1.6237
$M_{\mu\nu}$	400	200	$\frac{1}{2}$	1.1932
$M_{\mu,\,\nu+m}$	1000	200	0.2	0.6209
$M_{\nu,\,\mu+l}$	1200	400	$\frac{1}{3}$	0.9042

Thus, by (51)

$M = 0.001(12 \times 2.54)[1200(1.6237) + 400(1.1932) - 1000(0.6209) - 1200(0.9042)]$

$= 0.03048[1948.4 + 477.3 - 620.9 - 1085.0]$

$= 0.03048(719.8) = 21.939\,\mu\text{h}.$

This may be checked by formula (50). The calculated distances are, by (49),

$R_1 = 871.77,\qquad\qquad R_3 = 296.65,$

$R_2 = 963.33,\qquad\qquad R_4 = 871.77,$

$\tanh^{-1}\frac{m}{R_1+R_2} = \tanh^{-1}\frac{600}{1835.1} = 0.33941,$

$\tanh^{-1}\frac{l}{R_1+R_4} = \tanh^{-1}\frac{1000}{1743.5} = 0.65278,$

$\tanh^{-1}\frac{m}{R_3+R_4} = \tanh^{-1}\frac{600}{1168.4} = 0.56748,$

$\tanh^{-1}\frac{l}{R_2+R_3} = \tanh^{-1}\frac{1000}{1260.0} = 1.08120.$

54 CALCULATION OF MUTUAL INDUCTANCE AND SELF-INDUCTANCE

By formula (50)

$M = 0.002(0.7)(12 \times 2.54)[1200(0.33941) + 1000(0.65278) - 200(0.56748)$
$ -400(1.08120)]$
$ = 1.4(0.03048)[407.29 + 652.78 - 113.50 - 432.48]$
$ = 0.03048(719.7) = 21.936 \ \mu h.$

The results by the two methods differ by only about 1 part in 10,000.

Example 15: As a further example, we may suppose that instead of the inclination of the wires the given data consists of the lengths of the wires and the four measured distances between their ends. (See Fig. 20.)

Assume that

$$l = 2000 \text{ cm.}, \qquad m = 1200 \text{ cm.},$$
$$R_1 = 2020, \qquad R_2 = 2195,$$
$$R_3 = 500, \qquad R_4 = 1550.$$

Here the inclination ϵ and the distances μ and ν have to be calculated by formulas (46) to (48):

$$\alpha^2 = (1550)^2 - (500)^2 + (2195)^2 - (2020)^2 = 2.8900 \times 10^6,$$

$$2 \cos \epsilon = \frac{\alpha^2}{(2000)(1200)} = 1.2037,$$

so that

$$\cos \epsilon = 0.60185.$$

$R_2^2 - R_3^2 = 4.56(10^6), \qquad\qquad R_4^2 - R_3^2 = 2.16(10^6),$

$R_2^2 - R_3^2 - l^2 = 0.56(10^6), \qquad R_4^2 - R_3^2 - m^2 = 0.72(10^6),$

$$\frac{\nu}{m} = \frac{8(0.72) + 2.89(0.56)}{14.508}, \qquad \nu = 610.28.$$

$$\frac{\mu}{l} = \frac{2.88(0.56) + 2.89(0.72)}{14.508}, \qquad \mu = 509.18.$$

Also

$\nu + m = 1810.3, \qquad\qquad \mu + l = 2509.2,$

$R_1 + R_2 = 4215, \qquad\qquad R_1 + R_4 = 3570,$

$R_2 + R_3 = 2695, \qquad\qquad R_3 + R_4 = 2050.$

Therefore, by formula (50)

$$\frac{M}{0.002 \cos \epsilon} = 2509.2 \tanh^{-1} \frac{1200}{4215} + 1810.3 \tanh^{-1} \frac{2000}{3570}$$
$$- 509.18 \tanh^{-1} \frac{1200}{2050} - 610.28 \tanh^{-1} \frac{2000}{2695}$$
$$= 2509.2(0.29278) + 1810.3(0.63315)$$
$$- 509.18(0.67059) - 610.28(0.95515)$$
$$= 734.6 + 1146.2 - 341.4 - 582.9 = 956.5,$$

$M = 2 \times 0.60185(0.001)(956.5) = 1.1513 \ \mu h.$

For this form of problem, the summation formula (51) and Table 7 are much less convenient, since in general four double interpolations will have to be made in the table. In this especial example the details are:

	l_1	$\dfrac{m_1}{l_1}$	S_1
1st term	2509.2	0.72147	1.1880
2nd term	610.28	0.83433	1.2880
3rd term	2509.2	0.24322	0.5785
4th term	1810.3	0.28127	0.64275

and from (51)

$$M = 0.001[2980.9 + 786.1 - 1451.5 - 1158.2] = 1.1573\ \mu\text{h}.$$

This value is about ½ per cent higher than the more accurate value by the general formula. (Each of the four values of S_1 was interpolated for $\cos \epsilon = 0.60185$ and the appropriate value of $\dfrac{m_1}{l_1}$.)

If, therefore, the four distances between the ends are given, it is more convenient to use the direct formula (50) than to use the summation method and Table 7.

Mutual Inductance of Two Straight Filaments Placed in Any Desired Positions.

This is the most general case for straight filaments. Solutions have been given by F. F. Martens[32] and G. A. Campbell.[33] Although differing in form their formulas are equivalent.

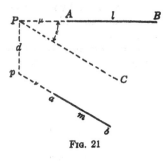

Fig. 21

In Fig. 21 are shown the two filaments AB and ab. Planes are passed through the filaments in such a way as to intersect at right angles. In the figure PC represents the line of intersection. From the point of intersection of the line PC and the direction of BA, produced, is drawn the line Pp perpendicular to BP and ba, produced. Pp is the common perpendicular to the filaments and its length d is the distance between the plane BPC and the parallel plane through filament ab. The angle BPC is the angle ϵ of inclination between the filaments.

Let l and m denote the lengths of the filaments AB and ab, respectively, and place $PA = \mu$, $pa = \nu$. Furthermore, designate the four distances between the ends of one filament and the ends of the other as follows:

$$Bb = R_1, \quad Ba = R_2, \quad Aa = R_3, \quad \text{and} \quad Ab = R_4.$$

The mutual inductance of the filaments AB and ab, which is sensibly the same as the mutual inductance of sections of wire with the filaments along their axes, is given by

$$\frac{M}{0.001\cos\epsilon} = 2\left[(\mu+l)\tanh^{-1}\frac{m}{R_1+R_2} + (\nu+m)\tanh^{-1}\frac{l}{R_1+R_4}\right.$$

$$\left. - \mu\tanh^{-1}\frac{m}{R_3+R_4} - \nu\tanh^{-1}\frac{l}{R_2+R_3}\right] - \frac{\Omega d}{\sin\epsilon}, \quad (52)$$

in which

$$\Omega = \tan^{-1}\left\{\frac{d^2\cos\epsilon + (\mu+l)(\nu+m)\sin^2\epsilon}{dR_1\sin\epsilon}\right\}$$

$$- \tan^{-1}\left\{\frac{d^2\cos\epsilon + (\mu+l)\nu\sin^2\epsilon}{dR_2\sin\epsilon}\right\}$$

$$+ \tan^{-1}\left\{\frac{d^2\cos\epsilon + \mu\nu\sin^2\epsilon}{dR_3\sin\epsilon}\right\}$$

$$- \tan^{-1}\left\{\frac{d^2\cos\epsilon + \mu(\nu+m)\sin^2\epsilon}{dR_4\sin\epsilon}\right\}. \quad (53)$$

Two cases have to be considered:

Case a. If the two wires are elements of existing circuits, the given data will be the measured lengths l, m of the filaments and the distances R_1, R_2, R_3, and R_4.

From these must be calculated the values of $\cos\epsilon$, μ, ν, and d for use in formula (52) by the relations

$$\cos\epsilon = \frac{\alpha^2}{2lm},$$

$$\alpha^2 = R_4^2 - R_3^2 + R_2^2 - R_1^2,$$

$$d^2 = R_3^2 - \mu^2 - \nu^2 + 2\mu\nu\cos\epsilon,$$

$$\frac{\mu}{l} = \frac{2m^2(R_2^2 - R_3^2 - l^2) + \alpha^2(R_4^2 - R_3^2 - m^2)}{4l^2m^2 - \alpha^4}, \quad (54)$$

$$\frac{\nu}{m} = \frac{2l^2(R_4^2 - R_3^2 - m^2) + \alpha^2(R_2^2 - R_3^2 - l^2)}{4l^2m^2 - \alpha^4}.$$

Such calculations will be of infrequent occurrence. A more common application of (52) is its employment in the derivation of the inductance of circuits of special form built up of straight filaments. For such the problem is presented in the form of the following case:

FILAMENTS INCLINED AT AN ANGLE TO EACH OTHER

Case b. The given data are l, m, μ, ν, and d. There is required the calculation of the distances R_1, R_2, R_3, and R_4 for use in formula (52) by the relations

$$R_1^2 = d^2 + (\mu + l)^2 + (\nu + m)^2 - 2(\mu + l)(\nu + m) \cos \epsilon,$$
$$R_2^2 = d^2 + (\mu + l)^2 + \nu^2 - 2\nu(\mu + l) \cos \epsilon,$$
$$R_3^2 = d^2 + \mu^2 + \nu^2 - 2\mu\nu \cos \epsilon,$$
$$R_4^2 = d^2 + \mu^2 + (\nu + m)^2 - 2\mu(\nu + m) \cos \epsilon.$$
(55)

The angles in the expression for Ω are to be chosen in the proper quadrants by noting that the denominators of the fractions that represent the tangents are always positive. If ϵ is in the second quadrant, the numerators may become negative and the angles are in the fourth quadrant, that is, the angle is to be considered negative in calculating Ω. In brief, in calculating the combination of the four angles in (53), an angle is to be considered as a positive acute angle if the numerator is positive, and as a negative acute angle if the numerator is negative.

Example 16: As an illustration of the use of (52), suppose measurements are made on two straight filaments with the results $l = 1000$, $m = 500$, $R_1 = 1570$, $R_2 = 1320$, $R_3 = 420$, $R_4 = 775$, all the distances being measured in centimeters.
Then $\alpha^2 = (775)^2 - (420)^2 + (1320)^2 - (1570)^2 = -0.2983(10^6)$, and $2lm = 10^6$, so that by (54) $\cos \epsilon = -0.2983$,

$$R_2^2 - R_3^2 - l^2 = 0.5660\ (10^6),$$
$$R_4^2 - R_3^2 - m_2 = 0.1742(10^6),$$
$$4l^2m^2 - \alpha^4 = 0.9110(10^{12}),$$
$$2l^2 = 2(10^6),$$
$$2m^2 = 0.5(10^6).$$

From these data and (54)

$$\frac{\mu}{l} = 0.2536, \text{ so that } \mu = 253.6,$$

$$\frac{\nu}{m} = 0.1971, \qquad \nu = 98.55,$$

$$d^2 = (420)^2 - (253.6)^2 - (98.55)^2 + 2(-0.2983)(253.6)(98.55) = 295.75.$$

To find the mutual inductance we have $(\mu + l) = 1253.6$, $(\nu + m) = 598.55$. The $\tanh^{-1}$ terms in (52) are

$$1253.6 \tanh^{-1} \frac{500}{2890} + 598.55 \tanh^{-1} \frac{1000}{2345} - 253.6 \tanh^{-1} \frac{500}{1195} - 98.55 \tanh^{-1} \frac{1000}{1740}$$

$$= 1253.6(0.17476) + 598.55(0.45554) - 253.6(0.44576) - 98.55(0.65453) = 314.19.$$

58   CALCULATION OF MUTUAL INDUCTANCE AND SELF-INDUCTANCE

To calculate the $\tan^{-1}$ terms we have $d^2 \cos \epsilon = -2.6091(10^4)$, $\log \sin \epsilon = \bar{1}.97976$, $\log d^2 \sin \epsilon = 2.45068$. The logarithms of the numerators and denominators for the four arguments are

log numerator	5.81789	4.93682	n3.52140	5.04997
log denominator	5.64638	5.57125	5.07394	5.33998
log tan	0.17151	$\bar{1}$.36557	$n\bar{2}$.44746	$\bar{1}$.70999
angle	56° 1!80	13° 3!84	−1° 36!30	27° 9!07

so that
$$\Omega = 56°\,1!80 - 13°\,3!84 - 1°\,36!30 - 27°\,9!07 = 14°\,12!59.$$

This angle is found to be 0.24801 radians and the term $\dfrac{\Omega d}{\sin \epsilon} = 70.01$. Formula (52), therefore, gives

$$M = -0.001(0.2983)[628.38 - 70.01]$$
$$= -0.1666 \ \mu\text{h}.$$

Chapter 8

CIRCUITS COMPOSED OF COMBINATIONS OF STRAIGHT WIRES

The inductance of circuits made up of straight elements, that is, of lengths of straight wire, whether parallel or inclined, joined in series, may be obtained by the summation method, using formulas already given. The self-inductance of each element is given by the formula for a straight wire of the appropriate cross section. The inductance of the whole circuit is the sum of these terms together with the sum of the mutual inductances of all the pairs of elements. The mutual inductances are found from the formulas given for parallel or inclined filaments. The mutual inductance of each pair of *wires* is sensibly the same as that of the *filaments* along the axes of the wires, whatever the wire cross sections, as long as the distance between the wires is large compared with the cross sectional dimensions.

If the circuit consists of many elements, the calculation will be tedious on account of the large number of terms to be included. Formulas for special cases derived by this method are therefore of importance.

General Formula for the Inductance of a Triangle of Round Wire. The sides of the triangle are a, b, c, and the radius of cross section of the wire ρ.

$$L = 0.002 \left[a \log_e \frac{2a}{\rho} + b \log_e \frac{2b}{\rho} + c \log_e \frac{2c}{\rho} \right.$$
$$- (b+c) \sinh^{-1} \frac{c^2 + b^2 - a^2}{V} - (a+b) \sinh^{-1} \frac{a^2 + b^2 - c^2}{V}$$
$$- (a+c) \sinh^{-1} \frac{a^2 + c^2 - b^2}{V} - (a+b+c)$$
$$\left. + \frac{\mu}{4}(a+b+c) \right], \tag{56}$$

where
$$V^2 = 2(a^2b^2 + a^2c^2 + b^2c^2) - a^4 - b^4 - c^4. \tag{57}$$

60 CALCULATION OF MUTUAL INDUCTANCE AND SELF-INDUCTANCE

The sinh^{-1} functions may be evaluated by a table of hyperbolic functions or calculated from the relation sinh$^{-1} x = \log_e (x + \sqrt{1 + x^2})$, making use of the Auxiliary Table 2. For copper wire and other wires of nonmagnetic material, $\mu = 1$.

Rectangle of Round Wire.[34] The sides of the rectangle are a and b and the radius of cross section ρ.

$$L = 0.004 \left[a \log_e \frac{2a}{\rho} + b \log_e \frac{2b}{\rho} + 2\sqrt{a^2 + b^2} - a \sinh^{-1} \frac{a}{b} - b \sinh^{-1} \frac{b}{a} - 2(a + b) + \frac{\mu}{4}(a + b) \right]. \quad (58)$$

For copper and other nonmagnetic materials, $\mu = 1$.

Regular Polygons of Round Wire.[35] The side of the polygon is s, and the radius of cross section ρ.

Equilateral Triangle:

$$L = 0.006s \left[\log_e \frac{s}{\rho} - 1.40546 + \frac{\mu}{4} \right]. \quad (59)$$

Square:

$$L = 0.008s \left[\log_e \frac{s}{\rho} - 0.77401 + \frac{\mu}{4} \right]. \quad (60)$$

Pentagon:

$$L = 0.010s \left[\log_e \frac{s}{\rho} - 0.40914 + \frac{\mu}{4} \right]. \quad (61)$$

Hexagon:

$$L = 0.012s \left[\log_e \frac{s}{\rho} - 0.15152 + \frac{\mu}{4} \right]. \quad (62)$$

Octagon:

$$L = 0.016s \left[\log_e \frac{s}{\rho} + 0.21198 + \frac{\mu}{4} \right]. \quad (63)$$

The term in each case involving the permeability μ of the wire material represents the internal linkages of the flux in the cross section of the wire. For the usual case of nonmagnetic material, $\mu = 1$, and this term combines with the constant in the formula.

General Formula for the Calculation of the Inductance of Any Plane Figure. If the inductance formulas of the preceding section are expressed in each case in terms of the perimeter l of the figure, they are all found to take the general form

$$L = 0.002l \left[\log_e \frac{2l}{\rho} - \alpha + \frac{\mu}{4} \right]. \quad (64)$$

CIRCUITS COMPOSED OF COMBINATIONS OF STRAIGHT WIRES

The constant α in any case is characteristic of the figure and of its shape factor. Table 8 shows values of α for certain important cases.

TABLE 8. VALUES OF α FOR CERTAIN PLANE FIGURES

Isosceles Triangles				Rectangles		Regular Polygons			
ϵ	α	ϵ	α	β	α	N	α	N	α
5°	4.884	90°	3.331	0.05	4.494	3	3.197	13	2.506
10°	4.152	100°	3.426	.10	3.905	4	2.854	14	2.500
20°	3.690	110°	3.546	.15	3.598	5	2.712	15	2.495
30°	3.424	120°	3.696	.20	3.404	6	2.636	16	2.492
40°	3.284	130°	3.875	.25	3.270	7	2.591	17	2.489
50°	3.217	140°	4.105	0.3	3.172	8	2.561	18	2.486
60°	3.197	150°	4.399	.4	3.041	9	2.542	19	2.484
70°	3.214	160°	4.813	.5	2.962	10	2.529	20	2.482
80°	3.260	170°	7.514	.6	2.913	11	2.519	21	2.481
90°	3.331			.7	2.882	12	2.513	22	2.480
				.8	2.865			23	2.478
				0.9	2.856			24	2.477
				1.0	2.854			∞	2.452

ϵ is the angle between the equal sides of the triangle.
β is the ratio of the sides of the rectangle.
N is the number of the sides of the polygon.

It was noted by Bashenoff [36] that the value of α is approximately equal to $\log_e \frac{l^2}{S}$ in which l is the perimeter of the figure and S is the area enclosed. Accordingly Bashenoff recommended the use of the general formula

$$L = 0.002l \left[\log_e \frac{2l}{\rho} - \left(2 \log_e \frac{l}{\sqrt{S}} + \phi \right) + \frac{\mu}{4} \right] \qquad (65)$$

for the calculation of the inductance of a circuit consisting of a wire bent so as to enclose any desired plane figure. The value of ϕ necessary in any complicated case may be considered with sufficient accuracy as equal to that of some simpler nearly equivalent figure.

In order that two figures may be equivalent it is necessary that:
1. Their perimeters be equal. This assures equality of the self-inductances of the elements of the circuit. This is the primary condition.
2. The mutual inductances of the pairs of elements be the same. This condition requires that corresponding elements shall be equidistant and of the same length.

62 CALCULATION OF MUTUAL INDUCTANCE AND SELF-INDUCTANCE

Evidently this can be true only if the figures are identical, but approximate equality of inductance will be attained if the figures are similar in shape, and enclose equal areas. These considerations led Bashenoff to suggest that ϕ should be chosen in any complicated case as equal to that of a simpler figure somewhat resembling it in shape and having the same ratio $l/\sqrt{S}$. Tests of this method show that it gives a fair degree of accuracy in such cases where the value may be checked by the much more laborious summation method. For cases where the boundary of the plane figure is an enclosed curve or oval, no other method is available. In Table 9 are given values of ϕ as a function of the ratio $l/\sqrt{S}$ for certain simple figures that will serve for calculations

TABLE 9. DATA FOR THE CALCULATION OF THE INDUCTANCE OF POLYGONS OF ROUND WIRE

	Isosceles Triangles						Rectangles	
ϵ	$l/\sqrt{S}$	ϕ	ϵ	$l/\sqrt{S}$	ϕ	β	$l/\sqrt{S}$	ψ
0°	∞	0.3069	90°	4.828	0.1818	0	∞	0
5°	9.999	.2788	100°	5.033	.1943	0.05	9.392	0.0140
10°	7.379	.2548	110°	5.308	.2090	.10	6.957	.0258
20°	5.676	.2172	120°	5.672	0.2249	.15	5.939	.0355
30°	5.035	0.1912	130°	6.160	.2417	.20	5.367	.0438
40°	4.734	.1747	140°	6.843	.2587	.25	5.000	0.0507
50°	4.597	.1657	150°	7.863	0.2750	0.3	4.747	.0565
60°	4.559	0.1630	160°	9.599	.2897	.4	4.427	.0653
70°	4.591	.1654	170°	13.549	.3012	.5	4.243	0.0714
80°	4.682	.1719	180°	∞	0.3069	.6	4.131	.0755
90°	4.828	0.1818				.7	4.064	.0782
						.8	4.025	.0798
						0.9	4.006	.0807
	Special Trapezium					1.0	4.000	0.0809

γ	$l/\sqrt{S}$	ϕ	γ	$l/\sqrt{S}$	ϕ	Regular Polygons		
0	∞	0.3069	0.35	4.564	0.1458	N	$l/\sqrt{S}$	ϕ
0.05	9.392	.2742	0.4	4.427	.1332			
.10	6.957	.2452	.5	4.245	.1133			
.15	5.939	.2196	.6	4.131	.0995	3	4.559	0.1629
.20	5.367	.1971	.7	4.064	.0903	4	4.000	.0809
0.25	5.000	0.1775	.8	4.025	.0846	5	3.812	.0354
.30	4.747	.1605	.9	4.006	.0817	6	3.722	+0.0077
0.35	4.564	0.1458	1.0	4.000	0.0809	8	3.641	−0.0237
						∞	3.545	−0.0794

CIRCUITS COMPOSED OF COMBINATIONS OF STRAIGHT WIRES 63

for more complicated figures. From the tabulated data a graph may be prepared making possible the interpolation of ϕ with all the accuracy of which the method is capable.

If the figure is a triangle of sides a, b, c the perimeter is $l = a + b + c$, and the area S may be calculated by the well-known formula

$$S = \sqrt{\frac{l}{2}\left(\frac{l}{2} - a\right)\left(\frac{l}{2} - b\right)\left(\frac{l}{2} - c\right)}.$$

Thence follows the ratio $l/\sqrt{S}$ and from the values of ϕ given for an isosceles triangle in Table 9 the required value of ϕ for use in formula (65) may be interpolated. Two cases need to be distinguished: the angle ϵ between the equal sides of the isosceles triangle may be either acute or obtuse. The value of ϕ holding for the same value of $l/\sqrt{S}$ is not very different for the two different cases, but it is a simple matter to calculate and plot the dimensions of the equivalent isosceles triangle and judge which solution most nearly represents the given triangle.

If a plot of ϵ be made as a function of $l/\sqrt{S}$, the value of ϵ corresponding to the given value of this ratio may be interpolated. Then $2a = \dfrac{l}{1 + \sin \dfrac{\epsilon}{2}}$ and $c = l - 2a$, or we may use the relation $a^2 = \dfrac{2S}{\sin \epsilon}$.

Example 17: Given a triangle of measured sides 83, 60, 42.4 m. The perimeter is $l = 185.4$ and the area $S = \sqrt{92.7(9.7)(32.7)(50.3)} = 1216.2$ sq. m. From these values $l/\sqrt{S} = 5.316$. The curve of ϕ for an isosceles triangle having this ratio gives $\phi = 0.204$ if ϵ is acute, and $\phi = 0.209$ if ϵ is obtuse. Interpolating the values of ϵ corresponding to the given ratio of $l/\sqrt{S}$, they are found to give the two possible triangles

$\epsilon = 24°4$, $\epsilon = 110°6$,

$a = 76.7$, $a = 50.9$,

$c = 32.4$. $c = 83.8$.

Plotting these triangles together with the given one, the obtuse angled triangle is seen to be the better equivalent. Accordingly the inductance for the given triangle, calculated for wire of radius $\rho = 0.2$ cm., is found by formula (65) to be

$$L = 0.002(18,540) \left[\log_e \frac{2(18,540)}{0.2} - 2 \log_e 5.316 - 0.209 + \frac{1}{4}\right]$$

$$= 37.08[12.130 - 3.341 - 0.209 + \tfrac{1}{4}]$$

$$= 37.08(8.830) = 327.4 \ \mu\text{h}.$$

Calculating this case directly by the exact formula (56), the correct value of the inductance is found to differ from this value by less than a part in 10,000. The much simpler approximate method is, therefore, entirely satisfactory.

64 CALCULATION OF MUTUAL INDUCTANCE AND SELF-INDUCTANCE

Circuits forming polygons of four or more sides, if they approximate regular polygons, will have a ratio $l/\sqrt{S}$ less than 4. The value of ϕ may be interpolated from the curve drawn through the values of ϕ given in Table 9 for polygons of different numbers of sides with $l/\sqrt{S}$ as abscissas. The area S is, of course, to be calculated by dividing the polygon into triangles, calculating the area of each of these, and by taking the sum of these values.

Example 18: A polygon $ABCDEA$ plotted with the data $AB = 61.0$ m., $BC = 45.4$, $CD = 33.2$, $DE = 50.0$, and $EA = 62.4$, with the further measurements $AC = 85.2$, $AD = 88.8$, is found to resemble somewhat a regular pentagon. The perimeter is $l = 252.0$ m. and the area $S = 4528.7$ sq. m. Accordingly $l/\sqrt{S} = 3.861$ (not very different from the value for a regular pentagon). The value of ϕ interpolated from Table 9 is 0.042. Using this in (65), the inductance for wire of $\rho = 0.2$ cm. comes out as 501.1 μh. The value of ϕ for the regular pentagon is 0.036. It is found that a regular pentagon of the same perimeter differs in inductance from the given irregular figure by only about 6 parts in 10,000, the regular figure having slightly the larger inductance.

Irregular polygons will often be found to have values of $l/\sqrt{S}$ greater than 4. If the shape of the given figure suggests a rectangle, the value of ϕ may be obtained from Table 9 or by interpolation from a curve plotted from these values. A curve plotted for the shape ratio β of the rectangle as a function of $l/\sqrt{S}$ will enable the value of β corresponding to the given value to be obtained, and thence the sides a and b of the equivalent rectangle may be calculated from the relations

$$a = \frac{l}{2(1+\beta)}, \quad b = \frac{S}{a}, \quad b = \beta a.$$

If the figure differs markedly from the rectangle, owing to the presence of an acute angle, for example, and the ratio $l/\sqrt{S}$ lies between the value 4 and the minimum value 4.559 for a triangle, the special trapezium shown in Fig. 22 may give a useful equivalent. Values of ϕ as a function of $l/\sqrt{S}$ are given for this figure in Table 9 and from the corresponding values of $\gamma = \frac{b}{a}$ the value of γ may be interpolated and the dimensions of the equivalent figure in a given case be calculated from the relations

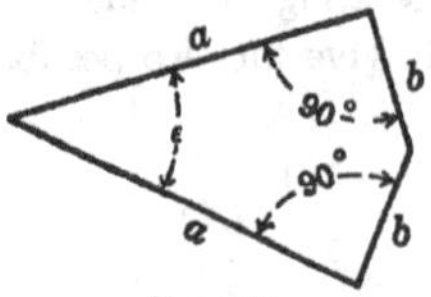

FIG. 22

$$a = \frac{l}{2(1+\gamma)}, \quad b = \frac{S}{a}, \quad \epsilon = 2\tan^{-1}\gamma.$$

Example 19: Bashenoff gave an example of a flattened pentagonal figure $ABCDEA$ for which the measurements were: $AB = 31.2$ m., $BC = 87.6$, $CD = 82.9$, $DE = 31.2$, $EA = 138.0$, with the further data that $AC = 95.16$ and $AD = 156.5$. The perimeter is $l = 370.9$ m. and the area $\dot{S} = 6658.5$ sq. m. The ratio is $l/\sqrt{S} = 4.546$.

When the figure is plotted it is seen to suggest an elongated rectangle. Data plotted from Table 9 for the rectangle indicates that the shape ratio of the rectangle having same value of $l/\sqrt{S}$ is $\beta = 0.36$, so that its sides are $a = \frac{370.9}{2(1.36)} = 136.3$, and $b = \frac{6658.5}{136.3} = 48.8$. The interpolated value of ϕ is 0.062 and the calculated value of L is 741.7 μh. This example was calculated also by the summation method, using formulas (7), (41), and (50), a laborious process. The result found was 3 parts in 1000 less. Evidently the Bashenoff method is accurate enough for practical purposes.

CIRCUITS COMPOSED OF COMBINATIONS OF STRAIGHT WIRES   65

Inductance of Circuits Enclosing Plane Curves. Circuits of wire bent so as to enclose oval areas and the like may be readily treated by the Bashenoff method. From the plotted curve may be found by planimeter or other instruments the perimeter and enclosed area. From these may be calculated the ratio $l/\sqrt{S}$. If this has a value less than 4, the value of ϕ may be interpolated from the curve for regular polygons. If the ratio is greater, the value may be obtained from the curve for rectangles or trapeziums. With the value of ϕ determined, the inductance is then given by formula (65).

Example 20: A wire of radius $\rho = 0.2$ is bent to form an ellipse. The semiaxes of the ellipse are as follows: major, $a = 100$ cm.; minor, $b = \dfrac{100}{\sqrt{2}}$. Then, since for an ellipse the eccentricity is given by $e^2 = \dfrac{a^2 - b^2}{a^2} = 1 - \dfrac{b^2}{a^2}$, the value here is $e = \dfrac{1}{\sqrt{2}}$.

The perimeter of an ellipse is $l = 4aE(e)$, where $E(e)$ is the complete elliptic integrate of second kind with modulus e. Tables of elliptic integrals, show that $E\left(\dfrac{1}{\sqrt{2}}\right) = 1.35064$. The area of an ellipse is $S = \pi ab$, so that for this case

$$\frac{l}{\sqrt{S}} = \frac{4aE(e)}{\sqrt{\pi ab}} = \frac{4(2)^{\frac{1}{4}}}{\sqrt{\pi}}(1.35064) = 3.625.$$

From Table 9 and the derived curve, $\phi = -0.034$ for this value, so that for the ellipse of wire formula (65) gives

$$L = 0.002l \left[\log_e \frac{2(400)(1.35064)}{0.2} - 2 \log_e 3.625 + 0.034 + \frac{1}{4} \right]$$

$$= 0.002l[6.019 + 0.034 + \tfrac{1}{4}] = 0.002l(6.303) = 6.81 \ \mu\text{h}.$$

For the same wire bent to form a circle we have by (64) and Table 8

$$L = 0.002l \left[\log_e \frac{2(400)(1.35064)}{0.2} - 2.452 + \frac{1}{4} \right]$$

$$= 0.002l(6.393) = 6.908 \ \mu\text{h}.$$

This illustrates the interesting fact that deforming a wire of fixed length from a circle to an ellipse of moderate eccentricity has a relatively small effect on the inductance. The decrease of inductance brought about by bringing nearer together part of the wire is partially offset by the increase of inductance occasioned by the increase of the distance between other parts of the circuit. The percentage change of the inductance would be smaller for the same value of eccentricity if the diameter of the wire were smaller.

Chapter 9

MUTUAL INDUCTANCE OF EQUAL, PARALLEL, COAXIAL POLYGONS OF WIRE

Let s = length of the side of the polygon,

d = distance between their planes.

Exact formulas are available for polygons of a few sides,[35, 37] and other cases could be treated without difficulty. These formulas are, however, long and tedious for computation, requiring the calculation of inverse hyperbolic and inverse trigonometrical functions. Routine calculations are much simplified by basing them on the tables for the mutual inductance of equal coaxial circular filaments, page 79. The procedure is as follows.

Calculate the diameter $2a$ of a circle that has the same perimeter as the given polygon. Making use of Table 16 or 17, obtain the value of f corresponding to the parameter $\dfrac{2a}{d}$ or $\dfrac{d}{2a}$, whichever is less than unity. The mutual inductance of the two equal circles is fa. The ratio F of the mutual inductance of the coaxial polygons to that of these coaxial circles is given in Table 10 as a function of $\dfrac{d}{s}$ or $\dfrac{s}{d}$. The summarized formulas for certain simple cases follow:

Equilateral Triangles:

$$\frac{2a}{d} = \frac{3}{\pi} \cdot \frac{s}{d}. \tag{66}$$

$$M = \frac{3s}{2\pi} fF. \tag{67}$$

Squares:

$$\frac{2a}{d} = \frac{4}{\pi} \cdot \frac{s}{d}. \tag{68}$$

$$M = \frac{4s}{2\pi} fF. \tag{69}$$

Hexagons:

$$\frac{2a}{d} = \frac{6}{\pi} \cdot \frac{s}{d}. \tag{70}$$

$$M = \frac{6s}{2\pi} fF. \tag{71}$$

Another possible method of treatment is to refer the mutual inductance of the polygons to that of coaxial equal circular filaments enclosing the same area. The resulting ratio F_1, which has to be tabulated, changes less regularly than F in Table 10 and cannot be so accurately interpolated.

TABLE 10. RATIOS FOR CALCULATING THE MUTUAL INDUCTANCE OF COAXIAL EQUAL POLYGONS

	Triangles		Squares		Hexagons			Triangles		Squares		Hexagons	
d/s	F	Diff.	F	Diff.	F	Diff.	d/s	F	Diff.	F	Diff.	F	Diff.
0	1.0000		1.0000		1.0000		0.75	0.4372		0.7085		0.8906	
0.05	0.7245	−605	0.8642	−280	0.9449	−99	.80	.4314	−58	.7035	−50	.8884	−22
.10	.6640	−423	.8362	−197	.9350	−67	.85	.4263	−51	.6988	−47	.8863	−21
.15	.6217	−327	.8165	−158	.9283	−52	.90	.4216	−47	.6941	−47	.8843	−20
.20	.5890	−266	.8007	−132	.9231	−43	0.95	.4175	−41	.6890	−42	.8823	−20
0.25	0.5624	−222	0.7875	−115	0.9188	−38	1.00	0.4138	−37	0.6861	−38	0.8802	−21
.30	.5402	−187	.7760	−102	.9150	−33	s/d						
.35	.5215	−161	.7658	−93	.9117	−30	1.0	0.4138	−72	0.6861	−78	0.8802	−41
.40	.5054	−140	.7565	−85	.9087	−30	0.9	.4066	−70	.6783	−82	.8761	−48
.45	.4914	−122	.7480	−78	.9057	−28	.8	.3996	−66	.6701	−88	.8713	−57
0.50	0.4792	−106	0.7402	−73	0.9029	−26	.7	.3930	−64	.6613	−88	.8656	−64
.55	.4686	−94	.7329	−67	.9003	−25	.6	.3866	−58	.6525	−86	.8592	−74
.60	.4592	−85	.7262	−62	.8978	−24	0.5	0.3808	−51	0.6439	−77	0.8518	−78
.65	.4507	−70	.7200	−60	.8954	−23	.4	.3757	−43	.6362	−73	.8440	−76
.70	.4437	−65	.7140	−55	.8931	−25	.3	.3714	−32	.6289	−68	.8364	−67
0.75	0.4372		0.7085		0.8906		.2	.3682	−20	.6221	−39	.8297	−54
							0.1	.3662	−7	.6182	−13	.8243	−18
							0	0.3655		0.6169		0.8225	

Polygons with a larger number of sides may also be treated by the use of circles having the same perimeter as the polygons. The following procedure leads to a moderate accuracy.

Generalizing formulas (66) to (71), they may be written in terms of the number of sides N of the polygon:

$$\frac{2a}{d} = \frac{N}{\pi} \cdot \frac{s}{d}, \tag{72}$$

$$M = \frac{Ns}{2\pi} fF. \tag{73}$$

Values of F may be obtained graphically by plotting the values in Table 8 against $\frac{1}{N}$, remembering that for the circle $\frac{1}{N} = 0$ and $F = 1$.

When the polygons are very close together, interpolation of F in Table 10 becomes inaccurate, and the following series formulas will prove accurate and convenient:[38]

Coaxial Triangles:

$$M = 0.006s \left[\log_e \frac{s}{d} - 1.4055 + 2.209 \frac{d}{s} - \frac{11}{12} \frac{d^2}{s^2} + \frac{203}{864} \frac{d^4}{s^4} - \cdots \right]. \tag{74}$$

Coaxial Squares:

$$M = 0.008s \left[\log_e \frac{s}{d} - 0.7740 + \frac{d}{s} - 0.0429 \frac{d^2}{s^2} - 0.109 \frac{d^4}{s^4} + \cdots \right]. \tag{75}$$

Coaxial Hexagons:

$$M = 0.012s \left[\log_e \frac{s}{d} - 0.15152 \right.$$
$$\left. + 0.3954 \frac{d}{s} + 0.1160 \frac{d^2}{s^2} - 0.052 \frac{d^4}{s^4} + \cdots \right]. \tag{76}$$

The preceding formulas apply strictly only to polygonal filaments. They may be used for polygons of round wire with negligible error and for wires of other cross sections also if the distance d is large compared with the cross sectional dimensions of the wire. If the polygons are composed of thick conductors and their planes are close together, the distance d in the series formulas should be replaced by the geometric mean distance of the cross sections of the conductors. For the simple cases of wire of rectangular cross section either Table 1 or 2 will be found useful. They give $\log_e R$ directly to replace $\log_e d$ in formulas (74), (75) or (76), and the geometric mean distance itself may be obtained from this using Auxiliary Table 1.

EQUAL, PARALLEL, COAXIAL POLYGONS OF WIRE

Example 21: Given two hexagons of round wire, the side of the hexagon being 20 cm., placed with their planes 25 cm. apart and with their sides parallel, calculate their mutual inductance by formulas (70) and (71).

For this case $\frac{s}{d} = \frac{20}{25} = 0.8$, and by (70) the equal circles of the same perimeter as the hexagons have a parameter $\frac{2a}{d} = \frac{6}{\pi}(0.8) = \frac{4.8}{\pi}$. Accordingly $\frac{d}{2a} = \frac{\pi}{4.8} = 0.6545$. To this value in Table 16 corresponds the value $f = 0.003225$. From Table 10, for $\frac{s}{d} = 0.8$, $F = 0.8713$, so that

$$M = \frac{120}{2\pi}(0.003225)(0.8713) = 0.05367 \ \mu h.$$

If the hexagons were moved nearer until the distance between their planes was only 2 cm., the new value of $\frac{d}{s} = 0.1$, and from Table 10, $F = 0.9350$. The parameter $\frac{d}{2a} = \frac{\pi}{60}$, and to this, from Table 16, corresponds $f = 0.029457$.

Thus,

$$M = \frac{120}{2\pi}(0.029457)(0.9350) = 0.5260 \ \mu h.$$

This value may be checked by formula (76). Substituting in this $\frac{d}{s} = 0.1$,

$$M = 0.012(20)[\log_e 10 - 0.15152 + 0.03954 + 0.00116]$$
$$= 0.24(2.19177) = 0.52602 \ \mu h,$$

agreeing with the result by the general method.

Suppose the two hexagons were replaced by two decagons of the same perimeter, spaced the same distance apart. The side s_1 would now be 12 cm. and $\frac{d}{s_1} = \frac{1}{6}$. Thus with $N = 10$ in (72), we would have the same value $\frac{d}{2a}$ as for the hexagons and, therefore, the same value of f.

Interpolating for $\frac{d}{s_1} = \frac{1}{6}$ in Table 10, we find for $N = 3$, $F = 0.6102$; for $N = 4$, $F = 0.8109$; and for $N = 6$, $F = 0.9265$. Plotting these values against $\frac{1}{N}$ with $F = 1$ for $\frac{1}{N} = 0$, the value $F = 0.97$ results for this example.

Thus the mutual inductance of the decagons is greater than that for the hexagons in the ratio of $F = 0.97$ to $F = 0.9350$. This gives a value $M = 0.5457 \ \mu h.$

Chapter 10

INDUCTANCE OF SINGLE-LAYER COILS ON RECTANGULAR WINDING FORMS

Formulas for Different Cases. The coil has N turns, each of which encloses a rectangular area a_1 by a cm. on a side, with a the greater. The pitch of the winding p and the axial length of the equivalent current sheet are related by the equation $b = Np$ (see Fig. 23). (This is approximately equal to the axial length of the coil.)

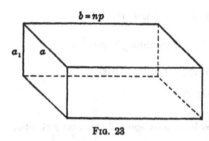

Fig. 23

To a close approximation, the inductance of the coil is equal to that of the rectangular current sheet with length b. The inductance of the rectangular current sheet was first derived by Niwa[30] in 1924. His general formula may be written in compact form as

$$L = 0.008 N^2 \frac{aa_1}{b} \left[\frac{1}{2} \frac{b}{a_1} \sinh^{-1} \frac{a}{b} + \frac{1}{2} \frac{b}{a} \sinh^{-1} \frac{a_1}{b} \right.$$

$$- \frac{1}{2}\left(1 - \frac{a_1^2}{b^2}\right) \frac{b}{a_1} \sinh^{-1} \frac{a}{b\sqrt{1 + \frac{a_1^2}{b^2}}}$$

$$- \frac{1}{2}\left(1 - \frac{a^2}{b^2}\right) \frac{b}{a} \sinh^{-1} \frac{a_1}{b\sqrt{1 + \frac{a^2}{b^2}}} - \frac{1}{2} \frac{a_1}{b} \sinh^{-1} \frac{a}{a_1}$$

$$\left. - \frac{1}{2} \frac{a}{b} \sinh^{-1} \frac{a_1}{a} + \left(\frac{\pi}{2} - \tan^{-1} \frac{aa_1}{b^2 \sqrt{1 + \frac{g^2}{b^2}}} \right) \right. \quad \text{(A)}$$

$$+ \frac{1}{3}\frac{b^2}{aa_1}\sqrt{1+\frac{g^2}{b^2}}\left(1-\frac{1}{2}\frac{g^2}{b^2}\right) + \frac{1}{3}\frac{b^2}{aa_1} - \frac{1}{3}\frac{b^2}{aa_1}\sqrt{1+\frac{a^2}{b^2}}\left(1-\frac{1}{2}\frac{a^2}{b^2}\right)$$

$$- \frac{1}{3}\frac{b^2}{aa_1}\sqrt{1+\frac{a_1{}^2}{b^2}}\left(1-\frac{1}{2}\frac{a_1{}^2}{b^2}\right) + \frac{1}{6}\frac{b}{aa_1}\left(\frac{g^3 - a^3 - a_1{}^3}{b^2}\right)\bigg]\,\mu\text{h},$$

in which $g^2 = a^2 + a_1{}^2$.

This expression is readily calculated by means of a table of hyperbolic functions. The antitangent term requires, in addition to a table of tangents, a standard table for converting from degrees to radians. The principal disadvantage of the formula lies in its length and, in some cases, in the near cancellation of terms of opposite signs. For the cases of very short solenoids and for long solenoids also, series developments have been given by Niwa,[40] which shorten the work of calculation.

It is convenient to express the inductance L_u of the current sheet in terms of the value L_∞ for an infinite solenoid having the same cross section so that $L_u = L_\infty F'$, the factor F' giving a measure of the effect of the ends. Expressed in this form

$$L_u = 0.004\pi N^2 \frac{aa_1}{b} F' \ \mu\text{h}. \tag{B}$$

For short solenoids Niwa writes

$$F' = \beta_1\gamma + \beta_1{}'\gamma \log_e \frac{1}{\gamma} + \beta_2\gamma^2 + \beta_3\gamma^3 - \beta_5\gamma^5 + \cdots, \tag{C}$$

in which

$$\gamma = \frac{b}{a} = \frac{\text{length of current sheet}}{\text{longer side of rectangle}},$$

and the coefficients are obtained from Table 11, due to Niwa, as functions of the ratio $k = \dfrac{a_1}{a}$.

This series converges only for values of γ smaller than k, so that, except for rectangles that depart little from squares, the range of (C) is small.

For long solenoids Niwa writes

$$F' = 1 - \alpha_1\epsilon + \alpha_2\epsilon^2 - \alpha_4\epsilon^4 + \alpha_6\epsilon^6 - \alpha_8\epsilon^8 + \cdots, \tag{D}$$

in which $\epsilon = \dfrac{a}{b}$, and the coefficients are functions of $k = \dfrac{a_1}{a}$. Their values may be taken from Table 12, which includes Niwa's values extended to include α_8 also. Formula (D) gives very accurate values and may be used with moderate accuracy even up to $\epsilon = 1$ and with all values of k.

It is necessary, therefore, for ordinary accuracy to use the general formula (A) only for short coils, where γ is greater than k, and also where k is small.

72 CALCULATION OF MUTUAL INDUCTANCE AND SELF-INDUCTANCE

TABLE 11. COEFFICIENTS, SHORT RECTANGULAR SOLENOID

$$\beta_1' = \frac{1}{\pi}\left(1 + \frac{1}{k}\right).$$

k	β_1	β_1'	β_2	β_3	β_5	β_7
1.00	0.4622	0.6366	0.2122	−0.0046	0.0046	− 0.0382
0.95	.4574	.6534	.2234	− .0046	.0053	
.90	.4512	.6720	.2358	− .0046	.0064	− .0525
.85	.4448	.6928	.2496	− .0042	.0080	
.80	.4364	.7162	.2653	− .0031	.0103	− .0838
0.75	0.4260	0.7427	0.2829	−0.0010	0.0141	
.70	.4132	.7730	.3032	+0.0026	.0198	− .1564
.65	.3971	.8080	.3265	.0085	.0291	
.60	.3767	.8488	.3537	.0179	.0432	− .3372
.55	.3500	.8970	.3858	.0331	.0711	
0.50	0.3151	0.9549	0.4244	0.0578	0.1183	− 0.7855
.40	+0.1836	1.1141	.5305	.1697	0.3898	− 2.403
.30	−0.0314	1.3359	0.7074	0.5433	2.0517	− 7.85
.20	−0.6409	1.9099	1.0610	2.3230	14.507	+15.51
0.10	−3.2309	3.5014	2.1220	22.548	497.36	14280

TABLE 12. VALUES OF COEFFICIENTS, RECTANGULAR SOLENOIDS

k	α_1	α_2	α_4	α_6	α_8
1.00	0.4732	0.1592	0.0265	0.0113	0.0102
0.95	.4612	.1512	.0240	.0097	.0085
.90	.4486	.1432	.0216	.0084	.0070
.85	.4356	.1353	.0194	.0072	.0058
.80	.4220	.1273	.0174	.0062	.0048
0.75	0.4077	0.1194	0.0155	0.0053	0.0040
.70	.3928	.1114	.0138	.0046	.0034
.65	.3773	.1035	.0123	.0040	.0028
.60	.3608	.0955	.0108	.0034	.0024
.55	.3435	.0875	.0095	.0029	.0020
0.50	0.3251	0.0796	0.0083	0.0025	0.0017
.45	.3056	.0716	.0072	.0022	.0015
.40	.2847	.0637	.0062	.0018	.0012
.35	.2623	.0557	.0052	.0016	.0011
.30	.2381	.0477	.0043	.0013	.0009
0.25	0.2117	0.0398	0.0035	0.0010	0.0007
.20	.1826	.0318	.0028	.0008	.0005
.15	.1499	.0239	.0020	.0006	.0004
.10	.1124	.0159	.0013	.0004	.0003
0.05	0.0669	0.0080	0.0007	0.0002	0.0001

SINGLE-LAYER COILS ON RECTANGULAR WINDING FORMS

Correction for Wire Insulating Space. The inductance of a *coil of wire* wound on a rectangular frame differs somewhat from that of the *equivalent current sheet*. Writing L_u for the inductance of the equivalent current sheet, the difference $\Delta L = L_u - L$ between L_u and the inductance L of the actual coil is given by

$$\Delta L = 0.002(a + a_1)N(G + H) \; \mu\text{h} \tag{E}$$

for a winding of round wire of bare diameter δ.

The values of G and H (see page 149) are to be obtained from Tables 38 and 39, for the given values of $\dfrac{\delta}{p}$ and N, respectively.

Example 22: A coil of 15 turns of wire 0.1 cm. in diameter is wound with a pitch of 0.2 cm. on a frame so that each turn encloses a rectangle 5 by 10 cm. Thus, $a_1 = 5$, $a = 10$, $p = 0.2$, $b = Np = 15(0.2) = 3$ cm. The parameters are, therefore, $\gamma = 0.3$, $k = 0.5$. Entering Table 11 with this value of k,

$\beta_1 = 0.3151,$ $\beta_3 = 0.0578,$ $\log_e \dfrac{1}{\gamma} = 1.2040,$

$\beta_1' = \dfrac{3}{\pi} = 0.9549,$ $\beta_5 = 0.1183,$

$\beta_2 = 0.4244,$ $\beta_7 = -0.785,$

so by the series (C)

$F' = 0.09453 + 0.34475 + 0.03820 + 0.00156 - 0.00029 - 0.00017$

$= 0.47858,$

and

$$L_u = 0.004\pi \, \frac{(10)(5)}{3} \, (15)^2 (0.47858)$$

$$= 225(0.10023) = 22.55 \; \mu\text{h}.$$

The exact formula (A), with the terms calculated in order, gives

$L_u = 0.008 \dfrac{(15)^2(10)(5)}{3} [0.57567 + 0.19257 + 0.69779 + 0.70112 - 1.20303$

$- 0.80202 + 0.60705 - 1.37622 + 0.06 + 0.95123 + 0.04535 + 0.30280]$

$= \dfrac{0.400(15)^2}{3} (0.75231) = 22.58 \; \mu\text{h}.$

For finding the correction for insulation we note that

$$\frac{\delta}{p} = \frac{0.1}{0.2} = 0.5, \quad N = 15,$$

and from Tables 38 and 39, $G = -0.1363$, $H = 0.2857$, so that $G + H = 0.1494$.

74 CALCULATION OF MUTUAL INDUCTANCE AND SELF-INDUCTANCE

Therefore, from formula (E)

$$\Delta L = 0.002(15)(15)(0.1494)$$
$$= 0.067 \, \mu h,$$

and the inductance of the coil by the exact formula is

$$L = 22.58 - 0.067 = 22.51 \, \mu h.$$

This correction is for many purposes of no importance.

Example 23: To illustrate the use of the series formula (D) for longer rectangular coils, there will be chosen a coil of 200 turns wound with a pitch of $p = 0.1$ cm. The rectangle formed by the turns has $a_1 = 7$, $a = 10$ cm., so that $k = 0.7$. The length of the equivalent current sheet is $b = 200(0.1) = 20$ cm., so that the parameter $\epsilon = \dfrac{a}{b} = 0.5$.

For $k = 0.7$, Table 12 yields

$$\alpha_1 = 0.3928, \qquad \alpha_6 = 0.0046,$$
$$\alpha_2 = 0.1114, \qquad \alpha_8 = 0.0034,$$
$$\alpha_4 = 0.0138,$$

so that by the series formula (D)

$$F' = 1 - 0.19640 + 0.02785 - 0.00086 + 0.00007 - 0.00001$$
$$= 0.8306,$$

and

$$L_u = 0.004\pi(200)^2 \frac{(10)(7)}{20} (0.8306)$$
$$= 1461.4 \, \mu h.$$

The exact formula (A) checks this to the last figure.

If the diameter of the bare wire is 0.75 of the pitch, then Tables 38 and 39 give $G = 0.2691$, $H = 0.3318$, so that $G + H = 0.6009$ and the correction is

$$\Delta L = 0.002(200)(17)(0.6009)$$
$$= 4.09 \, \mu h.$$

Therefore, $L = 1461.4 - 4.09 = 1457.3 \, \mu h$. The correction is only about 3 parts in 1000.

A test of formula (D) for $k = 0.7$ and $\epsilon = 1$ and the same values of a_1 and a as before gives $L_u = 0.0621 N^2$, whereas the value by the exact formula is $0.06227 N^2$. The agreement when $\epsilon = 1$ is still better for smaller values of k.

Coils on square forms may be calculated by the above formulas with $k = 1$. A simpler way is to treat them as a special case of polygonal solenoid (see page 171).

Part II

COILS AND OTHER CIRCUITS COMPOSED OF CIRCULAR ELEMENTS

Coils and circuits composed of circular elements are of great importance and especially those with coaxial arrangement of their turns. Such cases, for example, circular coils wound with rectangular cross sections, single-layer coils, coaxial solenoids and the like, all depend on certain basic ideal cases. The inductances of coils of many turns and coils with large cross sections may be obtained by integration and combination of the formulas for the basic cases.

For example, single-layer windings and coils of large axial dimension depend on the basic solenoid formula, while coils with many turns, wound in channels of rectangular cross section, depend upon the basic formula for the mutual inductance of coaxial circular filaments.

Chapter 11

MUTUAL INDUCTANCE OF COAXIAL CIRCULAR FILAMENTS

The circular filaments, Fig. 24, are supposed to have radii a and A, of which A is the larger, and the distance between their planes is denoted by d

The mutual inductance in this case is expressed in closed form in terms of complete elliptic integrals. Since the original formula given by Maxwell[41] numerous series developments have appeared, some applicable to circles near together, some applicable to distant circles, so that no less than a hundred formulas have been published,[42] and still others would be possible. By choice of a proper formula in any given case the mutual inductance may be calculated with an accuracy far beyond that necessary in practical problems.

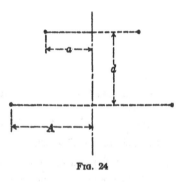

FIG. 24

For routine calculations, however, the tables here presented[43] give a simple solution of the problem, obviating the necessity of choice from among the existing formulas.

The mutual inductance depends upon the two parameters $\dfrac{a}{A}$ and $\dfrac{d}{A}$, while for pairs of circles having the same parameter values, the mutual inductance is proportional to the corresponding radius.

$$M = f\sqrt{Aa} = fA\sqrt{\frac{a}{A}}\ \mu\text{h},\tag{77}$$

in which f is to be obtained from Table 13 as a function of the variable

$$k'^2 = \frac{(A-a)^2 + d^2}{(A+a)^2 + d^2} = \frac{\left(1 - \dfrac{a}{A}\right)^2 + \dfrac{d^2}{A^2}}{\left(1 + \dfrac{a}{A}\right)^2 + \dfrac{d^2}{A^2}}.\tag{78}$$

78 CALCULATION OF MUTUAL INDUCTANCE AND SELF-INDUCTANCE

To realize the utmost accuracy of Table 13, third differences need to be taken into account. Generally, however, it will suffice to include second differences, and for rough values, linear interpolation will be satisfactory.

In Table 13 log f is given as well as f. This has the advantage of making possible a uniform percentage accuracy in the value of M for all parts of the table.

In extreme cases (very near circles and very distant circles), interpolation becomes difficult in Table 13. To provide for such cases Auxiliary Tables 14 and 15 (pages 81, 82) have been calculated. In Table 14 for circles very near together ($k'^2 \lessgtr 0.1$) advantage is taken of the fact that f is nearly proportional to log k'^2. For the other extreme ($k^2 = 1 - k'^2 \lessgtr 0.1$) Table 15 is provided, which makes use of the approximate proportionality of log f and log k^2 over this range. For direct calculation of k^2 the relation

$$k^2 = 1 - k'^2 = \frac{4aA}{(A+a)^2 + d^2} = \frac{4\frac{a}{A}}{\left(1+\frac{a}{A}\right)^2 + \frac{d^2}{A^2}} \qquad (79)$$

exists.

Example 24: Given circles of radii $a = 20$ and $A = 25$ cm., with their planes 10 cm. apart.

Here $\frac{a}{A} = 0.8$, $\frac{d}{A} = 0.4$, and $k'^2 = \frac{(1-0.8)^2 + (0.4)^2}{(1+0.8)^2 + (0.4)^2} = \frac{0.20}{3.40} = 0.058824$.

In Table 13, for $k'^2 = 0.050$, $f = 0.012026$, and the differences are $d_1 = -1009$, $d_2 = 171$. Interpolating, with $u = 0.8824$, $\frac{1-u}{2} = 0.059$, we find $f = 0.012026 + 0.8824\{-0.001009 - 0.059(0.000171)\} = 0.011126$. Similarly, there is interpolated the five-place logarithm, log $f = \bar{2}.04635$.

Thus, by (77),

$$M = 0.011126(25)\sqrt{0.8} = 0.24879 \ \mu h.$$

If only linear interpolation is used, there is found $f = 0.01136$, giving an inductance 9 parts in 10,000 greater.

These circles lie also within the bounds of Table 14, since log $k'^2 = \bar{2}.76956$. For log $k'^2 = \bar{2}.7$, the table gives $f = 0.012013$, with $d_1 = -1271$, $d_2 = 31$. Interpolating, with $u = 0.6956$, $f = 0.011126$, as before.

Example 25: If the dimensions are given in inches, it is necessary only to express the radius in centimeters. The shape ratios are independent of the units.

If $a = 2$ inches, $A = 5$ inches and $d = 4$ inches, for example, $\frac{a}{A} = 0.4$, $\frac{d}{A} = 0.8$, and $k'^2 = 0.38462$.

Interpolating from Table 13, $f = 0.0022884$, log $f = \bar{3}.35953$, and

$$M = 0.0022884(5 \times 2.54)\sqrt{0.4} = 0.018380 \ \mu h.$$

Since log $(5 \times 2.54) = 1.10380$, log $\sqrt{0.4} = \bar{1}.80103$, there is found log $M = \bar{2}.26436$, which gives $M = 0.018380$ also.

MUTUAL INDUCTANCE OF COAXIAL CIRCULAR FILAMENTS 79

TABLE 13. VALUES OF FACTOR f IN FORMULA (77)

$$M = f\sqrt{Aa}.$$

k'^2	f	Diff.	$\log f$	Diff.	k'^2	f	Diff.	$\log f$	Diff.
0.010	0.021474		$\bar{2}$.33191		0.260	0.003805		$\bar{3}$.58034	
		−4159		−9349			− 156		−1819
.020	.017315		.23842		.270	.003649		.56215	
		−2378		−6596			− 149		−1805
.030	.014937		.17246		.280	.003500		.54410	
		−1653		−4913			− 141		−1792
.040	.013284		.12333		.290	.003359		.52618	
		−1258		−4319			− 135		−1783
0.050	0.012026		$\bar{2}$.08014		0.300	0.003224		3.50835	
		−1009		−3807			− 129		−1773
.060	.011017		.04207		.310	.003095		.49062	
		− 838		−3437			− 124		−1767
.070	.010179		$\bar{2}$.00770		.320	.002971		.47295	
		− 715		−3162			− 118		−1760
.080	.009464		$\bar{3}$.97608		.330	.002853		.45535	
		− 621		−2946			− 113		−1757
.090	.008843		.94662		.340	.002740		.43778	
		− 546		−2772			− 108		−1754
0.100	0.008297		$\bar{3}$.91890		0.350	0.0026317		$\bar{3}$.42024	
		− 487		−2627			−1041		−1753
.110	.007810		.89263		.360	.0025276		.40271	
		− 439		−2509			−1000		−1753
.120	.007371		.86754		.370	.0024276		.38518	
		− 397		−2407			− 961		−1754
.130	.006974		.84347		.380	.0023315		.36764	
		− 363		−2321			− 924		−1756
.140	.006611		.82026		.390	.0022391		.35008	
		− 333		−2246			− 889		−1760
0.150	0.006278		$\bar{3}$.79780		0.400	0.0021502		$\bar{3}$.33248	
		− 308		−2181			− 856		−1765
.160	.005970		.77599		.410	.0020646		.31483	
		− 285		−2124			− 825		−1769
.170	.005685		.75475		.420	.0019821		.29712	
		− 265		−2074			− 795		−1778
.180	.005420		.73401		.430	.0019026		.27934	
		− 247		−2030			− 767		−1786
.190	.005173		.71371		.440	.0018259		.26148	
		− 232		−1991			− 740		−1796
0.200	0.004941		$\bar{3}$.69380		0.450	0.0017519		$\bar{3}$.24352	
		− 218		−1957			− 714		−1807
.210	.004723		.67423		.460	.0016805		.22545	
		− 205		−1926			− 689		−1819
.220	.004518		.65497		.470	.0016116		.20726	
		− 193		−1899			− 665		−1832
.230	.004325		.63598		.480	.0015451		.18894	
		− 183		−1875			− 643		−1846
.240	.004142		.61723		.490	.0014808		.17048	
		− 173		−1854			− 622		−1862
0.250	0.003969		$\bar{3}$.59869		0.500	0.0014186		$\bar{3}$.15186	
		− 164		−1835			− 601		−1879

CALCULATION OF MUTUAL INDUCTANCE AND SELF-INDUCTANCE

TABLE 13. VALUES OF FACTOR f IN FORMULA (77) (*Concluded*)

k'^2	f	Diff.	$\log f$	Diff.	k'^2	f	Diff.	$\log f$	Diff.
0.500	0.0014186		$\bar{3}$.15186		0.750	0.0003805		$\bar{4}$.58033	
		−601		−1879			−260		−3068
.510	.0013585		.13307		.760	.0003545		.54965	
		−581		−1898			−250		−3177
.520	.0013004		.11409		.770	.0003295		.51788	
		−561		−1917			−241		−3296
.530	.0012443		.09492		.780	.0003054		.48492	
		−543		−1939			−231		−3427
.540	.0011900		.07553		.790	.0002823		.45065	
		−526		−1962			−223		−3570
0.550	0.0011374		$\bar{3}$.05591		0.800	0.00025998		$\bar{4}$.41495	
		−509		−1987			−2139		−3730
.560	.0010865		.03604		.810	.00023859		.37765	
		−492		−2012			−2053		−3906
.570	.0010373		$\bar{3}$.01592		.820	.00021806		.33859	
		−476		−2041			−1966		−4105
.580	0.0009897		$\bar{4}$.99551		.830	.00019840		.29754	
		−461		−2071			−1881		−4326
.590	.0009436		.97480		.840	.00017959		.25428	
		−446		−2103			−1797		−4577
0.600	0.0008990		$\bar{4}$.95377		0.850	0.00016162		$\bar{4}$.20851	
		−432		−2137			−1712		−4867
.610	.0008558		.93240		.860	.00014450		.15986	
		−417		−2174			−1629		−5194
.620	.0008141		.91066		.870	.00012821		.10792	
		−405		−2213			−1545		−5577
.630	.0007736		.88853		.880	0.00011276		$\bar{4}$.05215	
		−391		−2254			−1461		−6028
.640	.0007345		.86599		.890	.00009815		$\bar{5}$.99187	
		−379		−2299			−1377		−6565
0.650	0.0006966		$\bar{4}$.84300		0.900	0.00008438		$\bar{5}$.92622	
		−366		−2346			−1292		
.660	.0006600		.81954		.910	.00007146		.85405	
		−354		−2398			−1206		
.670	.0006246		.79556		.920	.00005940		.77382	
		−343		−2451			−1116		
.680	.0005903		.77105		.930	.00004824		.68336	
		−332		−2510			−1026		
.690	.0005571		.74595		.940	.00003798		.57950	
		−320		−2573			−932		
0.700	0.0005251		$\bar{4}$.72022		0.950	0.00002866		$\bar{5}$.45732	
		−310		−2640			−831		
.710	.0004941		.69382		.960	.00002035		.30858	
		−299		−2714			−723		
.720	.0004642		.66668		.970	0.00001312		$\bar{5}$.11782	
		−289		−2791			−604		
.730	.0004353		.63877		.980	.00000708		$\bar{6}$.85035	
		−279		−2876			−459		
.740	.0004074		.61001		.990	0.00000249		$\bar{6}$.39551	
		−269		−2968			−249		
0.750	0.0003805		$\bar{4}$.58033		1.000	0			
		−260		−3068					

MUTUAL INDUCTANCE OF COAXIAL CIRCULAR FILAMENTS 81

TABLE 14. AUXILIARY TABLE FOR CIRCLES VERY CLOSE TOGETHER

$$k'^2 \lessgtr 0.1.$$

For still smaller values of k'^2 use the formula $f = 0.014468 \left(\log \frac{1}{k'^2} - 0.53307 \right).$

$\log k'^2$	f	Diff.	$\log k'^2$	f	Diff.	$\log k'^2$	f	Diff.
$\bar{6}.0$	0.079093		$\bar{4}.0$	0.050163		$\bar{2}.0$	0.021478	
		-1446			-1446			-1394
$\bar{6}.1$	.077647		$\bar{4}.1$	.048717		$\bar{2}.1$	.020084	
		-1447			-1445			-1384
$\bar{6}.2$	.076200		$\bar{4}.2$	.047272		$\bar{2}.2$	.018700	
		-1447			-1445			-1371
$\bar{6}.3$	.074753		$\bar{4}.3$	.045827		$\bar{2}.3$	.017329	
		-1447			-1445			-1357
$\bar{6}.4$	.073306		$\bar{4}.4$	.044382		$\bar{2}.4$	.015972	
		-1446			-1444			-1340
$\bar{6}.5$	0.071860		$\bar{4}.5$	0.042938		$\bar{2}.5$	0.014632	
		-1447			-1444			-1321
$\bar{6}.6$	.070413		$\bar{4}.6$	.041494		$\bar{2}.6$	.013311	
		-1447			-1443			-1298
$\bar{6}.7$	.068966		$\bar{4}.7$	.040051		$\bar{2}.7$	.012013	
		-1446			-1443			-1271
$\bar{6}.8$	.067520		$\bar{4}.8$	.038608		$\bar{2}.8$	0.010742	
		-1447			-1441			-1240
$\bar{6}.9$	.066073		$\bar{4}.9$	.037167		$\bar{2}.9$	.009502	
		-1446			-1440			-1205
$\bar{5}.0$	0.064626		$\bar{3}.0$	0.035727		$\bar{1}.0$	0.008297	
		-1447			-1439			
$\bar{5}.1$	.063180		$\bar{3}.1$	.034288				
		-1447			-1437			
$\bar{5}.2$	.061733		$\bar{3}.2$	.032851				
		-1446			-1435			
$\bar{5}.3$	.060287		$\bar{3}.3$	.031416				
		-1447			-1432			
$\bar{5}.4$	.058840		$\bar{3}.4$	.029984				
		-1446			-1430			
$\bar{5}.5$	0.057394		$\bar{3}.5$	0.028554				
		-1447			-1426			
$\bar{5}.6$	.055947		$\bar{3}.6$	.027128				
		-1447			-1421			
$\bar{5}.7$	.054500		$\bar{3}.7$	.025707				
		-1445			-1416			
$\bar{5}.8$	.053055		$\bar{3}.8$	.024291				
		-1446			-1410			
$\bar{5}.9$	.051609		$\bar{3}.9$	.022881				
		-1446			-1403			
$\bar{4}.0$	0.050163		$\bar{2}.0$	0.021478				

CALCULATION OF MUTUAL INDUCTANCE AND SELF-INDUCTANCE

TABLE 15. AUXILIARY TABLE FOR CIRCLES VERY FAR APART

$$k^2 \lessgtr 0.1.$$

For still smaller values of k^2 use the formula $\log f = \bar{3}.39224 + \frac{3}{2} \log k^2$.

$\log k^2$	$\log f$	Diff. d_1	Diff. d_2	$\log k^2$	$\log f$	Diff. d_1	Diff. d_2
$\bar{4}.0$	$\bar{9}.39227$			$\bar{3}.5$	$\bar{7}.64327$		
		15001				15027	
$\bar{4}.1$	.54228			$\bar{3}.6$	.79354		7
		15001				15034	
$\bar{4}.2$	.69229			$\bar{3}.7$	$\bar{7}.94388$		8
		15001				15042	
$\bar{4}.3$	.85230			$\bar{3}.8$	$\bar{6}.09430$		12
		15002				15054	
$\bar{4}.4$	$\bar{9}.99232$			$\bar{3}.9$	.24484		13
		15002				15067	
$\bar{4}.5$	$\bar{8}.14234$			$\bar{2}.0$	.39551		18
		15003				15085	
$\bar{4}.6$	.29237			$\bar{2}.1$	.54636		23
		15003				15108	
$\bar{4}.7$	.44240		1	$\bar{2}.2$	.69744		27
		15004				15135	
$\bar{4}.8$	.59244		2	$\bar{2}.3$	$\bar{6}.84879$		37
		15006				15172	
$\bar{4}.9$	.74250		1	$\bar{2}.4$	$\bar{5}.00051$		45
		15007				15217	
$\bar{3}.0$	$\bar{8}.89257$		1	$\bar{2}.5$	.15268		57
		15008				15274	
$\bar{3}.1$	$\bar{7}.04265$		3	$\bar{2}.6$	.30542		75
		15011				15349	
$\bar{3}.2$	.19276		2	$\bar{2}.7$	.45891		94
		15013				15443	
$\bar{3}.3$	.34289		4	$\bar{2}.8$	.61334		118
		15017				15565	
$\bar{3}.4$	.49306		4	$\bar{2}.9$	.76899		158
		15021				15723	
$\bar{3}.5$	$\bar{7}.64327$		6	$\bar{1}.0$	$\bar{5}.92622$		
		15027					

Special Case. Circles of Equal Radii. For circles of equal radii a, the tabulation of f may be made conveniently with respect to the parameter

$$\delta = \frac{\text{distance}}{\text{diameter}} = \frac{d}{2a} \text{ or } \Delta = \frac{\text{diameter}}{\text{distance}} = \frac{2a}{d},$$

whichever is less than unity.

Formula (77) becomes

$$M = fa \tag{80}$$

and f is to be taken from Table 16 or 17.

MUTUAL INDUCTANCE OF COAXIAL CIRCULAR FILAMENTS

TABLE 16. VALUES OF f FOR EQUAL CIRCLES NEAR TOGETHER

$$\delta = \frac{\text{distance}}{\text{diameter}} \leqq 1.$$

δ	f	Diff.	δ	f	Diff.	δ	f	Diff.	δ	f	Diff.
0.01	0.05016		0.26	0.010723		0.51	0.004800		0.76	0.0024659	
		−869			−383			−136			−599
.02	.04147		.27	.010340		.52	.004664		.77	.0024060	
		−508			−366			−132			−581
.03	.03639		.28	.009974		.53	.004532		.78	.0023479	
		−359			−347			−127			−563
.04	.03280		.29	.009627		.54	.004405		.79	.0022916	
		−277			−331			−122			−547
0.05	0.03003		0.30	0.009296		0.55	0.004283		0.80	0.0022369	
		−226			−314			−118			−531
.06	.02777		.31	.008980		.56	.004165		.81	.0021838	
		−189			−301			−114			−515
.07	.02588		.32	.008679		.57	.004051		.82	.0021323	
		−164			−289			−111			−500
.08	.02424		.33	.008390		.58	.003940		.83	.0020823	
		−143			−276			−106			−486
.09	.02281		.34	.008114		.59	.003834		.84	.0020337	
		−127			−264			−103			−472
0.10	0.021539		0.35	0.007850		0.60	0.003730		0.85	0.0019865	
		−1143			−253			−99			−458
.11	.020396		.36	.007597		.61	.003631		.86	.0019407	
		−1035			−243			−97			−445
.12	.019361		.37	.007354		.62	.003534		.87	.0018962	
		−944			−233			−93			−432
.13	.018417		.38	.007121		.63	.003441		.88	.0018530	
		−867			−223			−90			−421
.14	.017550		.39	.006898		.64	.003351		.89	.0018109	
		−800			−214			−88			−408
0.15	0.016750		0.40	0.006684		0.65	0.003263		0.90	0.0017701	
		−741			−207			−84			−397
.16	.016009		.41	.006477		.66	.003179		.91	.0017304	
		−690			−198			−82			−386
.17	.015319		.42	.006279		.67	.003097		.92	.0016918	
		−643			−190			−79			−376
.18	.014676		.43	.006089		.68	.003018		.93	.0016542	
		−603			−183			−77			−364
.19	.014073		.44	.005906		.69	.002941		.94	.0016178	
		−566			−176			−75			−356
0.20	0.013507		0.45	0.005730		0.70	0.002866		0.95	0.0015822	
		−532			−170			−72			−345
.21	.012975		.46	.005560		.71	.002794		.96	.0015477	
		−502			−164			−70			−336
.22	.012473		.47	.005396		.72	.002725		.97	.0015141	
		−473			−157			−68			−327
.23	.012000		.48	.005239		.73	.002657		.98	.0014814	
		−449			−152			−66			−318
.24	.011551		.49	.005087		.74	.002591		0.99	.0014496	
		−425			−146			−63			−310
0.25	0.011126		0.50	0.004941		0.75	0.002528		1.00	0.0014186	
		−403			−141			−62			

TABLE 17. VALUES OF f FOR EQUAL CIRCLES FAR APART

$$\Delta = \frac{\text{diameter}}{\text{distance}} \leqq 1.$$

Δ	f	Diff.	Δ	f	Diff.	Δ	f	Diff.	Δ	f	Diff.
1.00	0.0014186	−304	0.75	0.0007345	−235	0.50	0.00025999	−1377	0.25	0.00003683	−413
0.99	.0013882	−302	.74	.0007110	−231	.49	.00024622	−1335	.24	.00003270	−382
.98	.0013579	−300	.73	.0006879	−228	.48	.00023287	−1293	.23	.00002888	−353
.97	.0013279	−297	.72	.0006651	−224	.47	.00021994	−1251	.22	.00002535	−323
.96	.0012982	−296	.71	.0006427	−221	.46	.00020743	−1210	.21	.00002212	−296
0.95	0.0012686	−293	0.70	0.0006206	−217	0.45	0.00019533	−1168	0.20	0.000019165	−2687
.94	.0012393	−290	.69	.0005989	−214	.44	.00018365	−1126	.19	.000016478	−2429
.93	.0012103	−288	.68	.0005775	−210	.43	.00017239	−1085	.18	.000014049	−2184
.92	.0011814	−286	.67	.0005565	−206	.42	.00016154	−1044	.17	.000011865	−1949
.91	.0011529	−283	.66	.0005359	−202	.41	.00015109	−1003	.16	.000009916	−1827
0.90	0.0011246	−280	0.65	0.0005157	−198	0.40	0.00014106	−963	0.15	0.000008189	−1517
.89	.0010966	−278	.64	.0004959	−195	.39	.00013143	−922	.14	.000006672	−1319
.88	.0010688	−275	.63	.0004764	−191	.38	.00012221	−883	.13	.000005353	−1135
.87	.0010413	−272	.62	.0004573	−186	.37	.00011338	−843	.12	.000004218	−963
.86	.0010141	−270	.61	.0004387	−183	.36	0.00010495	−803	.11	.000003255	−806
0.85	0.0009871	−266	0.60	0.0004204	−179	0.35	0.00009692	−766	0.10	0.000002449	−661
.84	.0009605	−264	.59	.0004025	−175	.34	.00008926	−726	.09	.000001788	−531
.83	.0009341	−260	.58	.0003850	−170	.33	.00008200	−710	.08	.000001257	−414
.82	.0009081	−258	.57	.0003680	−167	.32	.00007510	−652	.07	.000000843	−311
.81	.0008823	−254	.56	.0003513	−163	.31	.00006858	−616	.06	.000000532	−224
0.80	0.0008569	−251	0.55	0.0003350	−158	0.30	0.00006242	−580	0.05	0.000000308	
.79	.0008318	−248	.54	.0003192	−154	.29	.00005662	−546			
.78	.0008070	−245	.53	.0003038	−150	.28	.00005116	−511			
.77	.0007825	−242	.52	.0002888	−146	.27	.00004605	−477			
.76	.0007583	−238	.51	.0002742	−142	.26	.00004128	−445			
0.75	0.0007345		0.50	0.0002600		0.25	0.00003683				

MUTUAL INDUCTANCE OF COAXIAL CIRCULAR FILAMENTS 85

Example 26: For two circles 4 inches in diameter, with their planes 2.5 inches apart, $\delta = \dfrac{d}{2a} = 0.625$, and from Table 16, we have, interpolating, including second differences, $f = 0.0034873$.

Thus
$$M = 0.0034873 \times (2.54 \times 2) = 0.017716 \, \mu h.$$

Table 18 gives values of f for values of $\log\left(\dfrac{\text{distance}}{\text{diameter}}\right) = \log \delta$ for equal circles so near together that Table 16 is not suitable, and for equal circles very far apart we have Auxiliary Table 19 (page 86) giving $\log f$ as a function of $\log\left(\dfrac{\text{diameter}}{\text{distance}}\right) = \log \Delta$.

TABLE 18. AUXILIARY TABLE FOR EQUAL CIRCLES VERY NEAR TOGETHER

$\delta \leqq 0.2.$

For still smaller values of δ use the formula $f = 0.028935 \left(\log \dfrac{1}{\delta} - 0.26654\right).$

log δ	f	Diff.	log δ	f	Diff.	log δ	f	Diff.
$\bar{4}.0$	0.10803		$\bar{3}.0$	0.07909		$\bar{2}.0$	0.05016	
		−289			−289			−289
$\bar{4}.1$	.10514		$\bar{3}.1$	.07620		$\bar{2}.1$	.04727	
		−289			−289			−289
$\bar{4}.2$	.10224		$\bar{3}.2$	.07331		$\bar{2}.2$	.04438	
		−289			−289			−288
$\bar{4}.3$	.09935		$\bar{3}.3$	.07041		$\bar{2}.3$	.04150	
		−289			−289			−289
$\bar{4}.4$	.09645		$\bar{3}.4$	.06752		$\bar{2}.4$	.03861	
		−289			−289			−288
$\bar{4}.5$	0.09356		$\bar{3}.5$	0.06463		$\bar{2}.5$	0.03573	
		−289			−289			−287
$\bar{4}.6$	.09067		$\bar{3}.6$	.06173		$\bar{2}.6$	.03286	
		−289			−289			−286
$\bar{4}.7$	.08777		$\bar{3}.7$	.05884		$\bar{2}.7$	.03000	
		−289			−289			−285
$\bar{4}.8$	.08488		$\bar{3}.8$	.05595		$\bar{2}.8$	.02715	
		−289			−289			−282
$\bar{4}.9$	.08199		$\bar{3}.9$	.05306		$\bar{2}.9$	.02433	
		−289			−289			−279
$\bar{3}.0$	0.07909		$\bar{2}.0$	0.05016		$\bar{1}.0$	0.02154	
								−274
						$\bar{1}.1$	.01880	
								−268
						$\bar{1}.2$	.01612	
								−259
						$\bar{1}.3$	0.01353	

CALCULATION OF MUTUAL INDUCTANCE AND SELF-INDUCTANCE

TABLE 19. AUXILIARY TABLE FOR EQUAL CIRCLES VERY FAR APART

$\Delta \gtrless 0.2$.

For still smaller values of Δ use the formula $\log f = \bar{3}.39224 + 3 \log \Delta$.

log Δ	log f	Diff.	log Δ	log f	Diff.
$\bar{2}.0$	$\bar{9}.39221$		$\bar{1}.00$	$\bar{6}.38900$	
		29998			14916
$\bar{2}.1$	$\bar{9}.69219$		$\bar{1}.05$	$\bar{6}.53816$	
		29997			14895
$\bar{2}.2$	$\bar{9}.99216$		$\bar{1}.10$	$\bar{6}.68711$	
		29995			14868
$\bar{2}.3$	$\bar{8}.29211$		$\bar{1}.15$	$\bar{6}.83579$	
		29992			14835
$\bar{2}.4$	$\bar{8}.59203$		$\bar{1}.20$	$\bar{6}.98414$	
		29988			14793
$\bar{2}.5$	$\bar{8}.89192$		$\bar{1}.25$	$\bar{5}.13207$	
		29980			14741
$\bar{2}.6$	$\bar{7}.19172$		$\bar{1}.30$	$\bar{5}.27948$	
		29970			
$\bar{2}.7$	$\bar{7}.49142$				
		29952			
$\bar{2}.8$	$\bar{7}.79094$				
		29924			
$\bar{2}.9$	$\bar{6}.09019$				
		29881			
$\bar{1}.0$	$\bar{6}.38900$				

The tables make clear how rapidly the mutual inductance increases as the distance between the circles is decreased. For circles near together, the mutual inductance varies nearly in proportion to the logarithm of the ratio of the largest to the smallest distance between the circumferences.

Figs. 46 and 47 (pages 189, 190), which are extensions of curves given by Curtis and Sparks,[44] are useful for purposes of orientation. In these the radius A of one of the circles is taken as unit distance. Each curve is the locus of the positions and radii of the circles that have a mutual inductance with the unit circle such that M in abhenrys divided by A is equal to the ratio to which the curve refers. Evidently the curves represent lines of flux, due to a current in the circle of radius A. For instance, the curve for $\dfrac{M}{A} = 5.0$ shows that circles having $\left(\dfrac{a}{A}, \dfrac{d}{A}\right)$ coordinates of (0.48, 0), (0.6, 0.5), (1, 0.99), and (2, 1.44) all have this same value of mutual inductance (0.005 μh) on a coaxial circle of radius 1 cm. It is evident also that no circle more distant than $1.47A$ can have a mutual inductance as large as this, nor can a coplanar circle of radius greater than $4.04A$ have as great a value of mutual inductance as this. In the case of this larger limiting circle, the flux lines corresponding to the ovals, designated by values of $\dfrac{M}{A}$, greater than 5.0 contribute nothing, since each flux line threads the circle in both directions.

MUTUAL INDUCTANCE OF COAXIAL CIRCULAR FILAMENTS

As an example of the use of Fig. 47, suppose it is desired to find the sizes and distance of two equal circles to have a mutual inductance of 0.050 μh. For equal circles of radius 2.5 cm. the value of $\frac{d}{A}$ to give this mutual inductance will be found on the curve 50 divided by 2.5, that is, 20 with $\frac{a}{A} = 1$. The value of $\frac{d}{A}$ is about 0.23, so that the value of d is 0.23 × 2.5 = 0.575 cm. Circles of 10 cm. radius are found from the curve $\frac{M}{A} = 5.0$ to require a separation of $d = 0.99 \times 10 = 9.9$ cm.

If a circle of radius 5 cm. is to have a mutual inductance of 0.030 μh with a larger circle of radius 10 cm., the curve $\frac{M}{A} = 3.0$ is used. For $\frac{a}{A} = \frac{1}{2}$, the value of $\frac{d}{A}$ is seen to be 0.60, so that $d = 6.0$.

Chapter 12

MUTUAL INDUCTANCE OF COAXIAL CIRCULAR COILS

Mutual Inductance of Coaxial Circles of Wire. The formulas in the preceding section apply exactly only to circular *filaments* of negligible cross section. However, these formulas give the mutual inductance of actual *circles* of *wire* placed coaxially with a high degree of accuracy if for the wires the filaments located at the centers of the cross sections of the wires are employed. This is also true if the separation of the planes of the wires is large, compared with the dimensions of the cross section of the wire, whatever the form of the cross section.

For circles of wire near together the formulas suffice also, if the wire has a circular cross section, for the reason that the principal term of the mutual inductance formula for coaxial circular filaments involves the logarithm of the ratio of the smaller radius to the smallest distance between their circumferences. This suggests that two circles of wire of appreciable cross section are approximately equivalent, as far as mutual inductance is concerned, to two circular filaments so situated that the shortest distance between them is equal to the geometric mean distance of the cross sections. Two equal circles of wire may therefore be replaced by two circles of the same mean radius separated by the geometric mean distance of their cross sections.

For circular cross sections the geometric mean distance is equal to the actual distance between centers, so that the assumption of their central filaments is sufficient, even when the planes of the wires are close together.

For equal rectangular cross sections, two circles of wire of equal mean radius may be replaced by circular filaments with this radius and with a distance between their planes equal to the g.m.d. R of the two rectangular cross sections. Tables 1 and 2 give $\log_e k$ in the formula $\log_e R = \log_e p + \log_e k$ where p = the distance between the centers of the rectangles. The value of R thus derived is to be used in place of d in calculating the mutual inductance of the equivalent coaxial circular filaments by Tables 16 to 19.

MUTUAL INDUCTANCE OF COAXIAL CIRCULAR COILS

Example 27: Given two equal circles of wire of rectangular cross section having axial and radial dimensions of 2 cm. and 1 cm. respectively. The distance between the planes of the central filaments is 5 cm. and the mean radius of each wire is 10 cm. Here in Table 1 we have to use $p = 5, C = 2, B = 1, \gamma = 0.4, \dfrac{1}{\Delta} = 0.5$. The table gives $\log_e k = -0.0102$, so that $\log k = -0.434 \times 0.0102 = -0.00443$. Thus $\log R = \log 5 - 0.00443 = 0.69454$, and R is found to be 4.949. The two wires are to be replaced by filaments of mean radius 10 cm. separated by a distance of 4.949 cm. between their planes. Thus in Table 16 we are to find f for $\delta = 0.4949$. The interpolated value is 0.005015. The value of δ for the central filaments is 0.5, to which corresponds $f = 0.004941$. Therefore the mutual inductance of the wires is greater than that of the central filaments by 74 parts in 4941 or about 15 parts in 1000.

The general case of circular wires of unequal rectangular cross sections or of unequal mean radii, or both, comes under the case of circular *coils* of rectangular cross section which follows.

Mutual Inductance of Coaxial Circular Coils of Rectangular Cross Section. The mutual inductance of the coils is to a first approximation

$$M = N_1 N_2 M_0, \qquad (81)$$

in which M_0 is the mutual inductance of the central filaments of the coils, and N_1 and N_2 are the numbers of turns on the coils. That is, formula (77) is to be used with a and A for the mean radii of the coils and for d the distance between the median planes of the coils.

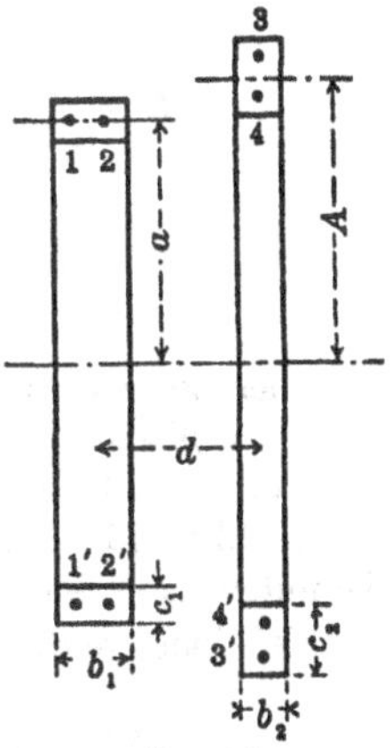

If the cross sectional dimensions are appreciable compared with the distance between the coils, the value of M calculated by (81) is too small, since the turns in the nearer portions of the coils contribute more to the total mutual inductance than is compensated for by the smaller contributions of the more distant turns.

A more accurate value of M is obtained by the use of Lyle's method,[12] which replaces each coil by two equivalent filaments with half of the turns supposed to be associated with each.

If the axial cross sectional dimension b_1 is greater than the radial c_1, Fig. 25, the equivalent filaments 1, 1' and 2, 2' have an equivalent radius r_1 slightly larger than the mean radius a, and the two filaments are located at an axial distance β on either side of the median plane. The defining equations for r_1 and β are

Fig. 25

$$r_1 = a\left(1 + \frac{1}{24}\frac{c_1^2}{a^2}\right),$$

$$\beta^2 = \frac{b_1^2 - c_1^2}{12}. \qquad (82)$$

CALCULATION OF MUTUAL INDUCTANCE AND SELF-INDUCTANCE

If the second coil has also its axial dimension b_2 greater than the radial c_2, the equivalent radius r_2 and the axial displacement of the equivalent filament are to be found from (82) with A in place of a, and the dimensions b_2, c_2 replacing b_1, c_1. If, however, the radial dimension c_2 is greater than the axial b_2, the coil is to be replaced by two coplanar circular filaments 3, 3' and 4, 4', half of the turns of the coil being assumed with each. The radii of the equivalent filaments are $r_2 + \delta$ and $r_2 - \delta$, respectively, where

$$r_2 = A\left(1 + \frac{1}{24}\frac{b_2{}^2}{A^2}\right),$$

$$\delta^2 = \frac{c_2{}^2 - b_2{}^2}{12}.$$
(83)

If coil 1 should also have a greater radial dimension c_1 than the axial b_1, the equivalent filaments are also coplanar with radii $r_1' + \delta_1$ and $r_1' - \delta_1$, the values of r_1' and δ_1 being obtained from (83) with a in place of A and with b_1, c_1 in place of b_2, c_2, respectively.

The mutual inductance of the coils is then given by the formula

$$M = \frac{N_1}{2} \cdot \frac{N_2}{2}(M_{13} + M_{14}) + \frac{N_1}{2} \cdot \frac{N_2}{2}(M_{23} + M_{24})$$

$$= N_1 N_2 \left(\frac{M_{13} + M_{14} + M_{23} + M_{24}}{4}\right),$$
(84)

or, otherwise expressed, the mutual inductance of the coils is obtained from (81) by replacing M_0 by the average of the mutual inductances of the four pairs of coaxial circular filaments, each one being calculated by (77) and Tables 13 to 15. The proper radii and separations to be used are shown in the following scheme for the coils of Fig. 25. Schemes for the other possible combinations are evident.

Filaments	Radii	Distance between Planes
1, 1' and 3, 3'	$r_1, r_2 + \delta$	$d + \beta$
1, 1' and 4, 4'	$r_1, r_2 - \delta$	$d + \beta$
2, 2' and 3, 3'	$r_1, r_2 + \delta$	$d - \beta$
2, 2' and 4, 4'	$r_1, r_2 - \delta$	$d - \beta$

Special Case. Equal Coils of Square Cross Section. An important case is that of equal coils of square cross section (see Fig. 26). Here β and δ are equal to zero and the equivalent filaments are only two, one for each coil.

MUTUAL INDUCTANCE OF COAXIAL CIRCULAR COILS

The modified radius of each is

$$r = a\left(1 + \frac{1}{24}\frac{c^2}{a^2}\right) = a\left[1 + \frac{1}{6}\left(\frac{c}{2a}\right)^2\right], \quad (85)$$

and the mutual inductance is

$$M = N^2 M_0 \quad (86)$$

in which M_0 is the mutual inductance of the coaxial circles of radius r and separation d.

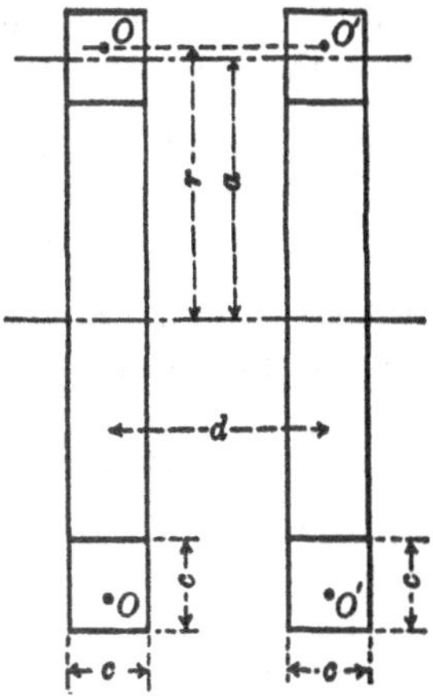

Example 28: Assume two coils of 200 turns each, mean radii $a = 10$ cm., separation between their median planes $d = 6$ cm., and side of the square cross section $c = 2.5$ cm.

For this case

$$r = a[1 + \tfrac{1}{24} \cdot \tfrac{1}{16}] = 10(1 + \tfrac{1}{384}) = 10.026 \text{ cm.}$$

The ratio

$$\frac{\text{distance}}{\text{diameter}} = \frac{d}{2r} = \frac{6}{20.052} = 0.29922.$$

Fig. 26

Interpolating from Table 16, $f = 0.009322$, and therefore from (80) and (86)

$$M = (200)^2 \times 10.026 \times 0.009322 = 3738.4 \ \mu\text{h} = 3.7384 \text{ mh.}$$

The formula (81) which employs the central filaments without correction uses Table 16 with $\dfrac{\text{distance}}{\text{diameter}} = \dfrac{6}{20} = 0.3$ and there is found $f = 0.009296$ so that $M = (200)^2 \times 10 \times 0.009296 = 3718.4 \ \mu\text{h} = 3.7184$ mh, or 0.53 per cent less.

Mutual Inductance of Brooks Coils. If the two equal coils have the shape proposed by Brooks [45] (essentially most economical coils, see p. 97) it is convenient to use tabulated values of the coefficient of coupling k in the equation

$$M = kL, \quad (87)$$

in which L is the inductance of each coil. Substituting for L from formula (94),

$$M = 0.016994 k a N^2, \quad (88)$$

in which $a =$ mean radius of each coil (Fig. 27), $N =$ number of turns of each coil and k is to be taken from Table 20 with the parameter $\alpha = \dfrac{d_0}{c}$ as argument.[46]

This ratio α expresses the distance between the windings as a fraction of the

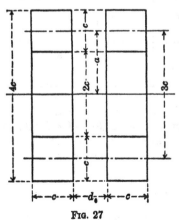

Fig. 27

92 CALCULATION OF MUTUAL INDUCTANCE AND SELF-INDUCTANCE

side of the winding channel, see Fig. 27. In this figure, which is drawn to scale, the mean diameter $2a$ of the windings and the side of the winding channel c have the Brooks ratio $\dfrac{c}{2a} = \dfrac{1}{3}$.

Evidently interpolation from Table 20 is not accurate for $\dfrac{1}{\alpha}$ less than 0.5. It is better in such cases to obtain the mutual inductance by the method of equivalent filaments, formulas (85), (86) and (80), pages 82 and 91.

TABLE 20. COUPLING COEFFICIENTS OF BROOKS COILS

α	k	d_1	d_2	$\dfrac{1}{\alpha}$	k	d_1	d_2
0	0.5036			1.0	0.19263		
		−511				−1738	
0.1	.4525		63	0.9	.17525		−171
		−448				−1909	
0.2	.4077		56	0.8	.15616		−181
		−392				−2090	
0.3	.3685		47	0.7	.13526		−187
		−345				−2277	
0.4	.3340		37	0.6	.11249		−165
		−308				−2442	
0.5	0.3032		34	0.5	0.08807		−102
		−274				−2544	
0.6	.2758		30	0.4	.06263		+ 51
		−244				−2493	
0.7	.2514		26	0.3	.03770		353
		−218				−2140	
0.8	.2296		23	0.2	.01630		809
		−195				−1331	
0.9	.2101		20	0.1	0.00299		1032
		−175				− 299	
1.0	0.1926			0	0		

Design of Equal Coaxial Coils of Square Cross Section. The approximate design may be based on the simple formula (81), $M = N^2 M_0$, in which M_0 is the mutual inductance of the central filaments that may be obtained from the diameter $2a$ and the separation d of the central filaments and Table 16, 17, 18, or 19.

The parameters $\gamma = \dfrac{c}{2a}$ and $\epsilon = \dfrac{d}{2a}$ are assumed to be given, together with the outside diameter of the wire δ. Then the number of turns and the mean diameter $2a$ of the coils are to be determined in order that the mutual inductance may have a desired value M in microhenrys.

Since $M_0 = fa$, Table 16 will give f for the assumed value of $\dfrac{d}{2a}$. The design formula is

$$(2a)^5 = \frac{2M}{f}\left(\frac{\delta}{\gamma}\right)^4 \tag{89}$$

Having calculated $(2a)$, the other data follow at once,

$$c = \gamma(2a), \quad d = \epsilon(2a), \quad N = \left(\frac{c}{\delta}\right)^2.$$

Example 29: To design a coil of 10 mh with a section ratio $\gamma = \frac{1}{4}$, space ratio $\epsilon = \frac{1}{2}$, to be wound with a wire of which 20 turns occupies 1 cm. of linear dimension.

For $\epsilon = \dfrac{d}{2a} = \dfrac{1}{2}$, Table 16 gives $f = 0.004941$.

Then $(2a)^5 = \dfrac{20{,}000}{0.004941}\left(\dfrac{0.05}{0.25}\right)^4$. From this

$2a = 5.785$ cm.,
$c = 1.446$ cm.,
$d = 2.892$ cm.,
$N = 28.92 \times 28.92 = 836$ turns.

$\log 2M = 4.30103$
$\log f = \overline{3}.69379$
$\log 2M/f = 6.60724$
$4 \log \dfrac{0.05}{0.25} = \overline{3}.20412$
$\log (2a)^5 = 3.81136$
$\log 2a = 0.76227.$

The coils could be wound with 29 turns per layer in a channel 1.45 cm. wide and with 29 layers lacking 5 turns. The inner diameter of the channel would be 4.34 cm., the outer diameter of the winding 7.23 cm., and the windings of the two coils would be separated by a clearance of 1.446 cm. between the channels.

Design of a Mutual Inductance Composed of Two Equal Brooks Coils.[45] This is evidently a special case of the foregoing with $\gamma = \dfrac{c}{2a} = \dfrac{1}{3}$. However, it is simpler to make use of Table 20 with a choice of a suitable spacing ratio to give a desired ratio of the mutual inductance to the self-inductance of the coils. In selecting the spacing ratio, mechanical considerations must be considered, since sufficient material must be left between the channels of the two coils to secure the necessary mechanical strength. From Table 20 it appears that, leaving a space between coils half of the width of the winding channels, a mutual inductance of about 0.3 of the value of the coil inductances may be attained. A good assumption for design purposes is that the ratio of mutual inductance to coil inductance shall be one quarter. This fixes the clearance between windings as 0.7 of the channel width approximately. The design of the coils then follows from the foregoing section with $\gamma = \frac{1}{3}$ and $\epsilon = 0.567$, or from formula (93) below, assuming each coil inductance to be made equal to four times the desired value of the mutual inductance.

Chapter 13

SELF-INDUCTANCE OF CIRCULAR COILS OF RECTANGULAR CROSS SECTION

Circular coils of rectangular cross section combine the advantages of large attainable inductance with simplicity of construction. By turning a simple channel in a circular disc of insulating material, a winding form is provided that enables a considerable number of turns to be wound in close proximity to one another. Thus each individual turn has a relatively large mutual inductance with the others, and large inductance of the whole coil results for the amount of wire used. Such a winding has, however, the disadvantage of a large distributed capacitance as compared with single-layer coils.

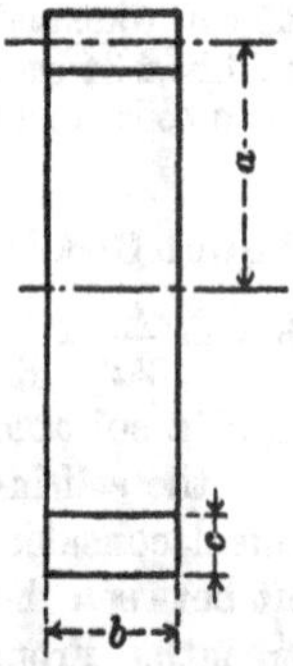

Nomenclature.

a = mean radius of the turns,
b = axial dimension of the cross section,
c = radial dimension of the cross section,
N = total number of turns,
n_b = number of turns per layer,
n_c = number of layers,
p_b = pitch of the winding in the layer = distance between centers of adjacent turns in the layer,
p_c = distance between centers of corresponding wires in consecutive layers.

Fig. 28

Then the relations exist that

$$b = n_b p_b, \quad c = n_c p_c, \quad N = n_b n_c. \tag{90}$$

For closely wound coils $p_b = p_c$, and if δ = the diameter of the covered wire

$$b = \delta n_b, \quad c = \delta n_c, \quad N = \frac{bc}{\delta^2}.$$

SELF-INDUCTANCE OF CIRCULAR COILS 95

The inductance is a function of the shape of the coil so that two shape ratios, such as $\frac{c}{2a}$ and $\frac{b}{c}$, are involved. Two coils, having the same shape ratios, have inductances that are proportional to the product aN^2.

The problem of calculating the inductance in the general case has received considerable attention.[47] Some of the known formulas apply to thin coils $\left(\frac{c}{2a} \text{ small, and } \frac{b}{c} \text{ large}\right)$; others apply to thick coils $\left(\frac{c}{2a} \text{ large, and } \frac{c}{b} \text{ large}\right)$. For the most part, calculations are not simple and in some cases the formulas are very complicated. To obviate the necessity of choice of a suitable formula in any given case, and to save the labor of a tedious calculation, tables will here be provided that reduce the computation to the simplest form. First, however, it is profitable to consider the simple and very important case of coils of square cross section ($b = c$).

Inductance of Circular Coils of Square Cross Section. Here the parameter $\frac{c}{2a}$ may have any desired value between 0 and 1, but $\frac{b}{c} = 1$. The inductance is given by

$$L = 0.001aN^2P_0' \quad \mu\text{h}, \tag{91}$$

in which P_0' is a function of $\frac{c}{2a}$ alone, and may be interpolated from Table 21.

The variation of P_0' with increasing $\frac{c}{2a}$, shown by the table, gives a picture of the decrease of the inductance as the turns are separated more and more to fill larger and larger cross sections, the mean radius a and the number of turns N being maintained constant.

For relatively small cross sections, so that $\frac{c}{2a} < 0.2$, interpolation in Table 21 becomes difficult. In such instances, it is about as easy to calculate P_0' directly by the formula [48]

$$P_0' = 4\pi \left[\frac{1}{2}\left\{1 + \frac{1}{6}\left(\frac{c}{2a}\right)^2\right\} \log_e \frac{8}{\left(\frac{c}{2a}\right)} - 0.84834 + 0.2041\left(\frac{c}{2a}\right)^2 \right]. \tag{92}$$

The logarithm may be obtained from a five-place table of common logarithms and the natural logarithm obtained from the common by Auxiliary Table 2 (p. 237).

CALCULATION OF MUTUAL INDUCTANCE AND SELF-INDUCTANCE

TABLE 21. VALUES OF CONSTANT P_0' IN FORMULA (91) FOR COILS OF SQUARE CROSS SECTION

$\frac{c}{2a}$	P_0'	d_1	$\frac{c}{2a}$	P_0'	d_1	$\frac{c}{2a}$	P_0'	d_1	$\frac{c}{2a}$	P_0'	d_1
0	∞		0.25	20.304		0.50	12.666		0.75	9.022	
					−460			−197			−102
0.01	60.277		.26	19.844		.51	12.469		.76	8.920	
					−440			−193			−100
.02	51.570		.27	19.404		.52	12.276		.77	8.820	
					−422			−186			−97
.03	46.480		.28	18.982		.53	12.090		.78	8.723	
					−405			−182			−95
.04	42.873		.29	18.577		.54	11.908		.79	8.628	
					−388			−176			−93
0.05	40.078		0.30	18.189		0.55	11.732		0.80	8.535	
					−375			−173			−90
.06	37.798		.31	17.814		.56	11.559		.81	8.445	
					−360			−167			−88
.07	35.873		.32	17.454		.57	11.392		.82	8.357	
					−348			−162			−86
.08	34.208		.33	17.107		.58	11.230		.83	8.271	
					−336			−158			−83
.09	32.743		.34	16.771		.59	11.072		.84	8.188	
					−323			−154			−82
0.10	31.436		0.35	16.448		0.60	10.918		0.85	8.106	
		−1180			−312			−151			−79
.11	30.256		.36	16.136		.61	10.767		.86	8.027	
		−1075			−302			−147			−77
.12	29.181		.37	15.834		.62	10.620		.87	7.950	
		−985			−292			−142			−76
.13	28.196		.38	15.542		.63	10.478		.88	7.874	
		−910			−282			−139			−73
.14	27.286		.39	15.260		.64	10.339		.89	7.801	
		−845			−274			−136			−71
0.15	26.441		0.40	14.986		0.65	10.203		0.90	7.730	
		−789			−265			−132			−70
.16	25.652		.41	14.721		.66	10.071		.91	7.660	
		−736			−257			−129			−68
.17	24.916		.42	14.464		.67	9.942		.92	7.592	
		−692			−248			−126			−66
.18	24.224		.43	14.216		.68	9.816		.93	7.526	
		−653			−242			−122			−64
.19	23.571		.44	13.974		.69	9.694		.94	7.462	
		−617			−235			−119			−62
0.20	22.954		0.45	13.739		0.70	9.575		0.95	7.400	
		−584			−227			−116			−61
.21	22.370		.46	13.512		.71	9.459		.96	7.339	
		−554			−221			−113			−59
.22	21.816		.47	13.291		.72	9.346		.97	7.280	
		−528			−214			−110			−58
.23	21.288		.48	13.077		.73	9.235		.98	7.222	
		−503			−209			−107			−56
.24	20.785		.49	12.868		.74	9.128		0.99	7.166	
		−481			−202			−105			−54
0.25	20.304		0.50	12.666		0.75	9.022		1.00	7.112	

SELF-INDUCTANCE OF CIRCULAR COILS

Example 30: Assume that four hundred turns of a wire that has such a diameter that it occupies 20 turns per centimeter is to be wound in a square cross section 1 cm. on a side, with a mean radius of the turns equal to 2.5 cm.

Here $a = 2.5$ cm., $c = 1$ cm., $\dfrac{c}{2a} = 0.2$, $N = 400$. From Table 21, $P_0' = 22.954$.

Therefore,
$$L = 0.001(2.5)(400)^2(22.954)$$
$$= 9181 \, \mu\text{h} = 9.181 \text{ mh}.$$

If, instead, 400 turns of the same wire were wound to a mean radius of 10 cm., $\dfrac{c}{2a} = \dfrac{1}{20}$ and $\left(\dfrac{c}{2a}\right)^2 = 0.0025$. In formula (92), $\left[1 + \dfrac{1}{6}\left(\dfrac{c}{2a}\right)^2\right] = 1.00042$, and $0.2041 \left(\dfrac{c}{2a}\right)^2 = 0.00051$.

$\dfrac{1}{2}\log\dfrac{8}{0.0025} = 1.75258$, and using Auxiliary Table 2 (page 237), the natural logarithm corresponding is 4.03546. Accordingly, from (92)

$$P_0' = 4\pi[(1.00042)(4.03546) - 0.84834 + 0.00051]$$
$$= 4\pi(3.1893) = 40.077.$$

The value given in Table 21 is 40.078.

From formula (91)
$$L = 0.001(10)(400)^2(40.077) = 64125 \, \mu\text{h}.$$
$$= 64.12 \text{ mh}.$$

Brooks Coils. Maxwell found [49] that for maximum inductance with a given length of a chosen wire, the mean diameter of the turns should be 3.7 times the dimension of the square cross section. This result, although often quoted, is only approximate. The more accurate formulas for the inductance now available show that the ratio lies quite close to $\dfrac{2a}{c} = 3$. This may be shown by expressing the inductance in terms of the length l of the wire and the diameter of the covered wire δ. Then

$$L = 0.0005 \, \frac{l^{5/3}\left(\dfrac{c}{2a}\right)^{3/2}}{(\pi)^{5/3}\delta^{3/2}} \, P_0'. \tag{93}$$

Differentiating this with respect to $\dfrac{c}{2a}$, with l and δ constant, the condition for maximum inductance is $\dfrac{1}{P_0'}\dfrac{dP_0'}{d\left(\dfrac{c}{2a}\right)} = -\dfrac{2}{3}\dfrac{1}{\left(\dfrac{c}{2a}\right)}$. Expressing P_0' in a Taylor's series for the critical region $\left(\dfrac{c}{2a} \text{ approximately } \dfrac{1}{3}\right)$ it is found that the maximum of inductance occurs with $\dfrac{c}{2a} = 0.337$, that is, $\dfrac{2a}{c} = 2.967$. The maximum is, however, very flat.

Accordingly, Brooks has proposed [46] that a coil for which $\dfrac{2a}{c} = 3$ is, for all practical purposes, one of the optimum form and has the advantage over that yielded by mathematical analysis of simplicity of the proportions. Such a coil offers, in fact, an inductance only 2 parts in 100,000 less than the maximum attainable with the wire in question. Fig. 29 shows such a coil, drawn to scale. The following simple relations exist between the principal dimensions:

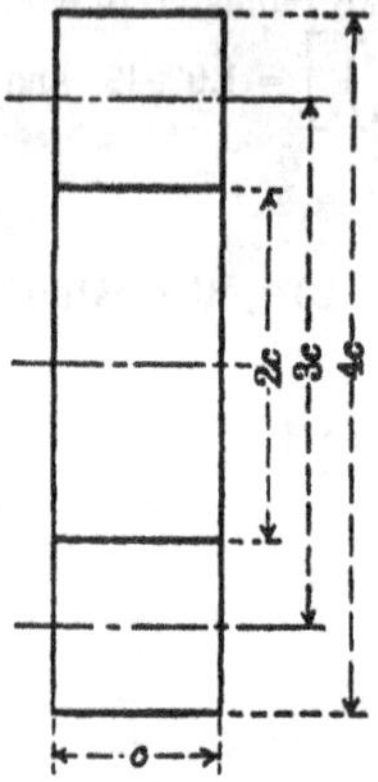

Side of square cross section = c,

Inner diameter of winding = $2c$,

Mean diameter of a turn = $3c = 2a$,

Outer diameter of winding = $4c$.

The inductance of a Brooks coil is given by the simple formula

$$L = 0.016994aN^2 \quad \mu\text{h}. \tag{94}$$

FIG. 29

Correction for Insulating Space. The preceding formulas all suppose that the current is uniformly distributed over the cross section. That is, the space occupied by insulating material is assumed to be negligible. This corresponds to the ideal condition of wire of square (or rectangular) cross section, the turns being kept from electrical contact by insulation of negligible thickness. In the case of the usual coil wound with wire, only thinly covered with insulating material, and with the turns crowded together in the winding channel, the correction for the insulating space is practically of very small importance. The unavoidable embedding of the wires as the coil is being wound tends to reduce the correction and to render its actual magnitude uncertain. Only in the case of wire with thick insulation or coils whose layers are purposely spaced to reduce coil capacitance will a correction be necessary.

Rosa has shown [50] that for an array of wires with equal spacing between wire centers, both between the layers and between the turns in the layer, the effect of insulating space is given by a correction ΔL, to be added to the inductance calculated by the foregoing formulas, of

$$\Delta L = 0.004\pi aN \left[\log_e \dfrac{p}{\delta_1} + 0.1381 + E \right], \tag{95}$$

SELF-INDUCTANCE OF CIRCULAR COILS 99

in which E = a constant, depending upon the number of layers and the number of turns in the layer,

p = pitch = distance between centers of adjacent wires,

δ_1 = diameter of the bare wire.

The following table gives the value of E for certain specific examples of square arrays of wire.

Turns	E
2 × 2	0.0169
4 × 4	0.0151
10 × 10	0.0171
20 × 20	0.0176
Infinite array	0.0181

Since, at most, ΔL is relatively small, it will suffice in general to assume $E = 0.017$ and this will give rise to a correction *factor* of

$$1 + \frac{\Delta L}{L} = 1 + \frac{0.739}{N}\left(\log_e \frac{p}{\delta_1} + 0.155\right). \quad (96)$$

Multiplying the value given by (94) by this factor, the inductance value, corrected for insulating space, is obtained.

Example 31: A Brooks coil is wound closely with wire of covered diameter 0.05 cm. and bare wire diameter 0.04 cm. There are 20 layers of wire of 20 turns per layer in the square cross section. Assuming an ideal square array of wires, $p = 0.05$ cm., $c = 20 \times 0.05 = 1$ cm., $2a = 3$, $N = 400$.

The uncorrected inductance is, by (94), $L = 0.016994 \times 1.5 \times (400)^2 = 4079$ μh $= 4.079$ mh.

For obtaining the correction factor, we have

$\frac{p}{\delta_1} = \frac{0.05}{0.04} = \frac{5}{4}$, $\frac{0.739}{N} = 0.00185$,

$\log_e \frac{5}{4} = 0.2231$, $\log_e \frac{5}{4} + 0.155 = 0.378$.

The correction factor is, therefore, $1 + 0.00185(0.378) = 1.00070$.

That is, the error arising from the neglect of the effect of insulating space is only 7 parts in 10,000 of the whole inductance. The effect of the embedding of the turns may easily amount to as much as this.

If, however, the same coil had been wound with wire of the same covered diameter 0.05 cm. but with such an insulation thickness that $\delta_1 = 0.02$ cm. only, the correction factor would amount to $1 + 0.00185(1.071) = 1.00198$, that is, three times as great. Even in that case its value is only 0.2 of 1 per cent, which is of small importance practically.

Design of a Brooks Coil to Obtain a Desired Inductance with a Chosen Size of Wire. Suppose the diameter of the covered wire to be δ cm. For cotton-covered wire, this may be obtained by measuring the length occupied by a counted number of turns wound on a mandrel.

100 CALCULATION OF MUTUAL INDUCTANCE AND SELF-INDUCTANCE

Since for the Brooks coil $\frac{c}{2a} = \frac{1}{3}$ and $N = \frac{c}{\delta} \cdot \frac{c}{\delta} = \frac{c^2}{\delta^2}$, formula (94) gives

so that
$$L = 25.49 \times 10^{-6} \frac{c^5}{\delta^4} \text{ mh},$$

$$c^5 = \frac{L\delta^4 10^6}{25.49} \quad (c \text{ and } \delta \text{ in cm.}, L \text{ in mh}), \tag{97}$$

$$c^5 = \frac{L\delta^4 10^6}{64.74} \quad (c \text{ and } \delta \text{ in inches}, L \text{ in mh}). \tag{98}$$

From the calculated value of c may be found the other principal dimensions of the coil. The number of turns per layer and the number of layers are each equal to $\frac{c}{\delta}$.

Example 32: To wind a Brooks coil of 10 mh with wire that is found to occupy 1.30 inches for a layer of 20 turns. That is, $\delta = 0.065$ inches. From (98)

$$c^5 = \frac{10(0.065)^4 10^6}{64.74} \text{ inches.}$$

$$\log \delta = \overline{2}.81291$$

$$\begin{array}{l} 4 \log \delta = \overline{5}.25164 \\ \log 10^7 = 7. \end{array}$$

$$\begin{array}{l} \text{sum} = 2.25164 \\ \log 64.74 = 1.81117 \end{array}$$

$$\log c^5 = 0.44047$$

$$\log c = 0.08809$$

$$\log \frac{c}{\delta} = 1.27518$$

$$\log \frac{c^2}{\delta^2} = 2.55036 = \log N.$$

Therefore,
$c = 1.225$ inches $=$ side of winding channel,
$2c = 2.45$ inches $=$ inner diameter,
$2a = 3c = 3.675$ inches $=$ mean diameter of the turns,
$4c = 4.90$ inches $=$ over-all diameter,
$N = 355$ turns.

The coil would have to consist of 19 turns per layer occupying a channel 1.235 inches wide and the 355 turns would require 18 complete layers plus 13 turns of a further layer. In actual practice, the inductance would be adjusted by measurement on a bridge, winding or unwinding a few turns, as found necessary.

SELF-INDUCTANCE OF CIRCULAR COILS

The length of wire required is

$$l = 2\pi Na = \pi(355)3.675 = 4098 \text{ inches} = 342 \text{ feet.}$$

The resistance would be about 0.84 ohm.

Design of a Brooks Coil to Obtain a Chosen Inductance and Time Constant. Brooks has given [51] design formulas and curves for this still more general design problem. These are based on round wires with negligible thickness of insulation. He shows how to apply these to actual insulated wire.

For a coil of $L_0 = 1$ mh and time constant $\tau = \dfrac{L_0}{R}$ in milliseconds, the side of the cross section c, the diameter of bare wire d_1, and the weight of copper w are for a Brooks coil wound with round wire with negligible insulation thickness, as follows:

Metric Units

$c = 9.142\tau^{1/2}$ mm.

$d_1 = 0.6336\tau^{5/8}$ mm.

$w = 0.05027\tau^{3/2}$ kg.

English Units

$c = 0.3599\tau^{1/2}$ inches

$d_1 = 0.02494\tau^{5/8}$ inches

$w = 0.1108\tau^{3/2}$ lb.

Plots of these data, which are straight lines on log-log paper, are here reproduced from Brooks's paper and are given in Figs. 30 and 31.

For a coil of any other inductance L, in millihenrys, use is made of the principle that for coils of the same absolute size the inductance is inversely proportional to the fourth power of the diameter of the wire over the insulation and, conversely, the diameter of the wire over the insulation is inversely proportional to the fourth root of the inductance. The factor $\dfrac{1}{L^{1/4}}$ s given in Fig. 32 for different values of inductance.

The process of design is as follows. Entering the curves with the desired time constant τ, first approximations c', d_1', w' are obtained. From the curve of $L^{1/4}$ the factor α corresponding to the actual desired inductance is taken, and the value $\alpha d_1'$ is the required size of wire. This value will, in general, lie between the diameter over insulation of two standard wire gages. Denoting the larger of these two insulated diameters by δ_1' and the corresponding thickness of insulation by t_1, calculate $\dfrac{\delta_1'}{\delta_1' - 2t_1}$, and $c'' = c'\left(\dfrac{\delta_1'}{\delta_1' - 2t_1}\right)$ will be a second approximation to the side of the square that will contain the required number of turns. For the assumed ratio $\dfrac{c}{2a} = \dfrac{1}{3}$ the value c'' means a new value of mean diameter and hence a new value of the number of turns. Accordingly the value δ'' corresponding to c'' is taken from the plot and a further approximation found. An example given by Brooks will make the process clearer.

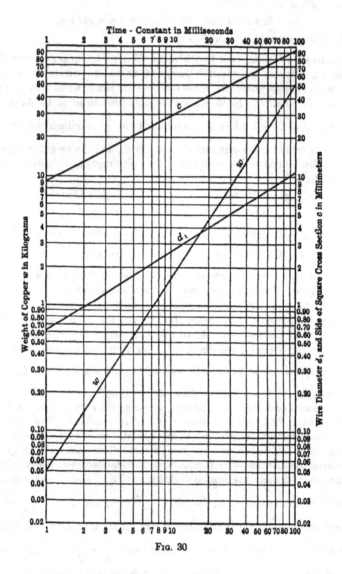

Fig. 30

SELF-INDUCTANCE OF CIRCULAR COILS

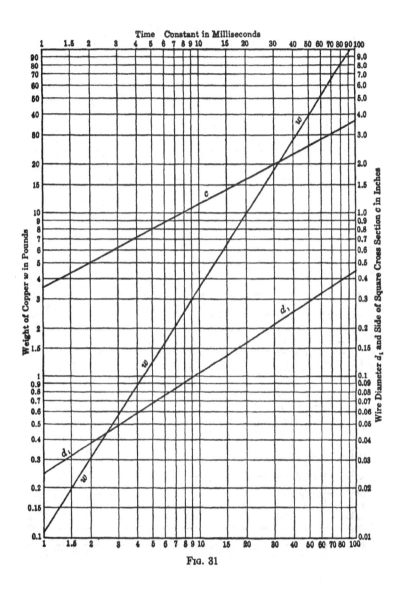

Fig. 31

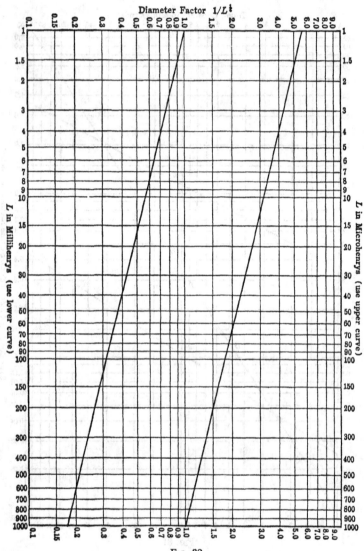

Fig. 32

SELF-INDUCTANCE OF CIRCULAR COILS 105

Example 33: To design a coil of 100 mh with a time constant of 3 msec.
For $\tau = 3$ msec., Figs. 30, 31 and 32 give $c' = 0.625$ inch, $d_1' = 0.0495$ inch, $w' = 0.575$ lb., and $\dfrac{1}{L^{1/4}} = 0.320$.

Hence $\delta' = 0.320 \times 0.0495$ inch $= 0.0158$ inch is the required size of wire of negligible insulation thickness. It is found that No. 27 wire has a covered diameter of $\delta_1' = 0.0176$ and $\delta_1' - 2t_1 = 0.0142$.
A second approximation is, therefore, $c'' = 0.625 \left(\dfrac{.0176}{.0142}\right) = 0.775$ inch. But for this value of c'' Fig. 31 gives $d_1'' = 0.0640$ inch, or for a 100 mh coil, $\delta'' = 0.320 \times 0.0640 = 0.0205$. Using a No. 25 wire for the next approximation, a value of δ''' results between No. 25 and No. 26, but nearer the latter.
Choosing No. 26 wire, $\delta = 0.0195$, the corresponding value of $d_1 = \dfrac{0.0195}{0.320} = 0.0610$, to which Fig. 31 gives a corresponding value $c = 0.741$ inch. Thus the mean radius is $a = \dfrac{3}{2}(0.741) = 1.111$ inch, and $N = \left(\dfrac{0.741}{0.0195}\right)^2 = 1444$ turns.
With these data a check calculation by (94) gives $L = 43.16 \times 10^{-6} \times 1.111 \times (1444)^2 = 99.98$ mh.
The length of the wire is $2\pi(1.111)1444 = 10{,}080$ inches $= 840$ feet.
The weight of copper for No. 26 wire is 0.840×0.769 or 0.646 lb., the resistance $R = 34.9$ and $\tau = \dfrac{99.98}{34.9} = 2.86$ msec.

Inductance of a Circular Coil with Rectangular Cross Section of Any Desired Proportions. For this general case two formulas are provided,[52] one useful especially for "thin coils," that is, coils of small radial thickness, and the other for "thick coils," that is, coils approximating a disc shape. Both formulas are useful for coils of intermediate shapes, and thus is afforded a valuable check on computations.

Thin coils (coils approximating solenoids):

$$L = 0.019739 \left(\frac{2a}{b}\right) N^2 a_{cm} K' \quad \text{(metric)}, \tag{99}$$

$$= 0.050138 \left(\frac{2a}{b}\right) N^2 a_{in} K' \quad \text{(English)}. \tag{99a}$$

The ratio $\dfrac{2a}{b}$ is independent of the choice of length unit. It is calculated as the quotient of the given parameters $\dfrac{c}{b}$ by $\dfrac{c}{2a}$.

The quantity K' is the difference $(K - k)$ of two quantities. Of these K is the constant in Nagaoka's formula for a solenoid and is tabulated as a function of $\dfrac{2a}{b}\left(\text{or } \dfrac{b}{2a}\right)$ in Table 36 (page 144). The quantity k takes into account the decrease of inductance due to the separation of the turns in the radial direction. It is obtained from Tables 22 and 23, where it is tabulated as a function of the two parameters $\dfrac{c}{2a}$ and $\dfrac{c}{b}\left(\text{or } \dfrac{b}{c}\right)$.

106 CALCULATION OF MUTUAL INDUCTANCE AND SELF-INDUCTANCE

TABLE 22. VALUES OF k FOR THIN, LONG COILS, FORMULA (99)

c/b	c/2a										c/b
	0	0.05	0.10	0.15	0.20	0.25	0.30	0.35	0.40	0.45	0.50	
0	0	0.0325	0.0633	0.0925	0.1200	0.1458	0.1700	0.1925	0.2133	0.2325	0.2500	0
0.05	0	.0316	.0621	.0911	.1186	.1445	.1687	.1913	.2123	.2317	.2494	0.05
0.10	0	.0308	.0608	.0896	.1170	.1428	.1671	.1898	.2109	.2304	.2484	0.10
0.15	0	.0300	.0594	.0879	.1151	.1409	.1651	.1879	.2091	.2288	.2470	0.15
0.20	0	.0293	.0581	.0861	.1131	.1388	.1630	.1858	.2071	.2270	.2453	0.20
0.25	0	0.0286	0.0569	0.0843	0.1109	0.1365	0.1607	0.1835	0.2048	0.2248	0.2432	0.25
0.30	0	.0280	.0557	.0826	.1085	.1342	.1583	.1810	.2024	.2224	.2410	0.30
0.35	0	.0274	.0546	.0810	.1069	.1319	.1558	.1785	.1999	.2199	.2386	0.35
0.40	0	.0269	.0535	.0796	.1051	.1297	.1533	.1759	.1973	.2174	.2361	0.40
0.45	0	.0264	.0525	.0782	.1033	.1276	.1510	.1732	.1945	.2146	.2335	0.45
0.50	0	0.0259	0.0516	0.0769	0.1016	0.1256	0.1487	0.1706	0.1916	0.2118	0.2308	0.50
0.55	0	.0254	.0507	.0755	.0999	.1236	.1464	.1682	.1890	.2091	.2280	0.55
0.60	0	.0250	.0498	.0742	.0982	.1216	.1442	.1658	.1866	.2065	.2252	0.60
0.65	0	.0246	.0490	.0730	.0966	.1197	.1421	.1635	.1842	.2039	.2225	0.65
0.70	0	.0242	.0482	.0719	.0952	.1179	.1400	.1613	.1818	.2014	.2199	0.70
0.75	0	0.0238	0.0474	0.0708	0.0937	0.1161	0.1380	0.1591	0.1794	0.1988	0.2173	0.75
0.80	0	.0234	.0467	.0697	.0923	.1144	.1360	.1569	.1770	.1963	.2147	0.80
0.85	0	.0230	.0460	.0687	.0910	.1128	.1341	.1548	.1747	.1939	.2121	0.85
0.90	0	.0227	.0453	.0677	.0897	.1113	.1323	.1527	.1725	.1915	.2096	0.90
0.95	0	.0224	.0447	.0667	.0884	.1097	.1305	.1507	.1704	.1892	.2071	0.95
1.00	0	0.0221	0.0441	0.0658	0.0872	0.1082	0.1288	0.1489	0.1683	0.1869	0.2047	1.00

c/b	c/2a										c/b
	0.50	0.55	0.60	0.65	0.70	0.75	0.80	0.85	0.90	0.95	1.00	
0	0.2500	0.2658	0.2800	0.2925	0.3033	0.3125	0.3200	0.3258	0.3300	0.3325	0.3333	0
0.05	.2494	.2655	.2800	.2928	.3040	.3135	.3213	.3275	.3321	.3351	.3363	0.05
0.10	.2484	.2648	.2795	.2926	.3040	.3138	.3221	.3287	.3337	.3371	.3388	0.10
0.15	.2470	.2636	.2786	.2920	.3037	.3139	.3225	.3294	.3349	.3386	.3408	0.15
0.20	.2453	.2621	.2773	.2910	.3031	.3136	.3225	.3298	.3356	.3397	.3423	0.20
0.25	0.2432	0.2603	0.2758	0.2897	0.3021	0.3129	0.3221	0.3298	0.3359	0.3404	0.3434	0.25
0.30	.2410	.2582	.2739	.2881	.3007	.3118	.3213	.3294	.3359	.3408	.3441	0.30
0.35	.2386	.2559	.2718	.2862	.2991	.3104	.3202	.3286	.3355	.3407	.3445	0.35
0.40	.2361	.2535	.2695	.2841	.2972	.3088	.3189	.3276	.3347	.3403	.3445	0.40
0.45	.2335	.2509	.2671	.2818	.2951	.3070	.3173	.3263	.3337	.3397	.3442	0.45
0.50	0.2308	0.2483	0.2645	0.2794	0.2929	0.3050	0.3156	0.3248	0.3326	0.3389	0.3436	0.50
0.55	.2280	.2456	.2619	.2769	.2906	.3028	.3137	.3231	.3311	.3377	.3428	0.55
0.60	.2252	.2428	.2592	.2743	.2881	.3005	.3116	.3213	.3295	.3363	.3417	0.60
0.65	.2225	.2400	.2564	.2716	.2856	.2981	.3093	.3192	.3277	.3348	.3404	0.65
0.70	.2199	.2373	.2536	.2689	.2830	.2957	.3070	.3171	.3258	.3331	.3390	0.70
0.75	0.2173	0.2346	0.2509	0.2662	0.2804	0.2932	0.3046	0.3149	0.3238	0.3313	0.3375	0.75
0.80	.2147	.2320	.2483	.2636	.2777	.2906	.3022	.3126	.3217	.3294	.3358	0.80
0.85	.2121	.2294	.2456	.2609	.2750	.2880	.2998	.3103	.3195	.3274	.3340	0.85
0.90	.2096	.2268	.2430	.2582	.2724	.2855	.2973	.3079	.3172	.3253	.3321	0.90
0.95	.2071	.2242	.2404	.2556	.2698	.2829	.2949	.3055	.3149	.3231	.3301	0.95
1.00	0.2047	0.2217	0.2378	0.2530	0.2672	0.2804	0.2929	0.3031	0.3126	0.3209	0.3281	1.00

SELF-INDUCTANCE OF CIRCULAR COILS

TABLE 23. VALUES OF k FOR SHORT, THICK COILS, FORMULA (99)

b/c	c/2a										b/c
	0	0.05	0.10	0.15	0.20	0.25	0.30	0.35	0.40	0.45	0.50	
0	0	0	0	0	0	0	0	0	0	0	0	0
0.05	0	0.0045	0.0090	0.0134	0.0178	0.0226	0.0269	0.0314	0.0359	0.0398	0.0436	0.05
0.10	0	.0077	.0153	.0229	.0304	.0379	.0453	.0527	.0601	.0674	.0745	0.10
0.15	0	.0098	.0194	.0291	.0387	.0482	.0576	.0669	.0763	.0855	.0944	0.15
0.20	0	.0114	.0228	.0341	.0454	.0566	.0677	.0786	.0894	.1000	.1105	0.20
0.25	0	0.0128	0.0256	0.0383	0.0510	0.0635	0.0759	0.0881	0.1001	0.1120	0.1237	0.25
0.30	0	.0140	.0280	.0419	.0557	.0694	.0829	.0962	.1093	.1221	.1347	0.30
0.35	0	.0151	.0301	.0451	.0598	.0745	.0889	.1031	.1172	.1309	.1443	0.35
0.40	0	.0160	.0319	.0478	.0634	.0789	.0942	.1092	.1240	.1385	.1526	0.40
0.45	0	.0168	.0335	.0501	.0666	.0828	.0988	.1146	.1300	.1452	.1599	0.45
0.50	0	0.0175	0.0349	0.0522	0.0694	0.0863	0.1030	0.1194	0.1354	0.1511	0.1663	0.50
0.55	0	.0182	.0362	.0541	.0720	.0895	.1068	.1237	.1403	.1563	.1720	0.55
0.60	0	.0188	.0374	.0559	.0743	.0924	.1102	.1276	.1446	.1611	.1772	0.60
0.65	0	.0193	.0385	.0576	.0764	.0950	.1132	.1311	.1485	.1655	.1819	0.65
0.70	0	.0198	.0395	.0591	.0784	.0974	.1160	.1343	.1521	.1694	.1861	0.70
0.75	0	0.0203	0.0404	0.0605	0.0802	0.0996	0.1186	0.1373	0.1554	0.1729	0.1899	0.75
0.80	0	.0207	.0413	.0617	.0818	.1016	.1210	.1400	.1584	.1762	.1934	0.80
0.85	0	.0211	.0421	.0629	.0833	.1035	.1232	.1424	.1611	.1792	.1966	0.85
0.90	0	.0215	.0428	.0639	.0847	.1052	.1252	.1447	.1637	.1820	.1995	0.90
0.95	0	.0218	.0435	.0649	.0860	.1068	.1271	.1469	.1661	.1846	.2022	0.95
1.00	0	0.0221	0.0441	0.0658	0.0872	0.1082	0.1288	0.1489	0.1683	0.1869	0.2047	1.00

b/c	c/2a										b/c
	0.50	0.55	0.60	0.65	0.70	0.75	0.80	0.85	0.90	0.95	1.00	
0	0	0	0	0	0	0	0	0	0	0	0	0
0.05	0.0436	0.0484	0.0529	0.0572	0.0613	0.0653	0.0692	0.0730	0.0767	0.0803	0.0839	0.05
0.10	.0745	.0816	.0885	.0953	.1020	.1085	.1149	.1211	.1272	.1331	.1388	0.10
0.15	.0944	.1023	.1120	.1204	.1287	.1368	.1447	.1523	.1597	.1668	.1736	0.15
0.20	.1105	.1208	.1308	.1406	.1501	.1594	.1684	.1771	.1854	.1933	.2009	0.20
0.25	0.1237	0.1350	0.1460	0.1568	0.1673	0.1775	0.1874	0.1968	0.2058	0.2142	0.2224	0.25
0.30	.1347	.1471	.1590	.1706	.1819	.1928	.2032	.2132	.2226	.2315	.2399	0.30
0.35	.1443	.1574	.1701	.1823	.1942	.2057	.2166	.2269	.2366	.2458	.2544	0.35
0.40	.1526	.1663	.1796	.1924	.2048	.2167	.2279	.2384	.2484	.2577	.2664	0.40
0.45	.1599	.1742	.1880	.2012	.2140	.2262	.2377	.2484	.2585	.2679	.2766	0.45
0.50	0.1663	0.1811	0.1953	0.2089	0.2220	0.2345	0.2462	0.2571	0.2672	0.2766	0.2852	0.50
0.55	.1720	.1872	.2018	.2157	.2291	.2418	.2536	.2645	.2748	.2842	.2926	0.55
0.60	.1772	.1927	.2076	.2218	.2354	.2482	.2601	.2712	.2814	.2907	.2990	0.60
0.65	.1819	.1977	.2128	.2272	.2409	.2539	.2659	.2770	.2872	.2964	.3046	0.65
0.70	.1861	.2022	.2175	.2321	.2459	.2589	.2710	.2821	.2923	.3014	.3095	0.70
0.75	0.1899	0.2062	0.2217	0.2365	0.2504	0.2634	0.2755	0.2866	0.2968	0.3058	0.3137	0.75
0.80	.1934	.2098	.2255	.2404	.2544	.2675	.2796	.2907	.3007	.3096	.3174	0.80
0.85	.1966	.2132	.2291	.2440	.2581	.2712	.2833	.2943	.3042	.3130	.3206	0.85
0.90	.1995	.2163	.2323	.2473	.2614	.2745	.2866	.2976	.3074	.3160	.3234	0.90
0.95	.2022	.2191	.2352	.2503	.2644	.2776	.2896	.3005	.3102	.3186	.3259	0.95
1.00	0.2047	0.2217	0.2378	0.2530	0.2672	0.2804	0.2924	0.3031	0.3126	0.3209	0.3281	1.00

108 CALCULATION OF MUTUAL INDUCTANCE AND SELF-INDUCTANCE

TABLE 24. VALUES OF F FOR DISC COILS, FORMULA (100)

b/c	\multicolumn{11}{c	}{c/2a}	b/c
	0	0.05	0.10	0.15	0.20	0.25	0.30	0.35	0.40	0.45	0.50	
0	1	1	1	1	1	1	1	1	1	1	1	0
0.05	1	0.9871	0.9843	0.9821	0.9801	0.9782	0.9763	0.9745	0.9728	0.9712	0.9696	0.05
0.10	1	.9749	.9695	.9651	.9612	.9575	.9541	.9507	.9474	.9442	.9412	0.10
0.15	1	.9634	.9555	.9491	.9434	.9381	.9332	.9283	.9237	.9189	.9145	0.15
0.20	1	.9524	.9422	.9339	.9266	.9197	.9133	.9070	.9010	.8951	.8894	0.20
0.25	1	0.9419	0.9294	0.9194	0.9105	0.9021	0.8943	0.8868	0.8795	0.8725	0.8656	0.25
0.30	1	.9318	.9172	.9059	.8950	.8853	.8762	.8675	.8591	.8510	.8431	0.30
0.35	1	.9221	.9054	.8920	.8802	.8692	.8589	.8491	.8396	.8305	.8217	0.35
0.40	1	.9128	.8941	.8792	.8660	.8537	.8423	.8314	.8210	.8110	.8013	0.40
0.45	1	.9038	.8832	.8668	.8523	.8389	.8264	.8145	.8032	.7923	.7819	0.45
0.50	1	0.8951	0.8727	0.8548	0.8391	0.8246	0.8111	0.7983	0.7861	0.7744	0.7632	0.50
0.55	1	.8867	.8625	.8434	.8264	.8108	.7964	.7827	.7697	.7573	.7454	0.55
0.60	1	.8786	.8527	.8322	.8141	.7975	.7822	.7677	.7540	.7409	.7284	0.60
0.65	1	.8707	.8432	.8214	.8022	.7847	.7686	.7533	.7389	.7251	.7121	0.65
0.70	1	.8630	.8339	.8109	.7907	.7723	.7554	.7394	.7243	.7100	.6965	0.70
0.75	1	0.8556	0.8249	0.8007	0.7796	0.7603	0.7426	0.7260	0.7103	0.6955	0.6816	0.75
0.80	1	.8484	.8162	.7908	.7688	.7487	.7303	.7130	.6968	.6816	.5672	0.80
0.85	1	.8413	.8077	.7813	.7584	.7374	.7184	.7005	.6838	.6681	.6533	0.85
0.90	1	.8345	.7995	.7720	.7482	.7265	.7068	.6884	.6712	.6551	.6399	0.90
0.95	1	.8279	.7914	.7629	.7383	.7159	.6956	.6767	.6591	.6426	.6271	0.95
1.00	1	0.8214	0.7837	0.7548	0.7287	0.7056	0.6848	0.6656	0.6474	0.6306	0.6148	1.00

b/c	\multicolumn{10}{c	}{c/2a}	b/c
	0.50	0.55	0.60	0.65	0.70	0.75	0.80	0.85	0.90	0.95	1.00	
0	1	1	1	1	1	1	1	1	1	1	1	0
0.05	0.9696	0.9679	0.9663	0.9648	0.9633	0.9618	0.9604	0.9591	0.9577	0.9562	0.9547	0.05
0.10	.9412	.9381	.9351	.9322	.9293	.9265	.9238	.9212	.9187	.9161	.9136	0.10
0.15	.9145	.9102	.9059	.9017	.8977	.8938	.8900	.8863	.8829	.8794	.8763	0.15
0.20	.8894	.8839	.8785	.8732	.8683	.8634	.8586	.8540	.8497	.8457	.8421	0.20
0.25	0.8656	0.8590	0.8526	0.8463	0.8405	0.8347	0.8292	0.8239	0.8190	0.8144	0.8103	0.25
0.30	.8431	.8355	.8282	.8210	.8143	.8078	.8017	.7959	.7905	.7854	.7807	0.30
0.35	.8217	.8132	.8051	.7971	.7897	.7825	.7759	.7697	.7638	.7583	.7532	0.35
0.40	.8013	.7920	.7831	.7745	.7664	.7587	.7515	.7450	.7388	.7329	.7274	0.40
0.45	.7819	.7718	.7622	.7531	.7443	.7360	.7285	.7217	.7152	.7090	.7032	0.45
0.50	0.7632	0.7526	0.7424	0.7327	0.7234	0.7146	0.7068	0.6996	0.6929	0.6865	0.6805	0.50
0.55	.7454	.7342	.7235	.7132	.7036	.6945	.6862	.6787	.6717	.6651	.6590	0.55
0.60	.7284	.7166	.7054	.6947	.6847	.6754	.6668	.6589	.6516	.6448	.6386	0.60
0.65	.7121	.6998	.6882	.6771	.6668	.6572	.6484	.6402	.6326	.6256	.6194	0.65
0.70	.6965	.6837	.6717	.6603	.6497	.6399	.6308	.6224	.6146	.6075	.6012	0.70
0.75	0.6816	0.6683	0.6560	0.6443	0.6334	0.6234	0.6140	0.6055	0.5976	0.5904	0.5839	0.75
0.80	.6672	.6536	.6409	.6290	.6179	.6076	.5981	.5894	.5815	.5742	.5676	0.80
0.85	.6533	.6394	.6264	.6143	.6030	.5926	.5829	.5742	.5662	.5589	.5521	0.85
0.90	.6399	.6257	.6124	.6002	.5888	.5782	.5684	.5597	.5517	.5443	.5375	0.90
0.95	.6271	.6126	.5991	.5867	.5752	.5644	.5545	.5459	.5378	.5303	.5236	0.95
1.00	0.6148	0.6001	0.5865	0.5738	0.5621	0.5512	0.5413	0.5324	0.5244	0.5170	0.5102	1.00

SELF-INDUCTANCE OF CIRCULAR COILS

TABLE 25. VALUES OF F FOR THIN, LONG COILS, FORMULA (100)

c/b	c/2a										c/b
	0	0.05	0.10	0.15	0.20	0.25	0.30	0.35	0.40	0.45	0.50	
0	0	0	0	0	0	0	0	0	0	0	0	0
0.05					0.1224	0.1064	0.0948	0.0856	0.0784	0.0725	0.0679	0.05
0.10		0.4002	0.3088	0.2560	.2197	.1942	.1747	.1595	.1472	.1370	.1287	0.10
0.15		.4845	.3942	.3388	.2977	.2669	.2426	.2232	.2074	.1944	.1831	0.15
0.20		.5439	.4600	.4038	.3608	.3276	.3007	.2787	.2604	.2451	.2317	0.20
0.25		0.5885	0.5103	0.4560	0.4131	0.3789	0.3506	0.3272	0.3073	0.2901	0.2754	0.25
0.30		.6247	.5514	.4993	.4571	.4227	.3940	.3695	.3487	.3305	.3147	0.30
0.35		.6540	.5853	.5354	.4946	.4609	.4320	.4071	.3856	.3667	.3503	0.35
0.40		.6786	.6140	.5663	.5269	.4939	.4654	.4407	.4188	.3995	.3825	0.40
0.45		.6996	.6386	.5932	.5551	.5231	.4951	.4707	.4488	.4292	.4118	0.45
0.50		0.7178	0.6600	0.6167	0.5801	0.5487	0.5215	0.4976	0.4759	0.4562	0.4385	0.50
0.55		.7337	.6789	.6675	.6023	.5718	.5453	.5218	.5003	.4807	.4631	0.55
0.60		.7478	.6957	.6561	.6222	.5929	.5668	.5436	.5224	.5031	.4856	0.60
0.65		.7604	.7107	.6728	.6401	.6119	.5863	.5636	.5426	.5235	.5063	0.65
0.70		.7717	.7242	.6877	.6563	.6289	.6041	.5818	.5612	.5426	.5253	0.70
0.75		0.7819	0.7364	0.7013	0.6710	0.6442	0.6204	0.5985	0.5785	0.5601	0.5430	0.75
0.80		.7912	.7475	.7138	.6845	.6585	.6353	.6140	.5944	.5762	.5594	0.80
0.85		.7997	.7577	.7253	.6969	.6717	.6491	.6282	.6091	.5912	.5747	0.85
0.90		.8075	.7671	.7358	.7083	.6839	.6619	.6415	.6227	.6052	.5890	0.90
0.95		.8147	.7757	.7454	.7189	.6952	.6738	.6540	.6354	.6183	.6023	0.95
1.00		0.8214	0.7837	0.7541	0.7287	0.7058	0.6848	0.6656	0.6474	0.6306	0.6148	1.00

c/b	c/2a										c/b
	0.50	0.55	0.60	0.65	0.70	0.75	0.80	0.85	0.90	0.95	1.00	
0	0	0	0	0	0	0	0	0	0	0	0	0
0.05	0.0679	0.0640	0.0606	0.0578	0.0554	0.0532	0.0512	0.0494	0.0480	0.0467	0.0455	0.05
0.10	.1287	.1215	.1155	.1102	.1056	.1015	.0980	.0950	.0924	.0900	.0878	0.10
0.15	.1831	.1734	.1652	.1581	.1517	.1461	.1412	.1372	.1335	.1301	.1271	0.15
0.20	.2317	.2202	.2103	.2017	.1940	.1872	.1813	.1761	.1715	.1673	.1636	0.20
0.25	0.2754	0.2625	0.2513	0.2415	0.2327	0.2250	0.2186	0.2121	0.2067	0.2019	0.1976	0.25
0.30	.3147	.3008	.2886	.2778	.2682	.2597	.2521	.2454	.2394	.2340	.2293	0.30
0.35	.3503	.3356	.3226	.3111	.3009	.2917	.2835	.2763	.2698	.2639	.2588	0.35
0.40	.3825	.3673	.3537	.3417	.3309	.3212	.3126	.3049	.2980	.2918	.2862	0.40
0.45	.4118	.3963	.3823	.3698	.3587	.3486	.3396	.3315	.3243	.3177	.3119	0.45
0.50	0.4385	0.4228	0.4086	0.3958	0.3843	0.3740	0.3647	0.3563	0.3488	0.3420	0.3360	0.50
0.55	.4631	.4472	.4328	.4198	.4081	.3976	.3881	.3795	.3717	.3648	.3586	0.55
0.60	.4856	.4697	.4552	.4421	.4302	.4195	.4098	.4011	.3932	.3861	.3797	0.60
0.65	.5063	.4905	.4760	.4628	.4507	.4399	.4301	.4213	.4133	.4060	.3995	0.65
0.70	.5253	.5097	.4953	.4820	.4699	.4590	.4491	.4402	.4321	.4248	.4182	0.70
0.75	0.5430	0.5274	0.5132	0.4999	0.4879	0.4769	0.4670	0.4580	0.4499	0.4425	0.4358	0.75
0.80	.5594	.5440	.5298	.5166	.5046	.4937	.4838	.4747	.4665	.4591	.4524	0.80
0.85	.5747	.5595	.5454	.5323	.5203	.5094	.4995	.4904	.4823	.4748	.4680	0.85
0.90	.5890	.5740	.5600	.5470	.5351	.5242	.5143	.5053	.4971	.4896	.4829	0.90
0.95	.6023	.5875	.5736	.5608	.5490	.5381	.5282	.5193	.5111	.5037	.4969	0.95
1.00	0.6148	0.6001	0.5865	0.5738	0.5621	0.5512	0.5413	0.5324	0.5244	0.5170	0.5102	1.00

110 CALCULATION OF MUTUAL INDUCTANCE AND SELF-INDUCTANCE

Thick coils (coils approximately pancake form):

$$L = 0.001 N^2 a_{cm} P' \text{ (metric units)}, \qquad (100)$$

$$= 0.00254 N^2 a_{in} P' \text{ (English units)}. \qquad (100a)$$

The quantity P' is the product of two factors P and F, of which the first is a function of $\dfrac{c}{2a}$ and applies to a coil of zero axial dimension, while F takes into account the reduction of inductance due to separating the turns in the axial direction. Values of P are to be obtained from Table 26 (page 113). The factor F is to be interpolated from Table 24 or 25; it depends upon two parameters, $\dfrac{c}{2a}$ and $\dfrac{b}{c}\left(\text{or } \dfrac{c}{b}\right)$.

Interpolation in Tables 22, 23 and 24, 25 (Double Interpolation). The quantities k and F in Tables 22-25 depend upon two parameters, so that to use the tables a double interpolation is, in general, necessary. Restricting the treatment to cases where differences higher than the second are neglected, a direct procedure would be to interpolate first with respect to one variable x for three different tabular values of the second variable y, these values being chosen to include the desired value of y, and then to interpolate with respect to y between these three interpolated values. This procedure may, however, be shortened and rendered very simple under the condition that no differences of higher order than the second are included.

Let u equal the fraction of the tabular interval of x, and v the corresponding fraction of the interval of y. The first and second differences with respect to x are denoted by d_x' and d_x'', and the corresponding quantities with respect to y by d_y' and d_y''. The first difference of d_x' with respect to changes of y is denoted by d_{xy}'' and this is equal to the difference d_{yx}'' of d_y' with respect to x.

The general interpolation formula for the desired function $f(x, y)$ at $x = x_0 + u w_x$ and $y = y_0 + v w_y$, in which w_x and w_y are the tabular intervals, is

$$f(x_0 + u w_x, y_0 + v w_y) = f(x_0, y_0) + u\left[d_x' - \frac{(1-u)}{2} d_x''\right]$$

$$+ v\left[d_y' - \frac{(1-v)}{2} d_y''\right] + uv\, d_{xy}''. \qquad (101)$$

The practical use of this formula is illustrated in the solution of the following example.

SELF-INDUCTANCE OF CIRCULAR COILS

Example 34: Let us interpolate the quantity F in Table 24 for the parameter values $\frac{c}{2a} = 0.43$ and $\frac{b}{c} = 0.66$. From the Table 24 we extract the following data:

$\frac{b}{c}$	$\frac{c}{2a} = 0.4$		0.45		0.50
0.65	0.7389	−138	0.7251	−130	0.7121
	−146		−151		
0.70	0.7243	−143	0.7100		
	−140				
0.75	0.7103				

Then $x_0 = 0.40$, $y_0 = 0.65$. The tabular intervals are $w_x = w_y = 0.05$, so that $u = \frac{0.43 - 0.40}{0.05} = 0.6$ and $v = \frac{0.66 - 0.65}{0.05} = 0.2$. Therefore, $\frac{1-u}{2} = 0.2$, $\frac{1-v}{2} = 0.4$, $uv = 0.12$.

The table shows that $F(x_0, y_0) = 0.7389$, and the differences are (in units in the fourth place):

$$d_x' = -138, \qquad d_y' = -146,$$
$$d_x'' = -130 - (-138) = +8, \qquad d_y'' = -140 - (-146) = +6,$$
$$d_{xy}'' = -143 - (-138) = -5, \qquad d_{yx}'' = -151 - (-146) = -5.$$

Formula (101) then gives
$$\frac{F(0.43, 0.66)}{0.0001} = 7389 + 0.6\{-138 - 0.2(8)\} + 0.2\{-146 - 0.4(6)\} + 0.12(-5)$$
$$= 7389 - 83.8 - 29.7 - 0.6 = 7275,$$

so that $F(0.43, 0.66) = 0.7275$.

If second differences had been neglected, the interpolation formula (101) would become
$$f(x_0 + uw_x, y_0 + vw_y) = f(x_0, y_0) + ud_x' + vd_y', \tag{102}$$

and for this example the approximate value $7389 + 0.6(-138) + 0.2(-146) = 7277$ would be found. The difference would be unimportant in many cases.

The two formulas (99) and (100) will now be used to solve a certain problem for which either may be used with good accuracy.

Example 35: To calculate the inductance of a coil for which $a = 3.41$ cm., $N = 220$, $\frac{c}{2a} = 0.209$, $\frac{b}{c} = \frac{4}{9}$.

From Table 23 is taken the values of k shown in the following scheme:

$\frac{b}{c}$	$\frac{c}{2a} = 0.20$		0.25		0.30
0.40	0.0634	155	0.0789	153	0.0942
	32		39		
0.45	0.0666	162	0.0828		
	28				
0.50	0.0694				

112 CALCULATION OF MUTUAL INDUCTANCE AND SELF-INDUCTANCE

$$u = 0.18, \quad \frac{1-u}{2} = 0.41, \quad v = \frac{8}{9}, \quad \frac{1-v}{2} = 0.056, \quad uv = 0.16.$$

$$d_z' = 155, \quad d_y' = 32, \quad d_z'' = -2, \quad d_y'' = -4, \quad d_{zy}'' = 7.$$

so that
$$\frac{k(0.209, \tfrac{4}{9})}{0.0001} = 634 + 28.0 + 28.6 + 1.1 = 692,$$

$$k(0.209, \tfrac{4}{9}) = 0.0692.$$

The ratio $\frac{b}{2a} = \frac{b}{c} \cdot \frac{c}{2a} = \frac{4}{9}$ (0.209) = 0.09289, and interpolating from Table 36 for this value there is found

$$K = 0.19319$$
$$k = 0.0692$$
$$K' = K - k = 0.1240,$$

and by (99),

$$L = 0.019739 \left(\frac{1}{0.09289}\right)(220)^2(0.1240)(3.41) = 4349 = 4.349 \text{ mh}.$$

Using the disc formula (100) we interpolate in Table 24, using the values in the following scheme:

$\dfrac{b}{c}$	$\dfrac{c}{2a} = 0.20$		0.25		0.30
0.40	0.8660	−123	0.8537	−114	0.8423
	−137		−148		
0.45	0.8523	−134	0.8389		
	−132				
0.50	0.8391				

Interpolating,
$$F = 0.8660 - 0.01220 - 0.00228 - 0.00017 = 0.85135.$$

Accordingly, using Table 26 with $\frac{c}{2a} = 0.209$ we find $P = 30.96$, and from formula (100)

$$L = 0.001(220)^2(3.41)(30.96)(0.85135)$$
$$= 4350 \ \mu\text{h} = 4.350 \text{ mh},$$

which closely checks the value by formula (99).

SELF-INDUCTANCE OF CIRCULAR COILS 113

TABLE 26. VALUES OF P FOR DISC COILS, FORMULAS (100) AND (100a)

c/2a	P	d_1	c/2a	P	d_1	c/2a	P	d_1	c/2a	P	d_1
0	∞		0.25	28.767		0.50	20.601		0.75	16.360	
					−477			−221			−125
0.01	69.008		.26	28.290		.51	20.381		.76	16.235	
		−8709			−458			−216			−123
.02	60.299		.27	27.832		.52	20.165		.77	16.112	
		−5093			−440			−210			−120
.03	55.206		.28	27.392		.53	19.955		.78	15.992	
		−3611			−424			−205			−118
.04	51.595		.29	26.968		.54	19.750		.79	15.874	
		−2801			−408			−200			−115
0.05	48.794		0.30	26.560		0.55	19.550		0.80	15.759	
		−2287			−394			−196			−113
.06	46.507		.31	26.166		.56	19.354		.81	15.646	
		−1933			−380			−192			−110
.07	44.574		.32	25.786		.57	19.162		.82	15.536	
		−1672			−368			−187			−108
.08	42.902		.33	25.418		.58	18.976		.83	15.428	
		−1474			−355			−183			−105
.09	41.428		.34	25.063		.59	18.793		.84	15.323	
		−1317			−344			−179			−103
0.10	40.111		0.35	24.719		0.60	18.614		0.85	15.220	
		−1191			−333			−174			−101
.11	38.920		.36	24.386		.61	18.440		.86	15.119	
		−1085			−323			−171			− 98
.12	37.835		.37	24.063		.62	18.269		.87	15.021	
		− 997			−313			−167			− 96
.13	36.838		.38	23.750		.63	18.102		.88	14.925	
		− 922			−304			−163			− 94
.14	35.916		.39	23.446		.64	17.939		.89	14.832	
		− 858			−296			−160			− 92
0.15	35.058		0.40	23.150		0.65	17.779		0.90	14.740	
		− 800			−287			−156			− 90
.16	34.258		.41	22.863		.66	17.623		.91	14.650	
		− 751			−279			−153			− 87
.17	33.507		.42	22.584		.67	17.470		.92	14.563	
		− 707			−271			−150			− 85
.18	32.800		.43	22.313		.68	17.320		.93	14.478	
		− 668			−264			−146			− 83
.19	32.132		.44	22.049		.69	17.174		.94	14.394	
		− 632			−257			−142			− 80
0.20	31.500		0.45	21.792		0.70	17.032		0.95	14.314	
		− 600			−251			−140			− 79
.21	30.900		.46	21.541		.71	16.891		.96	14.235	
		− 570			−244			−137			− 77
.22	30.330		.47	21.297		.72	16.754		.97	14.158	
		− 545			−238			−134			− 75
.23	29.785		.48	21.059		.73	16.620		.98	14.083	
		− 520			−232			−131			− 73
.24	29.265		.49	20.827		.74	16.489		0.99	14.010	
		− 498			−226			−129			− 71
0.25	28.767		0.50	20.601		0.75	16.360		1.00	13.939	

Chapter 14

MUTUAL INDUCTANCE OF A SOLENOID AND A COAXIAL CIRCULAR FILAMENT

Basic Case. Circle in the End Plane of the Solenoid. The basic case is that of a cylindrical current sheet and a coaxial circular filament in its end plane. By the principle of the interchange of lengths (page 133) the mutual inductance of a solenoid of radius A, length x and winding density n_1 on a coaxial circle of radius a in its end plane (Fig. 33a) is the same as the

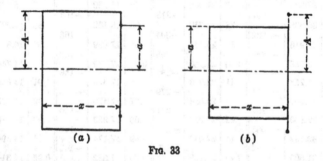

Fig. 33

mutual inductance of a solenoid of radius a, length x, and winding density n_1 on a coaxial circle of radius A, Fig. 33b. The mutual inductance may be obtained by the very accurate elliptic integral formula of Jones,[53] or the series formulas of Rosa,[54] Lorenz,[55] or Clem.[56] Calculations by these formulas are complicated and tedious. For routine calculations the results are found very simply by the formulas and tables that follow.

Let a and A be the two radii, A being always taken as the larger, $\alpha = \dfrac{a}{A}$, $x =$ the length of the solenoid, and the total number of turns N is related to the winding density by the equation $N = n_1 x$. The further shape

A SOLENOID AND A COAXIAL CIRCULAR FILAMENT

parameter $\rho^2 = \dfrac{A^2}{A^2 + x^2}$ will be introduced. Then,

$$M = 0.002\pi^2 a\alpha\rho N Q_0, \qquad (103)$$

in which Q_0 is a function of α and ρ^2, which may be interpolated from Table 27.

TABLE 27. VALUES OF Q_0 FOR MUTUAL INDUCTANCE SOLENOID AND CIRCLE, FORMULA (103)

ρ^2	$\alpha = 0$	0.05	0.10	0.15	0.20	0.25	0.30	0.35	0.40	0.45	0.50	ρ^2
0	1	1	1	1	1	1	1	1	1	1	1	0
0.05	1	1.0000	1.0000	1.0000	1.0001	1.0001	1.0001	1.0001	1.0002	1.0002	1.0002	0.05
.10	1	1.0000	1.0000	1.0001	1.0002	1.0002	1.0003	1.0004	1.0006	1.0008	1.0009	.10
.15	1	1.0000	1.0001	1.0002	1.0004	1.0006	1.0008	1.0010	1.0013	1.0016	1.0020	.15
.20	1	1.0000	1.0002	1.0004	1.0006	1.0010	1.0014	1.0018	1.0024	1.0030	1.0036	.20
0.25	1	1.0001	1.0003	1.0005	1.0009	1.0014	1.0021	1.0029	1.0038	1.0046	1.0057	0.25
.30	1	1.0001	1.0004	1.0007	1.0013	1.0021	1.0030	1.0041	1.0053	1.0067	1.0082	.30
.35	1	1.0001	1.0005	1.0010	1.0018	1.0028	1.0041	1.0056	1.0072	1.0091	1.0111	.35
.40	1	1.0002	1.0006	1.0014	1.0024	1.0037	1.0054	1.0072	1.0094	1.0119	1.0146	.40
.45	1	1.0002	1.0008	1.0017	1.0030	1.0047	1.0068	1.0092	1.0120	1.0152	1.0186	.45
0.50	1	1.0002	1.0010	1.0021	1.0037	1.0058	1.0084	1.0114	1.0148	1.0188	1.0231	0.50
.55	1	1.0003	1.0012	1.0026	1.0045	1.0071	1.0102	1.0138	1.0180	1.0228	1.0282	.55
.60	1	1.0003	1.0014	1.0031	1.0054	1.0084	1.0122	1.0166	1.0216	1.0274	1.0338	.60
.65	1	1.0004	1.0016	1.0036	1.0064	1.0100	1.0144	1.0196	1.0256	1.0325	1.0402	.65
.70	1	1.0004	1.0018	1.0042	1.0074	1.0116	1.0167	1.0228	1.0300	1.0382	1.0473	.70
0.75	1	1.0005	1.0021	1.0048	1.0085	1.0133	1.0193	1.0264	1.0348	1.0444	1.0552	0.75
.80	1	1.0006	1.0024	1.0054	1.0097	1.0152	1.0221	1.0303	1.0400	1.0512	1.0640	.80
.85	1	1.0006	1.0027	1.0061	1.0110	1.0173	1.0251	1.0345	1.0458	1.0588	1.0739	.85
.90	1	1.0007	1.0030	1.0069	1.0124	1.0195	1.0284	1.0392	1.0522	1.0673	1.0850	.90
0.95	1	1.0008	1.0034	1.0077	1.0139	1.0218	1.0320	1.0442	1.0592	1.0768	1.0976	0.95
1.00	1	1.0009	1.0038	1.0086	1.0154	1.0244	1.0358	1.0498	1.0668	1.0873	1.1117	1.00

ρ^2	$\alpha = 0.50$	0.55	0.60	0.65	0.70	0.75	0.80	0.85	0.90	0.95	$\alpha = 1.00$	ρ^2
0	1	1	1	1	1	1	1	1	1	1	1	0
0.05	1.0002	1.0002	1.0003	1.0004	1.0005	1.0005	1.0006	1.0007	1.0008	1.0008	1.0009	0.05
.10	1.0009	1.0011	1.0013	1.0016	1.0018	1.0020	1.0023	1.0026	1.0029	1.0032	1.0035	.10
.15	1.0020	1.0024	1.0029	1.0034	1.0040	1.0046	1.0051	1.0057	1.0063	1.0070	1.0077	.15
.20	1.0036	1.0044	1.0052	1.0061	1.0070	1.0080	1.0090	1.0101	1.0112	1.0123	1.0135	.20
0.25	1.0057	1.0068	1.0081	1.0094	1.0108	1.0124	1.0140	1.0156	1.0172	1.0190	1.0208	0.25
.30	1.0082	1.0098	1.0116	1.0136	1.0156	1.0178	1.0200	1.0224	1.0248	1.0273	1.0299	.30
.35	1.0111	1.0134	1.0158	1.0185	1.0213	1.0242	1.0272	1.0304	1.0337	1.0372	1.0406	.35
.40	1.0146	1.0176	1.0208	1.0242	1.0279	1.0317	1.0358	1.0400	1.0443	1.0488	1.0534	.40
.45	1.0186	1.0224	1.0264	1.0308	1.0355	1.0404	1.0456	1.0510	1.0566	1.0623	1.0683	.45
0.50	1.0231	1.0278	1.0330	1.0384	1.0444	1.0506	1.0571	1.0640	1.0710	1.0784	1.0858	0.50
.55	1.0282	1.0340	1.0404	1.0471	1.0544	1.0622	1.0704	1.0790	1.0878	1.0970	1.1064	.55
.60	1.0338	1.0410	1.0487	1.0570	1.0660	1.0756	1.0857	1.0964	1.1076	1.1189	1.1305	.60
.65	1.0402	1.0488	1.0582	1.0683	1.0792	1.0910	1.1036	1.1168	1.1304	1.1446	1.1593	.65
.70	1.0473	1.0575	1.0688	1.0812	1.0945	1.1089	1.1242	1.1405	1.1576	1.1754	1.1941	.70
0.75	1.0552	1.0674	1.0810	1.0960	1.1124	1.1301	1.1491	1.1693	1.1908	1.2134	1.2372	0.75
.80	1.0640	1.0785	1.0947	1.1126	1.1324	1.1545	1.1786	1.2044	1.2320	1.2612	1.2922	.80
.85	1.0739	1.0911	1.1106	1.1324	1.1568	1.1843	1.2150	1.2490	1.2864			.85
.90	1.0850	1.1054	1.1289	1.1558	1.1866	1.2220	1.2623	1.3080	1.3589	1.4150	1.4763	.90
0.95	1.0976	1.1216	1.1502	1.1842	1.2245							0.95
1.00	1.1117	1.1403	1.1752	1.2157	1.2733	1.3630	1.4978		1.7430		∞	1.00

116 CALCULATION OF MUTUAL INDUCTANCE AND SELF-INDUCTANCE

Evidently for the limiting case $\alpha = 1$, $x = 0$ and, therefore, $\rho^2 = 1$, we have two coincident coaxial circles, and the mutual inductance is logarithmically infinite. It is evident, therefore, that for short solenoids and circles nearly of the same radius (ρ^2 and α in the neighborhood of unity) the value of Q_0 cannot be accurately interpolated from the table. In such cases, this difficulty is avoided by the use of the following formula, which is based on an expression by Dwight.[57]

$$M = 0.002\pi NA\sqrt{\alpha R_0}\left[\log_\varepsilon\frac{16}{\gamma^2+\xi^2} - 2\frac{\gamma}{\xi}\tan^{-1}\frac{\xi}{\gamma} - 2\right], \quad (104)$$

in which appears the parameters $\gamma^2 = \frac{1}{4}\frac{(1-\alpha)^2}{\alpha}$, $\xi^2 = \frac{1}{4}\frac{(1-\rho^2)}{\alpha\rho^2}$, and the value of R_0 is to be taken from Table 28 for the given values of these parameters. The natural logarithm is obtained from the common logarithm by the use of Auxiliary Table 2 (page 237), and the antitangent is to be calculated in degrees and minutes by an ordinary trigonometrical table and converted to radians.

TABLE 28. VALUES OF R_0 FOR THE MUTUAL INDUCTANCE OF SOLENOID AND COAXIAL CIRCLE, FORMULA (104)

γ^2	$\xi^2 = 0$	0.01	0.02	0.03	0.04	0.05	$\xi^2 = 0.06$
0	1	1.0036	1.0071	1.0110	1.0149	1.0190	1.0232
0.005	1.0070	1.0116	1.0165	1.0213	1.0263	1.0314	1.0365
.010	1.0148	1.0204	1.0261	1.0318	1.0375	1.0433	1.0492
.015	1.0237	1.0299	1.0362	1.0424	1.0487	1.0551	1.0616
.020	1.0334	1.0400	1.0468	1.0535	1.0604	1.0672	1.0742
0.025	1.0438	1.0508	1.0580	1.0653	1.0726	1.0800	1.0875
.030	1.0548	1.0625	1.0702	1.0780	1.0858	1.0938	1.1018
.035	1.0666	1.0748	1.0830	1.0913	1.0996	1.1081	1.1168
0.040	1.0791	1.0878	1.0965	1.1055	1.1142	1.1232	1.1324

γ^2	$\xi^2 = 0.06$	0.07	0.08	0.09	0.10	0.11	$\xi^2 = 0.12$
0	1.0232	1.0274	1.0318	1.0362	1.0408	1.0454	1.0500
0.005	1.0365	1.0420	1.0471	1.0525	1.0580	1.0636	1.0693
.010	1.0492	1.0552	1.0613	1.0674	1.0737	1.0800	1.0866
.015	1.0616	1.0682	1.0749	1.0817	1.0886	1.0955	1.1026
.020	1.0742	1.0813	1.0886	1.0958	1.1033	1.1108	1.1185
0.025	1.0875	1.0951	1.1028	1.1107	1.1186	1.1268	1.1349
.030	1.1018	1.1100	1.1184	1.1268	1.1354	1.1440	1.1529
.035	1.1168	1.1255	1.1344	1.1434	1.1529	1.1619	1.1714
0.040	1.1324	1.1418	1.1512	1.1608	1.1706	1.1806	1.1908

A SOLENOID AND A COAXIAL CIRCULAR FILAMENT

These formulas apply strictly only to a cylindrical current sheet and a coaxial circular filament. However, they give very accurate values for a solenoid and a coaxial circular coil, provided the cross sectional dimensions of the latter are small in comparison with the radii. The value calculated by the preceding formulas must, of course, be multiplied by the number of turns N_2 on the circular coil.

General Case. Circle Not in the End Plane. In Figs. 34a and 34b are shown a solenoid and coaxial circle with a distance D between the plane

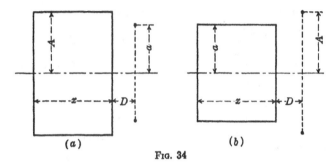

Fig. 34

of the circle and the end of the solenoid. Designating by M_{XD} the mutual inductance of the solenoid of length $(x + D)$ and the circle in its end plane (see figures), and M_D that of a solenoid of length D and the circle in its end plane, then $M_{XD} = M + M_D$, and the mutual inductance M of the solenoid of length x and the circle separated by a distance D is

$$M = M_{XD} - M_D. \qquad (105)$$

Both terms at the right are to be calculated by formula (103) or (104). This treatment applies equally to both of the cases shown in Figs. 34a and 34b.

Example 36: Given a solenoid of radius $A = 4$ inches and having 64 turns, wound with 32 turns of wire per inch of axial length, and a circular coil of 50 turns of mean radius 3 inches and negligible cross sectional dimensions, with its mean plane in the end plane of the solenoid.

The equivalent length x of the solenoid is $\dfrac{N}{n_1} = \dfrac{64}{32}$, or $x = 2$ inches.

$\alpha = \tfrac{3}{4} = 0.75$, $\qquad \rho^2 = \dfrac{(4)^2}{(4)^2 + (2)^2} = \dfrac{16}{20} = 0.8$,

$N = 64$, $\qquad N_2 = 50$.

From Table 27, $Q_0 = 1.1545$, and therefore, by (103)

$M = 0.002\pi^2(3 \times 2.54)(0.75)\sqrt{0.8}(64 \times 50)(1.1545)$

$\quad = 372.7 \ \mu\text{h}.$

118 CALCULATION OF MUTUAL INDUCTANCE AND SELF-INDUCTANCE

As an illustration of the use of formula (104) we may solve the same problem by that method also. The auxiliary parameters are

$$\gamma^2 = \frac{1}{4} \frac{(0.25)^2}{0.75} = 0.020833\cdots,$$

$$\xi^2 = \frac{1}{4} \frac{0.2}{(0.75)(0.8)} = 0.08333\cdots,$$

$$\gamma^2 + \xi^2 = 0.104167, \quad \frac{\gamma^2}{\xi^2} = \frac{1}{4}, \quad \frac{\gamma}{\xi} = \frac{1}{2}.$$

For the calculation of log and $\tan^{-1}$ terms we may systematize the work as follows:

$$\log 16 = 1.20412$$
$$\log(\gamma^2 + \xi^2) = \overline{1}.01773$$
$$\text{Diff.} = 2.18639$$

Converting by Auxiliary Table 2,

$$4.83543$$
$$.19802$$
$$90$$
$$\log_e \text{Diff.} = 5.03435$$

$$\tan^{-1}\frac{\xi}{\gamma} = 90° - \tan^{-1}\frac{\gamma}{\xi}$$
$$= 90° - \tan^{-1}\tfrac{1}{2}$$

$$90°$$
$$\tan^{-1}\tfrac{1}{2} = 26°\, 33'.91$$
$$\text{Diff.} = 63°\, 26'.09$$

Changing to radians,

$$1.099557$$
$$7563$$
$$26$$
$$\tan^{-1}\frac{\xi}{\gamma} = 1.10715 \text{ radians}$$

$$2\frac{\gamma}{\xi} = 1.$$

Therefore, by (104), the factor in the brackets is $5.03435 - 1.10715 - 2 = 1.92720$.
For the interpolation of the value of R_0 we take from Table 28 the extract shown and follow the example of page 111 for the details of the interpolation:

γ^2	$\xi^2 = 0.08$		0.09		0.10
0.020	1.0885₅	(73)	1.0958₅	(74.5)	1.1033
	(142.5)		(148.5)		
0.025	1.1028	(79)	1.1107		
	(155.5)				
0.030	1.1183₅				

$$d_\xi' = 73, \qquad\qquad d_\gamma' = 142.5,$$
$$d_\xi'' = 1.5, \qquad\qquad d_\gamma'' = 13,$$
$$d_{\xi\gamma}'' = 79 - 73, \qquad d_{\gamma\xi}'' = 148.5 - 142.5,$$
$$= 6. \qquad\qquad\qquad = 6.$$

The fractional interval of interpolation for ξ^2 is $u = \dfrac{1}{3}$, and for γ^2, $v = \dfrac{.000833}{.005} = \dfrac{1}{6}$.

Also $\dfrac{1-u}{2} = \dfrac{1}{3}$, $\dfrac{1-v}{2} = \dfrac{5}{12}$, and $uv = \dfrac{1}{18}$.

From these data

$$R_0 = 1.0885_5 + \tfrac{1}{3}[.0073 - \tfrac{1}{3}(.0001_5)] + \tfrac{1}{6}[.01425 - \tfrac{5}{12}(.0013)] + \tfrac{1}{18}(.0006)$$

$$= 1.0933,$$

and

$$M = 0.002\pi(4 \times 2.54)(64 \times 50)\sqrt{0.75}(1.92720)(1.0933)$$

$$= 372.7 \ \mu\text{h},$$

in agreement with the other formula.

Campbell Form of Mutual Inductance Standard. The assumption that the dimensions of the cross section of the circular coil which forms the secondary in the previous example shall be small, compared with the other dimensions, cannot always be conveniently managed. The N_2 turns are supposed to be situated at sensibly the same position as the circular filament through the center of the cross section, but actually some of the turns are nearer the center of the solenoid and others more distant than is the central filament. Likewise, some of the turns of the coil have a greater radius and others a smaller radius than the central filament. These differences will have effects which will not, in general, average out completely. In the case of the form of mutual inductance standard proposed by Campbell,[18] the residual errors due to the replacing of the coil by its central filament are negligibly small.

In the Campbell standard (Fig. 35) the primary consists of two equal solenoids of radius a and length x joined in series, placed coaxially with a distance $2x$ between their ends. In the median plane between the solenoids is located the secondary, which is a circular coil of such a radius A that the magnetic field is zero for all points of the central circular filament. Under this condition, the field strength varies little from zero over a region of appreciable dimensions. For values of the radius somewhat smaller or greater, the mutual inductance on a turn is only very slightly less than the maximum value which holds for the critical radius.

It is, therefore, possible to wind, for the secondary, a coil of a large number of turns in a channel of rectangular or square cross section, and the total

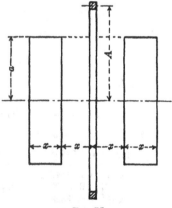

Fig. 35

120 CALCULATION OF MUTUAL INDUCTANCE AND SELF-INDUCTANCE

mutual inductance of the solenoids on the coil of N_2 turns is very closely equal to N_2 times the value for the central filament.

Example 37: Campbell gives for his standard the following data. The primary consists of two equal single-layer coils of 50 turns of radius $a = 10$ cm., wound in a screw thread with 10 turns per centimeter of axial length, so that $x = 5$ cm. The secondary consists of 1000 turns wound in a channel so as to have a cross section of about 1 cm. by 1 cm. Preliminary calculations showed that the central turn of the secondary should have a radius of $A = 14.6$ cm.

The mutual inductance of primary and secondary will be equal to twice the mutual inductance of one solenoid on the secondary. To calculate this we have to find the difference of the mutual inductances for the two cases [see (105)]:

I	II
$a = 10$	$a = 10$
$A = 14.6$	$A = 14.6$
length $= 2x = 10$	length $= x = 5$
$N_1 = 100$	$N_1 = 50$
$N_2 = 1000$	$N_2 = 1000$

For both cases $\alpha = \dfrac{10}{14.6} = 0.68493$,

$$\rho_1^2 = \frac{(14.6)^2}{(14.6)^2 + (10)^2} = 0.68069, \qquad \rho_2^2 = \frac{(14.6)^2}{(14.6)^2 + (5)^2} = 0.89502.$$

By formulas (105) and (103)

$$\frac{M}{2} = 0.002\pi^2(10)\alpha[(100 \times 1000)\rho_1 Q_0' - (50 \times 1000)\rho_2 Q_0''].$$

From Table 27 for the above values of α and ρ^2 there is found by interpolation $Q_0' = 1.0845$ and $Q_0'' = 1.1740$, so that

$$M = 0.002\pi^2(10)\alpha(100000)[2\rho_1 Q_0' - \rho_2 Q_0''].$$

The calculation may be arranged as follows:

$\log \rho_1^2 = \bar{1}.83294$	$\log \rho_2^2 = \bar{1}.95183$	$2\,\rho_1 Q_0' = 1.7895$
$\log \rho_1 = \bar{1}.91647$	$\log \rho_2 = \bar{1}.97592$	$\rho_2 Q_0'' = 1.1107$
$\log 2 = 0.30103$	$\log Q_0'' = 0.06967$	Diff. $= 0.6788$
$\log Q_0' = 0.03523$	0.04559	
0.25273		

Substituting this difference for the last factor there is found

$$M = 9.177 \text{ mh}.$$

Evidently the last figure is somewhat uncertain because of the subtraction of terms. Campbell,[16] using the very accurate and complicated formula of Jones[18] gave 9.1763 mh. It is of interest also to note that his calculations show that for a change of mean radius A from 14.3 to 14.7 cm. the mutual inductance was confined to the limits of

A SOLENOID AND A COAXIAL CIRCULAR FILAMENT   121

from 9.1754 to 9.1763 mh. Also a displacement of the secondary coil axially as much as 0.35 cm. reduced the value of M by less than 1 part in 10,000.

The calculation of the second term in the previous problem was also made by formula (104). The calculated values of the parameters are

$$\gamma^2 = 0.036233, \qquad \xi^2 = 0.042812,$$

$$\log_e\left(\frac{16}{\gamma^2 + \xi^2}\right) = 5.3103, \qquad \tan^{-1}\frac{\xi}{\gamma} = 47°\,23'23$$

$$= 0.82706 \text{ radians,}$$

and

$$\left[\log_e\left(\frac{16}{\gamma^2 + \xi^2}\right) - 2\frac{\gamma}{\xi}\tan^{-1}\frac{\xi}{\gamma} - 2\right] = 5.3103 - 1.5217 - 2$$

$$= 1.7886.$$

The interpolated value of R_0 from Table 28 is 1.1055 and thus $M_2 = 15.012$ mh. The first term in the previous problem comes out 24.194 mh, and for the standard

$$M = 24.194 - 15.012 = 9.182 \text{ mh.}$$

The agreement is satisfactory.

Chapter 15

MUTUAL INDUCTANCE OF COAXIAL SINGLE-LAYER COILS

Nomenclature. The two coils are supposed to have radius, axial length, and winding density, a, $2m_1$, and n_1, respectively, for one coil and the corresponding quantities A, $2m_2$, and n_2 for the other. The radius A is assumed to be the larger.

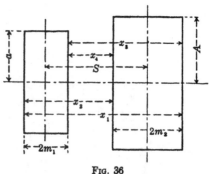

Fig. 36

The distance s between the centers of the coils may have any value. Thus one coil may be entirely inside the other, or partly inside, or entirely outside, depending on the dimensions and the spacing s.

The axial lengths are understood to be taken as equal to the distance p between centers of adjacent wires times the total number of turns on the coil. Thus

$$2m_1 = N_1 p_1 \quad \text{and} \quad 2m_2 = N_2 p_2.$$

Since the winding density is the reciprocal of the pitch, the further relations exist that

$$2m_1 n_1 = N_1 \quad \text{and} \quad 2m_2 n_2 = N_2.$$

The mutual inductance of coaxial cylindrical current sheets has been found by Nagaoka,[59] Terezawa,[60] and Olshausen.[61] Jones[62] has shown that these formulas give also the mutual inductance of a helical wire and a coaxial cylindrical current sheet. All these formulas make clear that in the most general case the mutual inductance is given as the sum of four integrals, two positive and two negative, which are functions of the four distances measured between the ends of one coil and the ends of the other. These distances, shown in Fig. 36, are

$$x_1 = s + (m_1 + m_2),$$
$$x_2 = s + (m_1 - m_2),$$
$$x_3 = s - (m_1 - m_2), \quad (106)$$
$$x_4 = s - (m_1 + m_2).$$

The values of the integrals in these absolute formulas depend upon elliptic integrals of all three kinds, and computations are tedious and complicated. To obtain more convenient expressions for numerical purposes, many series expansions of the elliptic integral expressions have been derived,[63] and it is possible in many cases, by choosing a suitable series formula, to calculate the mutual inductance without serious difficulty. Cases arise, however, where the absolute formulas have to be used because of poor convergence of the series expressions.

The general formula given below, used in conjunction with tables, avoids these difficulties, and applies in every case. Certain special cases are, however, simply treated also by special formulas which are given also to provide the possibility of check calculations in addition to the results by the general formula.

General Formula. The following general formula is based on one given by Clem.[64]

Having calculated the quantities x_n of equation (106), the four corresponding diagonals r_n are calculated by the relations

$$r_1 = \sqrt{A^2 + x_1^2}, \qquad r_3 = \sqrt{A^2 + x_3^2},$$
$$r_2 = \sqrt{A^2 + x_2^2}, \qquad r_4 = \sqrt{A^2 + x_4^2}, \quad (107)$$

and the mutual inductance of the coaxial solenoids is given by the general formula

$$M = 0.002\pi^2 a^2 n_1 n_2 [r_1 B_1 - r_2 B_2 - r_3 B_3 + r_4 B_4]. \quad (108)$$

The functions B_n in this expression depend upon the parameters

$$\rho_n^2 = \frac{A^2}{r_n^2} \quad \text{and} \quad \alpha = \frac{a}{A},$$

and are to be obtained [64a] by interpolation from Tables 29 and 30.

124 CALCULATION OF MUTUAL INDUCTANCE AND SELF-INDUCTANCE

TABLE 29. VALUES OF B_n AS FUNCTION OF α AND ρ^2, FORMULA (108)

α	$\rho^2 = 1.00$	0.95	0.90	0.85	0.80	0.75	0.70	0.65	0.60	0.55	$\rho^2 = 0.50$	α
1.00	0.84833	0.87727	0.89552	0.91020	0.92264	0.93345	0.94298	0.95144	0.95900	0.96576	0.97180	1.00
0.95	.86783	.89982	.90561	.91859	.92971	.93944	.94805	.95573	.96261	.96877	.97428	0.95
.90	.88418	.90175	.91531	.92666	.93655	.94524	.95298	.95990	.96612	.97169	.97668	.90
.85	.89870	.91296	.92456	.93444	.94314	.95085	.95774	.96393	.96951	.97452	.97901	.85
.80	.91176	.92344	.93329	.94185	.94944	.95622	.96231	.96781	.97276	.97723	.98124	.80
0.75	0.92356	0.93318	0.94150	0.94885	0.95542	0.96132	0.96668	0.97151	0.97588	0.97983	0.98338	0.75
.70	.93426	.94217	.94917	.95543	.96107	.96618	.97082	.97503	.97884	.98230	.98541	.70
.65	.94394	.95045	.95629	.96157	.96637	.97074	.97472	.97835	.98164	.98464	.98732	.65
.60	.95270	.95803	.96286	.96727	.97130	.97499	.97837	.98146	.98427	.98683	.98913	.60
.55	.96060	.96492	.96888	.97252	.97586	.97894	.98176	.98435	.98672	.98887	.99082	.55
0.50	0.96769	0.97115	0.97434	0.97730	0.98003	0.98256	0.98488	0.98702	0.98897	0.99076	0.99237	0.50
.45	.97400	.97673	.97927	.98163	.98382	.98584	.98772	.98945	.99103	.99248	.99379	.45
.40	.97958	.98169	.98366	.98550	.98721	.98880	.99028	.99164	.99289	.99404	.99508	.40
.35	.98444	.98603	.98751	.98890	.99020	.99142	.99254	.99358	.99454	.99542	.99622	.35
.30	.98862	.98976	.99084	.99186	.99280	.99369	.99451	.99527	.99598	.99662	.99721	.30
0.25	0.99212	0.99291	0.99365	0.99435	0.99500	0.99561	0.99618	0.99671	0.99720	0.099765	0.99806	0.25
.20	.99498	.99547	.99594	.99638	.99680	.99719	.99755	.99789	.99821	.99849	.99875	.20
.15	.99718	.99746	.99772	.99797	.99820	.99842	.99862	.99881	.99899	.99915	.99930	.15
.10	.99875	.99887	.99899	.99910	.99920	.99930	.99939	.99947	.99955	.99962	.99969	.10
.05	.99969	.99972	.99975	.99977	.99980	.99982	.99985	.99987	.99989	.99991	.99992	.05
0	1	1	1	1	1	1	1	1	1	1	1	0

MUTUAL INDUCTANCE OF COAXIAL SINGLE-LAYER COILS 125

TABLE 29. VALUES OF B_R AS FUNCTION OF α AND ρ^2, FORMULA (108) (*Concluded*)

α	$\rho^2 = 0.50$	0.45	0.40	0.35	0.30	0.25	0.20	0.15	0.10	0.05	$\rho^2 = 0$	α
1.00	0.97180	0.97718	0.98194	0.98612	0.98974	0.99282	0.99535	0.99735	0.99880	0.99969	1	1.00
0.95	.97428	.97919	.98354	.98736	.99066	.99346	.99577	.99759	.99891	.99972	1	0.95
.90	.97668	.98114	.98509	.98855	.99155	.99409	.99618	.99783	.99902	.99975	1	.90
.85	.97901	.98302	.98658	.98970	.99240	.99469	.99657	.99805	.99912	.99978	1	.85
.80	.98124	.98483	.98801	.99080	.99322	.99526	.99695	.98827	.99922	.99980	1	.80
0.75	0.98338	0.98656	0.98938	0.99185	0.99399	0.99581	0.99730	0.98847	0.99931	0.99983	1	0.75
.70	.98541	.98820	.99068	.99285	.99473	.99633	.99764	.98866	.99940	.99985	1	.70
.65	.98732	.98975	.99190	.99380	.99543	.99682	.99795	.98884	.99948	.99987	1	.65
.60	.98913	.99121	.99306	.99467	.99608	.99727	.99825	.99901	.99956	.99989	1	.60
.55	.99082	.99257	.99413	.99550	.99669	.99770	.99852	.99916	.99963	.99990	1	.55
0.50	0.99237	0.99383	0.99512	0.99626	0.99725	0.99809	0.98877	0.99931	0.99969	0.99992	1	0.50
.45	.99379	.99498	.99603	.99696	.99776	.99844	.99900	.99944	.99975	.99994	1	.45
.40	.99508	.99601	.99685	.99759	.99823	.98877	.99921	.99955	.99980	.99995	1	.40
.35	.99622	.99694	.99758	.99815	.99864	.99905	.99939	.99966	.99985	.99996	1	.35
.30	.99721	.99774	.99822	.99864	.99900	.99930	.99955	.99975	.99989	.99997	1	.30
0.25	0.99806	0.99843	0.98876	0.99905	0.99930	0.99952	0.99969	0.99982	0.99992	0.99998	1	0.25
.20	.99875	.99899	.99920	.99939	.99955	.99969	.99980	.99989	.99995	.99999	1	.20
.15	.99930	.99943	.99955	.99966	.99975	.99982	.99989	.99994	.99997	0.99999	1	.15
.10	.99969	.99975	.99980	.99985	.99989	.99992	.99995	.99997	.99999	1.00000	1	.10
0.05	0.99992	0.99994	0.99995	0.99996	0.99997	0.99998	0.99999	0.99999	1.00000	1.00000	1	0.05
0	1	1	1	1	1	1	1	1	1	1	1	0

It is to be noted that the values of the parameters are always included in the range of from zero to unity. Evidently the $\rho_n{}^2$ give the squares of the sine of the angle subtended by the radius of the larger coil at the various axial distances x_n.

TABLE 30. VALUES OF B_n. AUXILIARY TABLE FOR LARGE α AND ρ^2

α	$\rho^2 = 1.00$	0.99	0.98	0.97	0.96	0.95	α
1.00	0.84883	0.85698	0.86298	0.86820	0.87292	0.87727	1.00
0.99	.85294	.86035	.86606	.87107	.87562	.87982	0.99
.98	.85686	.86366	.86910	.87391	.87829	.88236	.98
.97	.86063	.86693	.87210	.87672	.88094	.88487	.97
.96	.86428	.87014	.87506	.87949	.88356	.88736	.96
0.95	0.86783	0.87329	0.87798	0.88223	0.88615	0.88982	0.95
.94	.87127	.87639	.88086	.88494	.88872	.89226	.94
.93	.87462	.87944	.88370	.88761	.89125	.89468	.93
.92	.87788	.88242	.88649	.89024	.89375	.89706	.92
.91	.88107	.88536	.88924	.89285	.89622	.89942	.91
0.90	0.88418	0.88824	0.89195	0.89541	0.89866	0.90175	0.90

α	$\rho^2 = 0.95$	0.94	0.93	0.92	0.91	$\rho^2 = 0.90$	α
1.00	0.87727	0.88133	0.88515	0.88877	0.89222	0.89552	1.00
0.99	.87982	.88376	.88747	.89100	.89436	.89757	0.99
.98	.88236	.88617	.88978	.89320	.89647	.89960	.98
.97	.88487	.88857	.89207	.89539	.89858	.90162	.97
.96	.88736	.89094	.89433	.89757	.90066	.90362	.96
0.95	0.88982	0.89329	0.89658	0.89972	0.90273	0.90561	0.95
.94	.89226	.89562	.89881	.90186	.90478	.90759	.94
.93	.89468	.89792	.90102	.90397	.90681	.90954	.93
.92	.89706	.90020	.90320	.90607	.90883	.91148	.92
.91	.89942	.90246	.90536	.90815	.91082	.91340	.91
0.90	0.90175	0.90469	0.90750	0.91020	0.91280	0.91531	0.90

The values of the B_n will usually have to be obtained by double interpolation in the tables for the two parameters. The method illustrated on page 111 should be used. In some cases second order differences may be neglected. The values of B_n in the tables are carried to a greater number of figures than appears at first sight necessary, for the reason that such an accuracy is required in unfavorable cases in order that the difference of the terms in formula (108) may be obtained with sufficient accuracy.

MUTUAL INDUCTANCE OF COAXIAL SINGLE-LAYER COILS   127

Example 38:

Given $a = 20$, $2m_1 = 10$, $s = 10$,

$A = 25$, $2m_2 = 16$, $n_1 = 10$,

$n_2 = 20$.

From these data is found $\alpha = 0.8$.

$x_1 = 23$, $x_2 = 7$, $x_3 = 13$, $x_4 = -3$,

$r_1^2 = 625 + 529$ $r_2^2 = 625 + 49$ $r_3^2 = 625 + 169$ $r_4^2 = 625 + 9$

$= 1154$, $= 674$, $= 794$, $= 634$,

$\rho_1^2 = \frac{625}{1154}$, $\rho_2^2 = \frac{625}{674}$, $\rho_3^2 = \frac{625}{794}$, $\rho_4^2 = \frac{625}{634}$,

log A^2 = 2.79588   2.79588   2.79588   2.79588
log r_n^2 = 3.06221   2.82866   2.89982   2.80209

log ρ_n^2 = $\overline{1}$.73367 $\overline{1}$.96722 $\overline{1}$.89606 $\overline{1}$.99379

$\rho_1^2 = 0.54159$, $\rho_2^2 = 0.92730$, $\rho_3^2 = 0.78715$, $\rho_4^2 = 0.98580$.

Interpolating in Table 29, there result

$B_1 = 0.97794$, $B_2 = 0.928135$, $B_3 = 0.95126$, $B_4 = 0.91489$,

log r_n = 1.531105   1.41433   1.44991   1.401045
log B_n = $\overline{1}$.99031 $\overline{1}$.96761 $\overline{1}$.97830 $\overline{1}$.961365

log $r_n B_n$ = 1.521415   1.38194   1.42821   1.36241

$r_1 B_1 = 33.221$, $r_2 B_2 = 24.096$, $r_3 B_3 = 26.805$, $r_4 B_4 = 23.036$.

$r_1 B_1 + r_4 B_4 = 56.257$
$r_2 B_2 + r_3 B_3 = \underline{50.901}$

Diff. = 5.356.

$a^2 n_1 n_2 = 80,000$

log $a^2 n_1 n_2 = 4.90309$
log $.002\pi^2 = \overline{2}.29533$
log Diff. = 0.72884

log M = 3.92726

$M = 8458\ \mu\text{h}$

$= 8.458\ \text{mh}$.

Thus the calculation is readily systematized and the principal labor is that of interpolating the values of the B_n in the tables.

It will be noticed that the mutual inductance comes out as the difference of the sums of pairs of terms that must be calculated with a greater degree of accuracy than is attainable in the desired value of the mutual inductance. This difficulty is inherent in the nature of the problem and may be avoided only in certain special cases where the values of the dimensions are such as to make possible the use of a series expansion of the general formula. Such special cases will next be illustrated.

128 CALCULATION OF MUTUAL INDUCTANCE AND SELF-INDUCTANCE

Concentric Coaxial Coils. This is a most important practical case. Here $s = 0$, $x_1 = m_1 + m_2$, $x_2 = m_1 - m_2$ and, furthermore, $x_4 = -x_1$, $x_3 = -x_2$. Since, however, the x_n enter only as squares in the expressions for the diagonals, $r_4 = r_1$ and $r_3 = r_2$. Formula (108) becomes, therefore,

$$M = 0.004\pi^2 a^2 n_1 n_2 [r_1 B_1 - r_2 B_2]. \tag{109}$$

This is the general formula for the concentric case. If, further, the concentric coils have the same axial length, $x_2 = x_3 = 0$, $r_2 = r_3 = A$, and $\rho_2{}^2 = \rho_3{}^2 = 1$, so that (109) becomes, for coils of the same length,

$$M = 0.004\pi^2 a^2 n_1 n_2 [r_1 B_1 - A B_2]. \tag{110}$$

Example 39: For calibration purposes in the laboratory frequent use has been made of mutual inductance standards consisting of a long solenoid as primary with a short coil wound about the center to serve as secondary. To illustrate the use of formula (109) for such a standard, suppose that we have two coils for which

$$a = 3, \qquad 2m_1 = 50, \qquad n_1 = 10,$$

$$A = 4, \qquad 2m_2 = 4, \qquad n_2 = 50,$$

so that the total number of turns on the two coils are, respectively, $N_1 = 500$ and $N_2 = 200$. Here $a^2 n_1 n_2 = 4500$ and $\alpha = 0.75$,

$$x_1 = 27, \qquad\qquad x_2 = 23,$$

$$r_1{}^2 = 16 + 729 = 745, \qquad r_2{}^2 = 16 + 529 = 545,$$

$$\rho_1{}^2 = \tfrac{16}{745} = 0.02148, \qquad \rho_2{}^2 = \tfrac{16}{545} = 0.02936.$$

Interpolating in Table 29, with $\alpha = 0.75$,

$$B_1 = 0.99997, \quad B_2 = 0.99994,$$

which give

$$r_1 B_1 = 27.294$$

$$r_2 B_2 = \underline{23.344}$$

$$\text{Diff.} = 3.950,$$

and

$$M = 0.004\pi^2 (4500)(3.950)$$

$$= 701.7 \mu h = 0.7017 \text{ mh.}$$

When the inner of two concentric solenoids has the shorter length, a series formula is useful for checking the value calculated from the general formula (109). This expression, derived from one given by Searle and Airey,[65] reads

$$M = 0.002 \frac{\pi^2 a^2 N_1 N_2}{\rho} \left[1 - \frac{1}{2} \frac{A^2}{\rho^2} \cdot \frac{\delta^2}{\rho^2} \left\{ \lambda_2 + \lambda_4 \xi_2 \frac{\delta^2}{\rho^2} + \lambda_6 \xi_4 \frac{\delta^4}{\rho^4} + \lambda_8 \xi_6 \frac{\delta^6}{\rho^6} + \cdots \right\} \right], \tag{111}$$

MUTUAL INDUCTANCE OF COAXIAL SINGLE-LAYER COILS

in which $\delta^2 = a^2 + m_1^2$, $\rho^2 = A^2 + m_2^2$, and the coefficients are as follows:

$$\lambda_2 = 1 - \frac{7}{4}\frac{a^2}{\delta^2},$$

$$\lambda_4 = 1 - \frac{9}{2}\frac{a^2}{\delta^2} + \frac{33}{8}\frac{a^4}{\delta^4},$$

$$\lambda_6 = 1 - \frac{33}{4}\frac{a^2}{\delta^2} + \frac{143}{8}\frac{a^4}{\delta^4} - \frac{715}{64}\frac{a^6}{\delta^6}.$$

The coefficients ξ_{2n} are the same functions of $\frac{A^2}{\rho^2}$ as λ_{2n} are of $\frac{a^2}{\delta^2}$. Values of these functions are given in Tables 31, 32, 33, and 34.

Evidently formula (111) converges more rapidly, assuming a given shape ratio of the outer coil, the smaller the ratio $\frac{\delta^2}{\rho^2}$, that is, the smaller the diagonal of the inner coil as compared with the diagonal of the outer coil. It is to be noted that the total numbers of turns on the coils enter in formula (111) instead of the winding densities that appear in formula (109).

TABLE 31. VALUES OF POLYNOMIAL $\lambda_2(\gamma^2)$ AS FUNCTION OF γ^2

$$\lambda_2(\gamma^2) = 1 - \tfrac{7}{4}\gamma^2$$

γ^2	λ_2	d_1	γ^2	λ_2	d_1	γ^2	λ_2	d_1	γ^2	λ_2	d_1
0	1.0000		0.25	0.5625		0.50	0.1250		0.75	−0.3125	
		−175			−175			−175			−175
0.01	0.9825		.26	.5450		.51	.1075		.76	− .3300	
.02	.9650		.27	.5275		.52	.0900		.77	− .3475	
.03	.9475		.28	.5100		.53	.0725		.78	− .3650	
.04	.9300		.29	.4925		.54	.0550		.79	− .3825	
0.05	0.9125		0.30	0.4750		0.55	0.0375		0.80	−0.4000	
		−175			−175			−175			−175
.06	.8950		.31	.4575		.56	.0200		.81	− .4175	
.07	.8775		.32	.4400		.57	+ .0025		.82	− .4350	
.08	.8600		.33	.4225		.58	− .0150		.83	− .4525	
.09	.8425		.34	.4050		.59	− .0325		.84	− .4700	
0.10	0.8250		0.35	0.3875		0.60	−0.0500		0.85	−0.4875	
		−175			−175			−175			−175
.11	.8075		.36	.3700		.61	− .0675		.86	− .5050	
.12	.7900		.37	.3525		.62	− .0850		.87	− .5225	
.13	.7725		.38	.3350		.63	− .1025		.88	− .5400	
.14	.7550		.39	.3175		.64	− .1200		.89	− .5575	
0.15	0.7375		0.40	0.3000		0.65	−0.1375		0.90	−0.5750	
		−175			−175			−175			−175
.16	.7200		.41	.2825		.66	− .1550		.91	− .5925	
.17	.7025		.42	.2650		.67	− .1725		.92	− .6100	
.18	.6850		.43	.2475		.68	− .1900		.93	− .6275	
.19	.6675		.44	.2300		.69	− .2075		.94	− .6450	
0.20	0.6500		0.45	0.2125		0.70	−0.2250		0.95	−0.6625	
		−175			−175			−175			−175
.21	.6325		.46	.1950		.71	− .2425		.96	− .6800	
.22	.6150		.47	.1775		.72	− .2600		.97	− .6975	
.23	.5975		.48	.1600		.73	− .2775		.98	− .7150	
.24	.5800		.49	.1425		.74	− .2950		0.99	− .7325	
0.25	0.5625		0.50	0.1250		0.75	−0.3125		1.00	−0.7500	

130 CALCULATION OF MUTUAL INDUCTANCE AND SELF-INDUCTANCE

TABLE 32. VALUES OF POLYNOMIAL $\lambda_4(\gamma^2)$ AS FUNCTION OF γ^2

Second differences have constant value $+8$

$$\lambda_4(\gamma^2) = 1 - \tfrac{9}{2}\gamma^2 + \tfrac{33}{8}\gamma^4$$

γ^2	λ_4	d_1	γ^2	λ_4	d_1	γ^2	λ_4	d_1	γ^2	λ_4	d_1
0	1.0000		0.25	0.1328		0.50	−0.2188		0.75	−0.0547	
		−446			−240			−33			173
0.01	0.9554		.26	.1088		.51	−.2221		.76	−.0374	
		−438			−231			−25			181
.02	.9116		.27	.0857		.52	−.2246		.77	−.0193	
		−429			−223			−17			189
.03	.8687		.28	.0634		.53	−.2263		.78	−.0004	
		−421			−215			−9			198
.04	.8266		.29	.0419		.54	−.2271		.79	+.0194	
		−413			−207			−1			206
0.05	0.7853		0.30	0.0212		0.55	−0.2272		0.80	+0.0400	
		−405			−198			+8			214
.06	.7448		.31	+.0014		.56	−.2264		.81	.0614	
		−396			−190			16			222
.07	.7052		.32	−.0176		.57	−.2248		.82	.0836	
		−388			−182			24			231
.08	.6664		.33	−.0358		.58	−.2224		.83	.1067	
		−380			−174			33			239
.09	.6284		.34	−.0532		.59	−.2191		.84	.1306	
		−372			−165			41			247
0.10	0.5912		0.35	−0.0697		0.60	−0.2150		0.85	0.1553	
		−363			−157			49			255
.11	.5549		.36	−.0854		.61	−.2101		.86	.1808	
		−355			−149			57			264
.12	.5194		.37	−.1003		.62	−.2044		.87	.2072	
		−347			−141			66			272
.13	.4847		.38	−.1144		.63	−.1978		.88	.2344	
		−339			−132			74			280
.14	.4508		.39	−.1276		.64	−.1904		.89	.2624	
		−330			−124			82			288
0.15	0.4178		0.40	−0.1400		0.65	−0.1822		0.90	0.2912	
		−322			−116			90			297
.16	.3856		.41	−.1516		.66	−.1732		.91	.3209	
		−314			−108			99			305
.17	.3542		.42	−.1624		.67	−.1633		.92	.3514	
		−306			−99			107			313
.18	.3236		.43	−.1723		.68	−.1526		.93	.3827	
		−297			−91			115			321
.19	.2939		.44	−.1814		.69	−.1411		.94	.4148	
		−289			−83			123			330
0.20	0.2650		0.45	−0.1897		0.70	−0.1288		0.95	0.4478	
		−281			−75			132			338
.21	.2369		.46	−.1972		.71	−.1156		.96	.4816	
		−273			−66			140			346
.22	.2096		.47	−.2038		.72	−.1016		.97	.5162	
		−264			−58			148			354
.23	.1832		.48	−.2096		.73	−.0868		.98	.5516	
		−256			−50			156			363
.24	.1576		.49	−.2146		.74	−.0712		0.99	.5879	
		−248			−42			165			371
0.25	0.1328		0.50	−0.2188		0.75	−0.0547		1.00	0.6250	

MUTUAL INDUCTANCE OF COAXIAL SINGLE-LAYER COILS 131

TABLE 33. VALUES OF POLYNOMIAL $\lambda_6(\gamma^2)$ AS FUNCTION OF γ^2

Third difference less than 1 unit in last place

$$\lambda_6(\gamma^2) = 1 - \tfrac{33}{4}\gamma^2 + \tfrac{143}{8}\gamma^4 - \tfrac{715}{64}\gamma^6$$

γ^2	λ_6	d_1	d_2	γ^2	λ_6	d_1	d_2	γ^2	λ_6	d_1	d_2	γ^2	λ_6	d_1	d_2
0	1.0000			0.25	−0.1199			0.50	−0.0527			0.75	0.1540		
		−807				−131				126				−36	
0.01	0.9193		35	.26	−.1330		18	.51	−.0402		+2	.76	.1504		−15
		−772				−113				127				−52	
.02	.8421		34	.27	−.1443		18	.52	−.0274		+1	.77	.1452		−16
		−738				−95				128				−67	
.03	.7683		34	.28	−.1538		17	.53	−.0146		0	.78	.1385		−16
		−704				−78				128				−84	
.04	.6979		33	.29	−.1617		16	.54	−.0018		0	.79	.1301		−17
		−671				−62				128				−101	
0.05	0.6308		32	0.30	−0.1679		16	0.55	+0.0110		−1	0.80	0.1200		−18
		−638				−46				127				−119	
.06	.5669		32	.31	−.1725		15	.56	.0236		−2	.81	.1081		−18
		−607				−32				125				−138	
.07	.5063		31	.32	−.1757		14	.57	.0361		−2	.82	.0943		−19
		−576				−17				122				−157	
.08	.4487		30	.33	−.1774		14	.58	.0484		−3	.83	.0787		−20
		−545				−4				119				−177	
.09	.3941		30	.34	−.1778		13	.59	.0603		−4	.84	.0610		−20
		−516				+9				116				−197	
0.10	0.3426		29	0.35	−0.1768		12	0.60	0.0719		−4	0.85	0.0413		−21
		−487				22				111				−218	
.11	.2939		28	.36	−.1746		12	.61	.0830		−5	.86	+.0195		−22
		−458				33				106				−240	
.12	.2481		28	.37	−.1713		11	.62	.0936		−6	.87	−.0046		−23
		−430				44				100				−263	
.13	.2050		27	.38	−.1669		10	.63	.1036		−6	.88	−.0309		−23
		−404				55				94				−286	
.14	.1647		26	.39	−.1614		10	.64	.1130		−7	.89	−.0595		−24
		−377				64				86				−310	
0.15	0.1270		26	0.40	−0.1550		9	0.65	0.1216		−8	0.90	−0.0905		−25
		−351				73				79				−335	
.16	.0918		25	.41	−.1477		8	.66	.1295		−8	.91	−.1240		−25
		−326				81				70				−360	
.17	.0592		24	.42	−.1396		8	.67	.1365		−9	.92	−.1600		−26
		−302				89				61				−386	
.18	.0290		24	.43	−.1307		7	.68	.1426		−10	.93	−.1985		−27
		−278				96				51				−412	
.19	+.0012		23	.44	−.1211		6	.69	.1477		−10	.94	−.2398		−27
		−255				102				41				−440	
0.20	−0.0244		22	0.45	−0.1108		6	0.70	0.1518		−11	0.95	−0.2838		−28
		−233				108				30				−468	
.21	−.0477		22	.46	−.1001		5	.71	.1548		−12	.96	−.3305		−29
		−211				113				18				−496	
.22	−.0688		21	.47	−.0888		4	.72	.1565		−12	.97	−.3801		−29
		−190				117				+5				−526	
.23	−.0878		20	.48	−.0771		4	.73	.1570		−13	.98	−.4327		−30
		−170				120				−8				−556	
.24	−.1048		20	.49	−.0651		3	.74	.1562		−14	0.99	−.4882		−31
		−150				123				−22				−586	
0.25	−0.1199		19	0.50	−0.0527		+2	0.75	0.1540		−14	1.00	−0.5469		

132 CALCULATION OF MUTUAL INDUCTANCE AND SELF-INDUCTANCE

TABLE 34. VALUES OF POLYNOMIAL $\lambda_8(\gamma^2)$ AS FUNCTION OF γ^2

Fourth differences less than one unit in last place

$$\lambda_8(\gamma^2) = 1 - 13\gamma^2 + \tfrac{195}{4}\gamma^4 - \tfrac{1105}{16}\gamma^6 + \tfrac{4199}{128}\gamma^8$$

γ^2	λ_8	d_1	d_2	γ^2	λ_8	d_1	d_2	γ^2	λ_8	d_1	d_2	γ^2	λ_8	d_1	d_2
0	1.0000			0.25	−0.1541				0.50	0.1050			0.75	−0.0843		
		−1252				56					30				−102	
0.01	0.8748		93	.26	−.1484		16	.51	.1080		−11	.76	−.0944		10
		−1159				73					18				−92	
.02	.7589		89	.27	−.1412		14	.52	.1098		−12	.77	−.1036		12
		−1069				87					+7				−80	
.03	.6520		86	.28	−.1324		12	.53	.1105		−12	.78	−.1116		14
		−984				100					−5				−66	
.04	.5536		82	.29	−.1225		10	.54	.1100		−12	.79	−.1182		16
		−902				110					−16				−50	
0.05	0.4634		78	0.30	−0.1115		9	0.55	0.1084		−11	0.80	−0.1232		18
		−824				119					−27				−32	
.06	.3810		74	.31	−.0996		7	.56	.1057		−11	.81	−.1264		20
		−750				126					−38				−12	
.07	.3060		70	.32	−.0871		5	.57	.1018		−11	.82	−.1276		22
		−680				131					−49				+10	
.08	.2380		67	.33	−.0740		4	.58	.0969		−10	.83	−.1266		25
		−613				134					−60				35	
.09	.1767		63	.34	−.0606		2	.59	.0909		−10	.84	−.1231		27
		−550				136					−70				62	
0.10	0.1217		60	0.35	−0.0469		+1	0.60	0.0840		−9	0.85	−0.1169		30
		−490				137					−79				92	
.11	.0728		57	.36	−.0332		−1	.61	.0761		−9	.86	−.1077		32
		−433				136					−88				124	
.12	+.0295		53	.37	−.0195		−2	.62	.0673		−8	.87	−.0953		35
		−380				135					−96				159	
.13	−.0085		50	.38	−.0061		−3	.63	.0577		−7	.88	−.0794		38
		−329				132					−103				197	
.14	−.0414		47	.39	+.0071		−4	.64	.0474		−6	.89	−.0597		40
		−282				127					−110				238	
0.15	−0.0696		44	0.40	0.0198		−5	0.65	0.0364		−5	0.90	−0.0359		43
		−238				122					−115				281	
.16	−.0934		41	.41	.0320		−6	.66	.0249		−4	.91	−.0078		46
		−196				116					−120				327	
.17	−.1130		38	.42	.0436		−7	.67	.0129		−3	.92	+.0249		50
		−158				109					−123				377	
.18	−.1288		36	.43	.0544		−8	.68	+.0006		−2	.93	.0626		53
		−122				101					−125				429	
.19	−.1411		33	.44	.0645		−9	.69	−.0119		−1	.94	.1055		56
		−89				92					−126				485	
0.20	−0.1500		30	0.45	0.0738		−9	0.70	−0.0245		0	0.95	0.1540		59
		−59				83					−126				544	
.21	−.1559		28	.46	.0820		−10	.71	−.0371		+2	.96	.2085		62
		−31				73					−124				607	
.22	−.1590		25	.47	.0894		−10	.72	−.0496		3	.97	.2691		66
		−6				63					−121				672	
.23	−.1596		23	.48	.0956		−11	.73	−.0617		5	.98	.3364		69
		+17				52					−116				742	
.24	−.1579		21	.49	.1009		−11	.74	−.0733		6	0.99	.4106		73
		+38				41					−110				814	
0.25	−0.1541		18	0.50	0.1050		−11	0.75	−0.0843		8	1.00	0.4922		

MUTUAL INDUCTANCE OF COAXIAL SINGLE-LAYER COILS

Example 40: The use of formula (111) may be illustrated by the solution of the following problem:

$a = 5$, $2m_1 = 4$, $n_1 = 20$,
$A = 10$, $2m_2 = 16$, $n_2 = 10$.

From these dimensions

$$N_1 = 4 \times 20 = 80, \quad N_2 = 160, \quad \delta^2 = 29, \quad \text{and} \quad \rho^2 = 164.$$

Accordingly,

$$\frac{a^2}{\delta^2} = \frac{25}{29}, \quad \frac{A^2}{\rho^2} = \frac{100}{164}, \quad \text{and} \quad \frac{\delta^2}{\rho^2} = \frac{29}{164}.$$

From Tables 31 and 32, or by direct calculation, there are found

$$\lambda_2 = -\tfrac{59}{116}, \quad \lambda_4 = 0.1862, \quad \text{and} \quad \xi_2 = -0.0671.$$

The series terms are then

$$\lambda_2 = -0.5086 \qquad a^2 N_1 N_2 = 320{,}000,$$

$$\lambda_4 \xi_2 \frac{\delta^2}{\rho^2} = -0.0023 \qquad \frac{A^2}{\rho^2} \cdot \frac{\delta^2}{\rho^2} \text{ (Sum)} = -0.02755,$$

$$\text{Sum} = -0.5109,$$

so that by (111),

$$M = \frac{0.002\pi^2(320000)}{\sqrt{164}} (1.02755) = 506.8\,\mu\text{h}.$$

By the general formula (109)

$r_1^2 = 200$, $r_2^2 = 136$, $\alpha = 0.5$, $r_1 B_1 = 14.034$
$\rho_1^2 = \tfrac{1}{2}$, $\rho_2^2 = 0.73530$, $r_2 B_2 = \underline{11.466}$
$B_1 = 0.99237$, $B_2 = 0.98318$, $a^2 n_1 n_2 = 5000$,   Diff. = 2.568.

$M = 0.004\pi^2(5000)(2.568) = 506.9\,\mu\text{h}$, agreeing with the result by formula (111).

The problem of the long solenoid with a short coil outside considered in example 39 cannot, as it stands, be solved by the formula (111), since the condition that $\frac{\delta^2}{\rho^2}$ shall be small is not satisfied. It is, however, possible to reduce this problem quite simply to an equivalent pair of coils that come under the conditions assumed for formula (111). To do this we make use of the principle of interchange of lengths.

Principle of Interchange of Lengths. Referring to equations (106) and (107), it is evident that an interchange of m_1 and m_2 does not affect the distances x_1 and x_4, while x_2 and x_3 are merely interchanged, one for the other. Consequently r_1 and r_4 are unchanged and r_2 and r_3 merely change places. As a result the mutual inductance is unchanged.

Generalizing, the mutual inductance of two coils of radius a, length $2m_1$, and radius A, length $2m_2$, (Fig. 37a) is the same as that of the pair in Fig. 37b, where the coil of radius a has length $2m_2$ and the coil of radius A a length $2m_1$, provided that the winding density n_1 is transferred to remain associated with the length $2m_1$ and the winding density n_2 with the length $2m_2$.

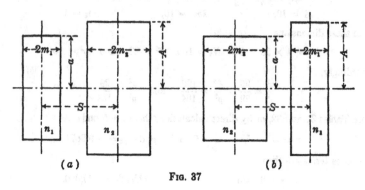

Fig. 37

This principle was first pointed out for concentric solenoids,[66] but it applies generally and not only to single-layer coils but to thick coils as well, provided that in changing the length of a coil, its inner and outer radii are retained.

Example 41: Applying the principle of interchange of lengths to the coils of example 39, the coil dimensions of the equivalent pair are

$$a = 3, \quad 2m_1 = 4, \quad n_1 = 50, \quad N_1 N_2 = 200 \times 500,$$
$$A = 4, \quad 2m_2 = 50, \quad n_2 = 10, \quad = 100{,}000.$$

We have now the case of a short coil inside a much larger one and the dimensions are favorable for the use of formula (111):

$$\delta^2 = 9 + 4 = 13, \qquad \rho^2 = 16 + 625 = 641,$$

$$\frac{\delta^2}{\rho^2} = \frac{13}{641}, \qquad \frac{A^2}{\rho^2} = \frac{16}{641}, \qquad \frac{a^2}{\delta^2} = \frac{9}{13},$$

$$\lambda_2 = -0.2115, \qquad \lambda_4 = -0.1384, \qquad \xi_2 = 0.9563.$$

Consequently the series terms give $-0.2115 - 0.0027 = -0.2142$. This value multiplied by

$$\frac{1}{2} \frac{A^2}{\rho^2} \cdot \frac{\delta^2}{\rho^2} = -0.00005,$$

and the mutual inductance is

$$0.002\pi^2 \frac{(900000)}{\sqrt{641}} (1.00005) = 701.7 \; \mu\text{h},$$

agreeing with the result of example 39.

Loosely Coupled Coils. When the separation of the solenoids is considerable, compared with their dimensions, it is difficult to obtain satisfactory

MUTUAL INDUCTANCE OF COAXIAL SINGLE-LAYER COILS 135

accuracy in the value of the mutual inductance, since in formula (108) the position and negative terms nearly cancel. To improve the accuracy, we may write

$$B_1 = 1 - \gamma_1, \quad B_2 = 1 - \gamma_2, \quad B_3 = 1 - \gamma_3, \quad B_4 = 1 - \gamma_4,$$

the γ's being the small amounts by which the B's, interpolated from the tables as usual, differ from unity.

Equation (108) then becomes

$$M = 0.002 \ \pi^2 a^2 n_1 n_2 [(r_1 + r_4) - (r_2 + r_3) \\ - (r_1\gamma_1 + r_4\gamma_4) + (r_2\gamma_2 + r_3\gamma_3)]. \quad (112)$$

The values of the diagonals r_1, r_2, r_3, and r_4 may be found accurate to as many figures as is desired by arithmetical methods or a calculating machine. The accuracy of the mutual inductance will be limited by the uncertainties of the smaller terms that involve the γ's, and these depend upon the accuracy of the tables.

Example 42: Two loosely coupled coils [47] have the dimensions

$a = 4.435,$ $2m_1 = 27.38,$ $n_1 = 0.7296,$ $s = 31.165.$
$A = 6.44,$ $2m_2 = 20.55,$ $n_2 = 2.737,$

From these values are found

$x_1 = 55.13,$ $x_2 = 34.58,$ $x_3 = 27.75,$ $x_4 = 7.2,$
$r_1^2 = 3080.79,$ $r_2^2 = 1237.25,$ $r_3^2 = 811.54,$ $r_4^2 = 93.314,$
$\rho_1^2 = 0.013462,$ $\rho_2^2 = 0.033521,$ $\rho_3^2 = 0.051105,$ $\rho_4^2 = 0.44445,$

and by interpolation in Table 29 with $\alpha = 0.68866$ there result

$B_1 = 0.99999,$ $B_2 = 0.999935,$ $B_3 = 0.99985,$ $B_4 = 0.98884,$
$\gamma_1 = 0.00001,$ $\gamma_2 = 0.000065,$ $\gamma_3 = 0.00015,$ $\gamma_4 = 0.01116,$
$r_1\gamma_1 = 0.00056,$ $r_2\gamma_2 = 0.00229,$ $r_3\gamma_3 = 0.00427,$ $r_4\gamma_4 = 0.10780,$

so that $-(r_1\gamma_1 + r_4\gamma_4) = -0.10836$, $(r_2\gamma_2 + r_3\gamma_3) = 0.00656$, and the sum of the correction terms is -0.10180.

If we carry out the calculation of the r's to include the fifth decimal place

$r_1 =$ 9.65990 $r_2 =$ 28.48747
$r_4 =$ 55.50487 $r_3 =$ 35.17457
$r_1 + r_4 =$ 65.16477 $r_2 + r_3 =$ 63.66204
$r_2 + r_3 =$ 63.66204
Diff. = 1.50273
Corr. terms = -0.10180
Sum = 1.4009.

136 CALCULATION OF MUTUAL INDUCTANCE AND SELF-INDUCTANCE

The value of the correction terms is somewhat uncertain, so that this result is uncertain by a few units in the last figure. The value of the mutual inductance corresponding to this value is, by (112),

$$M = 0.002\pi^2(4.435)^2(0.7296)(2.737)(1.4009)$$
$$= 1.0862 \ \mu\text{h}.$$

In order to get this degree of accuracy using formula (109) it would be necessary to calculate the $r_n B_n$ accurate to about a part in a million, which is a greater degree of accuracy than holds for the values in Tables 29 and 30. The true value of the mutual inductance in this case, as found by the most accurate absolute formulas,[67] is about 1.0865 μh.

This difficulty inherent in calculations for loosely coupled coils is, of course, more serious, the looser the coupling is. For extreme cases, however, a series formula may be employed [68] that converges more rapidly the farther the coils are apart compared with their dimensions.

$$M = \frac{0.002\pi^2 a^2 A^2 N_1 N_2 s}{q_1{}^2 q_2{}^2} \left[1 + 2\left(\frac{\delta s}{q_1 q_2}\right)^2 \left(\lambda_2 - \frac{3}{4}\frac{A^2}{\delta^2}\right)\left(1 + \frac{m_2{}^2}{s^2}\right) \right.$$
$$+ 3\left(\frac{\delta s}{q_1 q_2}\right)^4 \left(\lambda_4 - \frac{5}{2}\frac{A^2}{\delta^2}\lambda_2 + \frac{5}{8}\frac{A^4}{\delta^4}\right)\left(1 + \frac{10}{3}\frac{m_2{}^2}{s^2} + \frac{m_2{}^4}{s^4}\right)$$
$$+ 4\left(\frac{\delta s}{q_1 q_2}\right)^6 \left(\lambda_6 - \frac{21}{4}\lambda_4 \frac{A^2}{\delta^2} + \frac{35}{8}\lambda_2 \frac{A^4}{\delta^4} - \frac{35}{64}\frac{A^6}{\delta^6}\right)\left(1 + 7\frac{m_2{}^2}{s^2} + 7\frac{m_2{}^4}{s^4} + \frac{m_2{}^6}{s^6}\right)$$
$$\left. + \cdots \right]. \quad (113)$$

In this formula

$$q_1 = s - m_2, \qquad q_2 = s + m_2, \qquad \delta^2 = a^2 + m_1{}^2,$$
$$\lambda_2 = 1 - \frac{7}{4}\frac{a^2}{\delta^2}, \qquad \lambda_4 = 1 - \frac{9}{2}\frac{a^2}{\delta^2} + \frac{33}{8}\frac{a^4}{\delta^4}, \qquad (114)$$
$$\lambda_6 = 1 - \frac{33}{4}\frac{a^2}{\delta^2} + \frac{143}{8}\frac{a^4}{\delta^4} - \frac{715}{64}\frac{a^6}{\delta^6}.$$

The convergence of formula (113) is more rapid, the smaller the ratio $\left(\frac{\delta s}{q_1 q_2}\right)^2$, that is, in general, the greater the separation s when compared with a, m_1 and m_2.

Example 43: The use of formula (113) may be illustrated by applying it to loosely coupled coils for which

$$a = 2, \qquad 2m_1 = 10, \qquad s = 18, \qquad n_1 = n_2 = 10,$$
$$A = 3, \qquad 2m_2 = 6,$$

so that $N_1 = 100$, $N_2 = 60$.
From these data
$$q_1 = 15, \qquad \delta^2 = 29,$$
$$q_2 = 21,$$

MUTUAL INDUCTANCE OF COAXIAL SINGLE-LAYER COILS

so that

$$\frac{\delta^2}{q_1^2} \cdot \frac{s^2}{q_2^2} = \frac{29}{225} \cdot \frac{324}{441} = 0.09469,$$

$$\lambda_2 = 0.75862, \quad \lambda_4 = 0.4578, \quad \lambda_6 = 0.1728,$$

$$M = \frac{7776\pi^2}{(315)^2} [1 + 0.10236 - 0.00208 - 0.00110]$$

$$= 0.85017 \ \mu\text{h}.$$

The principle of interchange of lengths may be employed to obtain a better convergence. This gives rise to coils, equivalent to this case for which

$$a = 2, \qquad 2m_1 = 6, \qquad s = 18,$$
$$A = 3, \qquad 2m_2 = 10,$$

with $N_1 = 60$, $N_2 = 100$. These dimensions give

$$q_1 = 13, \quad q_2 = 23, \quad \delta^2 = 13, \quad \left(\frac{\delta s}{q_1 q_2}\right)^2 = 0.04711,$$

$$\lambda_2 = 0.46154, \quad \lambda_4 = 0.00595, \quad \lambda_6 = -0.1710,$$

$$M = \frac{7776\pi^2}{(299)^2} [1 - 0.005855 - 0.004150 + 0.000392]$$

$$= 0.85019 \ \mu\text{h}.$$

Formula (113) is somewhat tedious to use, but it gives a satisfactory accuracy where the general formula (108) is least precise. The Tables 31–34 will be found useful in obtaining the λ's in (113).

Coaxial Coils of Equal Radii. This is an important practical case on account of the convenience with which two coupled coils may be obtained as single-layer windings on the same cylindrical winding form.

In calculating the mutual inductance by the general formula (108), the B functions are all obtained by interpolation from Table 30 or 35 with $\alpha = 1$. In the special case that the coils are of equal winding length, $2m_1 = 2m_2$, and the distances $x_2 = x_3 = s$, $x_1 = s + 2m_1$, $x_4 = s - 2m_1$.

A useful check on the calculation by the general formula is obtained by computing the mutual inductance from the self-inductance formulas [69] for single-layer coils, pages 143, 153.

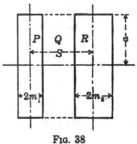

Fig. 38

In Fig. 38, supposing a single-layer coil Q of radius a and length $s - (m_1 + m_2)$ to be placed between the given coils P and R, just filling the space, then the mutual inductance M of the coils P and R may be obtained from the formula

$$2M = (L_{PQR} + L_Q) - (L_{PQ} + L_{QR}). \tag{115}$$

138 CALCULATION OF MUTUAL INDUCTANCE AND SELF-INDUCTANCE

In this equation L_{PQR} is the self-inductance of a single-layer coil made up of P, Q, and R in series, and similarly L_{PQ} is the inductance of P and Q considered as a single coil, etc. Each term of (115) is calculated by formula (118) and Table 36 or 37 of pages 144, 146.

TABLE 35. VALUES OF B_n FOR COILS OF EQUAL RADII ($\alpha = 1$)

ρ^2	B_n	ρ^2	B_n	ρ^2	B_n	ρ^2	B_n
0	1	0.25	0.992815	0.50	0.971802	0.75	0.933448
0.01	0.999987	.26	.992244	.51	.970649	.76	.931397
.02	.999950	.27	.991650	.52	.969469	.77	.929294
.03	.999889	.28	.991035	.53	.968262	.78	.927135
.04	.999804	.29	.990399	.54	.967027	.79	.924918
0.05	0.999695	0.30	0.989742	0.55	0.965763	0.80	0.922639
.06	.999562	.31	.989062	.56	.964471	.81	.920297
.07	.999407	.32	.988360	.57	.963149	.82	.917886
.08	.999228	.33	.987637	.58	.961798	.83	.915403
.09	.999026	.34	.986891	.59	.960416	.84	.912843
0.10	0.998802	0.35	0.986123	0.60	0.959002	0.85	0.910202
.11	.998556	.36	.985332	.61	.957558	.86	.907472
.12	.998287	.37	.984520	.62	.956080	.87	.904648
.13	.997996	.38	.983684	.63	.954570	.88	.901721
.14	.997684	.39	.982826	.64	.953024	.89	.898683
0.15	0.997349	0.40	0.981944	0.65	0.951443	0.90	0.895522
.16	.996992	.41	.981039	.66	.949826	.91	.892225
.17	.996614	.42	.980110	.67	.948172	.92	.888774
.18	.996214	.43	.979158	.68	.946480	.93	.885151
.19	.995793	.44	.978182	.69	.944748	.94	.881327
0.20	0.995351	0.45	0.977181	0.70	0.942975	0.95	0.877266
.21	.994886	.46	.976156	.71	.941161	.96	.872917
.22	.994401	.47	.975106	.72	.939302	.97	.868201
.23	.993894	.48	.974031	.73	.937398	.98	.862983
.24	.993366	.49	.972930	.74	.935448	0.99	.856980
0.25	0.992815	0.50	0.971802	0.75	0.933448	1.00	0.848826

Since the radii are all the same, but the axial lengths b differ, the formula (115) may be written

$$M = 0.002\pi^2 a^2 n_1 n_2 [(b_{PQR}K_{PQR} + b_Q K_Q) - (b_{PQ}K_{PQ} + b_{QR}K_{QR})], \quad (116)$$

in which the subscripts refer to the sections and the K's are the Nagaoka constants taken from Table 36 or 37 for the values of $\dfrac{2a}{b}\left(\text{or }\dfrac{b}{2a}\right)$ which apply to the different combinations of section represented in the terms of (116).

MUTUAL INDUCTANCE OF COAXIAL SINGLE-LAYER COILS   139

Example 44: Two coaxial coils of radius 5 cm. and axial lengths 1 cm. are separated by an axial distance of 1 cm. The winding densities of the coils are $n_1 = n_2 = 10$. Since $s = 2$, the four distances in (108) are

$$x_1 = 3, \qquad x_2 = x_3 = 2, \qquad x_4 = 1,$$
$$r_1^2 = 34, \qquad r_2^2 = r_3^2 = 29, \qquad r_4^2 = 26,$$
$$\rho_1^2 = 0.73529, \qquad \rho_2^2 = \rho_3^2 = 0.86207, \qquad \rho_4^2 = 0.96154.$$

Interpolating for these values of ρ^2 and with $\alpha = 1$ in Table 35, there are found

$$B_1 = 0.93637, \qquad B_2 = 0.90690, \qquad B_4 = 0.87222,$$

so that

$$r_1 B_1 = 5.45994 \qquad\qquad r_2 B_2 = r_3 B_3 = 4.883784$$
$$r_4 B_4 = \underline{4.44746}$$
$$\text{Sum} = 9.90740$$
$$2 r_2 B_2 = \underline{9.76757}$$
$$\text{Diff.} = 0.13983,$$

and

$$M = 0.002\pi^2(25)(100)(0.13983) = 6.900_3 \ \mu\text{h}.$$

Using the self-inductance formula (116) we have $2a = 10$ for each coil section. The lengths and shape ratios are as follows:

$$L_{PQR} \qquad\qquad L_{PQ} = L_{QR} \qquad\qquad L_Q$$
$$b_{PQR} = 3 \qquad\qquad b_{PQ} = b_{QR} = 2 \qquad\qquad b_Q = 1$$
$$\frac{b}{2a} = 0.3 \qquad\qquad \frac{b}{2a} = 0.2 \qquad\qquad \frac{b}{2a} = 0.1,$$

and for these values of $\dfrac{b}{2a}$, Table 36 gives

$$K_{PQR} = 0.405269, \quad K_{PQ} = K_{QR} = 0.319825, \quad K_Q = 0.203324,$$

so that

$$b_{PQR} K_{PQR} = 1.215807$$
$$b_Q K_Q = \underline{0.203324}$$
$$\text{Sum} = 1.419131$$
$$2 b_{PQ} K_{PQ} = \underline{1.279300}$$
$$\text{Diff.} = 0.139831$$

and, by (116),

$$M = 0.002\pi^2(2500)(0.139831) = 6.9003 \ \mu\text{h},$$

agreeing with the value found by the general formula (108). However, the separate terms do not need to be known so accurately in this latter case as those in the general formula.

Concentric Coils of Equal Length. The general formula (110) gives satisfactory accuracy in this case except when the coil lengths b are small, compared with the coil (nearly equal) diameters. The modified Searle and

CALCULATION OF MUTUAL INDUCTANCE AND SELF-INDUCTANCE

Airey formula (111) does not converge well for this case. Such arrangements do not occur often in practice except with adjacent layers of a multilayer coil whose layers are spaced to reduce coil capacitance.

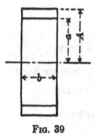

Fig. 39

To calculate the mutual inductance in this special case a series expansion of a formula by Dwight [70] may be used. Setting

$$\delta^2 = \frac{(A-a)^2}{4Aa}, \quad \xi^2 = \frac{b^2}{4Aa},$$

and assuming that, not only are these quantities small, but that $\gamma^2 = \dfrac{\delta^2}{\xi^2}$ is small also,

$$M = 0.008\pi(Aa)^{3/2}\xi^2 n_1 n_2 \left[\left\{m + \frac{n\xi^2}{8} - \frac{p\xi^4}{64} + \frac{5}{1024}\xi^6 + \cdots\right\}\log_e\frac{16}{\xi^2}\right.$$

$$-v + \frac{w\xi^2}{16} + \frac{y\xi^4}{48} - \frac{109}{12{,}288}\xi^6 + \cdots - 2\pi u\gamma + \gamma^2\left(t\log_e\frac{1}{\gamma^2} + 3\right)$$

$$\left.+\frac{1}{6}\gamma^4 - \frac{1}{30}\gamma^6 + \cdots\right], \tag{117}$$

in which $m = 1 + \frac{3}{4}\delta^2 - \frac{15}{64}\delta^4 + \frac{35}{256}\delta^6,$

$n = 1 - \frac{5}{8}\delta^2 + \frac{35}{64}\delta^4,$

$v = 1 - \frac{7}{4}\delta^2,$

$t = 1 + \frac{3}{8}\delta^2 - \frac{5}{64}\delta^4 + \frac{35}{1024}\delta^6,$

$u = 1 + \dfrac{\delta^2}{2} - \dfrac{\delta^4}{8} + \dfrac{\delta^6}{16},$ \hfill (117a)

$v = 1 - \dfrac{9}{8}\delta^2 - \dfrac{\delta^4}{16} + \dfrac{325}{3072}\delta^6,$

$w = 1 + \frac{3}{2}\delta^2 - \frac{895}{384}\delta^4,$

$y = 1 - \frac{149}{64}\delta^2.$

This formula is, of course, of rather limited range of application since it converges rapidly only for very short coils of equal length and of very nearly equal radii. It will be required only in cases where the general formula (110) is difficult to use and where a check on the results is desirable.

Example 45: An example of the numerical relations involved which render the use of (117) advantageous is that offered by two concentric coils 1 cm. long with radii $a = 4.9$ and $A = 5.0$. It will be assumed that each coil has 10 turns, so that $n_1 = n_2 = 10$. Then $b = 1$.

MUTUAL INDUCTANCE OF COAXIAL SINGLE-LAYER COILS 141

Then, using the general formula (109) and Table 30, with $\alpha = 0.38$,

$$x_1 = 1, \quad x_4 = -1, \qquad x_2 = x_3 = 0,$$
$$r_1^2 = 26, \qquad r_2^2 = 25,$$
$$\rho_1^2 = 0.96154, \qquad \rho_2^2 = 1,$$
$$B_1 = 0.87764, \qquad B_2 = 0.85686,$$

so that
$$r_1 B_1 = 4.47510$$
$$r_2 B_2 = 4.28430$$
$$\overline{\text{Diff.} = 0.19080}$$

Therefore,
$$M = 0.004\pi^2(2500)(0.19080) = 18.086\ \mu\text{h}.$$

Applying the series formula (117) we have $A - a = 0.1$, $4aA = 98$.

$$\delta^2 = \frac{0.01}{98} = 0.00010204,$$

$$\xi^2 = \tfrac{1}{98}, \quad \gamma^2 = 0.01.$$

Since δ^4 is negligible, the constants in (117a) are easy to calculate:

$$m = 1.000076, \qquad u = 1.000051,$$
$$n = 0.999943, \qquad w = 1.000153,$$
$$t = 1.000044, \qquad y = 0.999765,$$
$$v = 0.999885,$$

$$\log_e \frac{16}{\xi^2} = 7.3575, \qquad \log_e \frac{1}{\gamma^2} = 4.6053.$$

Accordingly the terms in the brackets in equation (117) are

$$\{1.000076 + 0.001275 - 0.000016\}(7.3575) - 0.999885 + 0.000638 + 0.000022$$

and
$$- 0.2\pi(1.000051) + 0.01(4.6053 + 3) + \tfrac{1}{8}(.0001) = 5.8158,$$

$$M = 0.008\pi(24.5)^{3/2}(\tfrac{1}{98})^2(100)(5.8158) = 18.087\ \mu\text{h},$$

agreeing with the value found by the general formula (108). Even in this rather extreme case the general formula gives a precision of better than 5 parts in 10,000, so that the more complicated formula is of value only for a check.

Chapter 16

SINGLE-LAYER COILS ON CYLINDRICAL WINDING FORMS

Basic Current Sheet Formulas. Probably no other type of coil is so widely used as a simple helical winding, such as is obtained by winding a single layer of wire on a cylindrical form. Such coils are most often wound with thin insulated wire with their turns close together, but more open windings of bare wire or copper strip are common in high voltage transmitters. In such cases the wire is sufficiently stiff to hold its position without the necessity of a winding form.

Single-layer coils have the advantage, not only of simple and inexpensive construction, but the effective capacitance of the winding is small. For this reason, they are especially useful in high frequency circuits, except where larger inductances are required.

The calculation of the inductance of a single-layer coil is based on formulas for a cylindrical current sheet, that is, a winding where the current flows around the axis of a cylinder in a layer of infinitesimal radial thickness on the surface of the cylinder. Except in the case of very open helical windings, the inductance of a single-layer coil is closely equal to that of a cylindrical current sheet having the same number of turns N as the coil, the same mean radius a, and a length b equal to the number of turns in the coil times the distance between centers of adjacent wires.

An exact formula for the inductance of a cylindrical current sheet was first found in 1879 by Lorenz,[71] who integrated the expression for the mutual inductance of two equal coaxial circular filaments twice over the length of the current sheet. Lorenz's formula is in elliptic integrals and involves both positive and negative terms. Formulas in the form of converging series, more convenient for purposes of numerical computation, have been derived by a number of authors.[72] By the choice of a suitable series formula, a result for any given case may be obtained with a greater degree of accuracy than is required in practice. However, the necessity for the selection of the appropriate formula is avoided and much time is saved if tables are available that render possible the use of a single general formula.

For this purpose we may adopt Nagaoka's formula,[73] which bases the calculation on the well-known formula for the inductance of a cylindrical cur-

SINGLE-LAYER COILS ON CYLINDRICAL WINDING FORMS

rent sheet of infinite length and applies a correction to take account of the effect of the ends.

Nagaoka's formula is
$$L = 0.004\pi^2 a^2 b n^2 K$$
$$= 0.002\pi^2 a \left(\frac{2a}{b}\right) N^2 K, \tag{118}$$

in which n is the winding density in turns per centimeter of axial length and K is the factor that takes account of the effect of the ends. Nagaoka gave [74] a table of values of K as a function of the shape ratio $\frac{2a}{b} = \frac{\text{diameter}}{\text{length}}$, which is reproduced here. For relatively short coils, b less than $2a$, it is convenient to tabulate K for such coils as a function of $\frac{b}{2a}$. This has been done in Table 36. Table 37 includes Nagaoka's values of K for long coils. From this table very accurate values of K may be interpolated, sufficing for the calculation of the inductance of standard coils intended for the most precise work.

For very short coils, interpolation in Table 36 becomes uncertain and it is better to derive K directly from the following series formula: [75]

$$K = \frac{2\beta}{\pi}\left[\left(\log_e \frac{4}{\beta} - \frac{1}{2}\right) + \frac{\beta^2}{8}\left(\log_e \frac{4}{\beta} + \frac{1}{8}\right) \right.$$
$$\left. - \frac{\beta^4}{64}\left(\log_e \frac{4}{\beta} - \frac{2}{3}\right) + \frac{5}{1024}\beta^6\left(\log_e \frac{4}{\beta} - \frac{109}{120}\right) - \cdots\right], \tag{119}$$

in which $\beta = \frac{b}{2a}$. For values of β as large as $\frac{1}{4}$ three terms will suffice for an accuracy better than 1 part in 1000.

Inductance of Ring Conductor. A limiting case of formula (119) is that of the inductance of an annular conductor whose cross section is a line of length b, that is, a single turn of very thin metal tape of mean radius a and width b. The formula for this case is

$$L = 0.004\pi a \left[\log_e \frac{8a}{b} - \frac{1}{2}\right] \tag{119a}$$

Another important case is that of a single turn of round wire of radius of cross section ρ bent in a circle of mean radius a. This may be treated by formula (80) and the formula for the geometric mean distance of a circle. The inductance is

$$L = 0.004\pi a \left[\log_e \frac{8a}{\rho} - 1.75\right] \tag{119b}$$

If the cross section of the ring is rectangular, the inductance may be calculated by formula (99) or (100) together with the Tables 22–25.

144 CALCULATION OF MUTUAL INDUCTANCE AND SELF-INDUCTANCE

TABLE 36. VALUES OF K FOR SHORT SINGLE-LAYER COILS, FORMULA (118)

$b/2a$	K	d_1	d_2	$b/2a$	K	d_1	d_2
0	0			0.25	0.365432		
		34960				8386	
0.01	0.034960		−8822	.26	.373818		−219
		26138				8168	
.02	.061098		−3329	.27	.381986		−209
		22809				7958	
.03	.083907		−2154	.28	.389944		−200
		20655				7758	
.04	.104562		−1602	.29	.397703		−192
		19053				7566	
0.05	0.123615		−1273	0.30	0.405269		−184
		17780				7382	
.06	.141395		−1056	.31	.412650		−176
		16724				7205	
.07	.158119		−901	.32	.419856		−170
		15823				7035	
.08	.173942		−784	.33	.426890		−164
		15038				6871	
.09	.188980		−696	.34	.433762		−158
		14343				6713	
0.10	0.203324		−623	0.35	0.440474		−152
		13720				6562	
.11	.217044		−564	.36	.447036		−147
		13156				6414	
.12	.230200		−514	.37	.453450		−141
		12642				6274	
.13	.242842		−473	.38	.459724		−137
		12169				6136	
.14	.255011		−436	.39	.465860		−132
		11732				6004	
0.15	0.266744		−406	0.40	0.471865		−128
		11327				5877	
.16	.278070		−378	.41	.477742		−124
		10948				5754	
.17	.289019		−354	.42	.483496		−120
		10595				5634	
.18	.299614		−333	.43	.489129		−116
		10262				5518	
.19	.309876		−313	.44	.494646		−112
		9949				5405	
0.20	0.319825		−295	0.45	0.500052		−108
		9654				5296	
.21	.329479		−280	.46	.505348		−105
		9374				5191	
.22	.338852		−266	.47	.510539		−102
		9108				5089	
.23	.347960		−252	.48	.515628		−100
		8856				4989	
.24	.356816		−240	.49	.520617		−96
		8616				4893	
0.25	0.365432		−229	0.50	0.525510		−93

SINGLE-LAYER COILS ON CYLINDRICAL WINDING FORMS

TABLE 36. VALUES OF K FOR SHORT SINGLE-LAYER COILS, FORMULA (118) (*Concluded*)

$b/2a$	K	d_1	d_2	$b/2a$	K	d_1	d_2
0.50	0.525510			0.75	0.623011		
		4800				3111	
.51	.530310		−92	.76	.626122		−48
		4708				3063	
.52	.535018		−89	.77	.629185		−48
		4619				3016	
.53	.539637		−86	.78	.632200		−46
		4534				2970	
.54	.544170		−84	.79	.635170		−46
		4450				2924	
0.55	0.548620		−81	0.80	0.638094		−44
		4368				2880	
.56	.552988		−78	.81	.640974		−43
		4290				2837	
.57	.557278		−77	.82	.643811		−43
		4212				2794	
.58	.561491		−76	.83	.646605		−41
		4137				2753	
.59	.565628		−74	.84	.649358		−41
		4063				2712	
0.60	0.569691		−71	0.85	0.652070		−39
		3992				2673	
.61	.573683		−68	.86	.654743		−39
		3923				2634	
.62	.577606		−68	.87	.657376		−38
		3856				2596	
.63	.581462		−66	.88	.659972		−37
		3790				2560	
.64	.585252		−65	.89	.662532		−37
		3724				2522	
0.65	0.588976		−63	0.90	0.665054		−36
		3662				2486	
.66	.592638		−61	.91	.667540		−34
		3601				2452	
.67	.596239		−60	.92	.669991		−34
		3541				2417	
.68	.599780		−58	.93	.672408		−33
		3483				2384	
.69	.603263		−57	.94	.674792		−33
		3426				2350	
0.70	0.606689		−56	0.95	0.677142		−32
		3370				2318	
.71	.610060		−54	.96	.679460		−32
		3316				2286	
.72	.613376		−54	.97	.681747		−31
		3263				2256	
.73	.616639		−52	.98	.684003		−31
		3211				2225	
.74	.619850		−50	0.99	.686228		−30
		3161				2195	
0.75	0.623011		−50	1.00	0.688423		

146 CALCULATION OF MUTUAL INDUCTANCE AND SELF-INDUCTANCE

TABLE 37. VALUES OF K FOR LONG SINGLE-LAYER COILS, FORMULA (118)

$2a/b$	K	d_1	d_2	$2a/b$	K	d_1	d_2
0	1.000000			0.25	0.901649		
		−4231				−3616	
0.01	0.995769		+24	.26	.898033		+23
		−4207				−3593	
.02	.991562		26	.27	.894440		24
		−4181				−3569	
.03	.987381		24	.28	.890871		23
		−4157				−3546	
.04	.983224		25	.29	.887325		24
		−4132				−3522	
0.05	0.979092		+25	0.30	0.883803		+24
		−4107				−3498	
.06	.974985		25	.31	.880305		22
		−4082				−3476	
.07	.970903		26	.32	.876829		24
		−4056				−3452	
.08	.966847		24	.33	.873377		23
		−4032				−3429	
.09	.962815		24	.34	.869948		23
		−4008				−3406	
0.10	0.958807		+26	0.35	0.866542		+22
		−3982				−3384	
.11	.954825		25	.36	.863158		24
		−3957				−3360	
.12	.950868		24	.37	.859799		23
		−3933				−3338	
.13	.946935		23	.38	.856461		23
		−3910				−3315	
.14	.943025		26	.39	.853146		22
		−3884				−3293	
0.15	0.939141		+26	0.40	0.849853		+23
		−3857				−3270	
.16	.935284		23	.41	.846583		22
		−3834				−3248	
.17	.931450		23	.42	.843335		23
		−3811				−3225	
.18	.927639		26	.43	.840110		21
		−3785				−3204	
.19	.923854		24	.44	.836906		21
		−3761				−3183	
0.20	0.920093		+23	0.45	0.833723		+23
		−3737				−3160	
.21	.916356		24	.46	.830563		21
		−3713				−3139	
.22	.912643		24	.47	.827424		22
		−3689				−3117	
.23	.908954		25	.48	.824307		21
		−3664				−3096	
.24	.905290		23	.49	.821211		+21
		−3641				−3075	
0.25	0.901649		+25	0.50	0.818136		

SINGLE-LAYER COILS ON CYLINDRICAL WINDING FORMS

TABLE 37. VALUES OF K FOR LONG SINGLE-LAYER COILS, FORMULA (118) (*Concluded*)

$2a/b$	K	d_1	d_2	$2a/b$	K	d_1	d_2
0.50	0.818136			0.75	0.747762		+18
		−3054				−2571	
.51	.815082		+21	.76	.745191		17
		−3033				−2554	
.52	.812049		21	.77	.742637		17
		−3012				−2537	
.53	.809037		21	.78	.740100		18
		−2991				−2519	
.54	.806046		20	.79	.737581		17
		−2971				−2502	
0.55	0.803075		+21	0.80	0.735079		+16
		−2950				−2486	
.56	.800125		20	.81	.732593		19
		−2930				−2467	
.57	.797195		20	.82	.730126		16
		−2910				−2451	
.58	.794285		20	.83	.727675		16
		−2890				−2435	
.59	.791395		20	.84	.725240		16
		−2870				−2419	
0.60	0.788525		+20	0.85	0.722821		+17
		−2850				−2402	
.61	.785675		19	.86	.720419		16
		−2831				−2386	
.62	.782844		19	.87	.718033		16
		−2812				−2370	
.63	.780032		20	.88	.715663		15
		−2792				−2355	
.64	.777240		19	.89	.713308		16
		−2773				−2339	
0.65	0.774467		+19	0.90	0.710969		+17
		−2754				−2322	
.66	.771713		19	.91	.708647		14
		−2735				−2308	
.67	.768978		19	.92	.706339		16
		−2716				−2292	
.68	.766262		19	.93	.704047		15
		−2697				−2277	
.69	.763565		18	.94	.701770		16
		−2679				−2261	
0.70	0.760886		+18	0.95	0.699509		+14
		−2661				−2247	
.71	.758225		18	.96	.697262		15
		−2643				−2232	
.72	.755582		19	.97	.695030		15
		−2624				−2217	
.73	.752958		17	.98	.692813		15
		−2607				−2202	
.74	.750351		19	0.99	.690611		14
		−2589				−2188	
0.75	0.747762		+18	1.00	0.688423		

CALCULATION OF MUTUAL INDUCTANCE AND SELF-INDUCTANCE

TABLE 38. CORRECTION TERM G IN FORMULAS (120) AND (135)

For small values of $\frac{\delta}{p}$ it is more accurate to calculate G from the defining equation

$$G = \frac{5}{4} - \log_e 2\frac{p}{\delta}$$

δ/p	G	Diff.	δ/p	G	Diff.	δ/p	G	Diff.	δ/p	G	Diff.
1.00	0.5568		0.75	0.2691		0.50	−0.1363		0.25	−0.8294	
		−100			−134			−202			−408
0.99	.5468		.74	.2557		.49	− .1565		.24	− .8702	
		−101			−136			−206			−426
.98	.5367		.73	.2421		.48	− .1771		.23	− .9128	
		−103			−138			−211			−445
.97	.5264		.72	.2283		.47	− .1982		.22	−0.9573	
		−104			−140			−215			−465
.96	.5160		.71	.2143		.46	− .2197		.21	−1.0038	
		−105			−142			−219			−488
0.95	0.5055		0.70	0.2001		0.45	−0.2416		0.20	−1.0526	
		−106			−144			−225			−513
.94	.4949		.69	.1857		.44	− .2641		.19	−1.1039	
		−107			−146			−230			−541
.93	.4842		.68	.1711		.43	− .2871		.18	−1.1580	
		−108			−148			−235			−571
.92	.4734		.67	.1563		.42	− .3106		.17	−1.2151	
		−109			−150			−241			−606
.91	.4625		.66	.1413		.41	− .3347		.16	−1.2757	
		−110			−152			−247			−645
0.90	0.4515		0.65	0.1261		0.40	−0.3594		0.15	−1.3402	
		−112			−155			−253			−690
.89	.4403		.64	.1106		.39	− .3847		.14	−1.4092	
		−113			−157			−260			−741
.88	.4290		.63	.0949		.38	− .4107		.13	−1.4833	
		−114			−160			−267			−801
.87	.4176		.62	.0789		.37	− .4374		.12	−1.5634	
		−116			−163			−274			−870
.86	.4060		.61	.0626		.36	− .4648		.11	−1.6504	
		−117			−166			−281			−953
0.85	0.3943		0.60	0.0460		0.35	−0.4929		0.10	−1.7457	
		−118			−168			−290			−1054
.84	.3825		.59	.0292		.34	− .5219		.09	−1.8511	
		−120			−171			−299			−1178
.83	.3705		.58	+ .0121		.33	− .5518		.08	−1.9689	
		−121			−174			−308			−1335
.82	.3584		.57	− .0053		.32	− .5826		.07	−2.1024	
		−123			−177			−317			−1541
.81	.3461		.56	− .0230		.31	− .6143		.06	−2.2565	
		−124			−180			−328			−1824
0.80	0.3337		0.55	−0.0410		0.30	− .6471		0.05	−2.4389	
		−126			−184			−339			
.79	.3211		.54	− .0594		.29	− .6810		.04	−2.6620	
		−127			−187			−351			
.78	.3084		.53	− .0781		.28	− .7161		.03	−2.9497	
		−129			−190			−364			
.77	.2955		.52	− .0971		.27	− .7525		.02	−3.3551	
		−131			−194			−377			
.76	.2824		.51	− .1165		.26	− .7902		0.01	−4.0483	
		−133			−198			−392			
0.75	0.2691		0.50	−0.1363		0.25	−0.8294		0	∞	

SINGLE-LAYER COILS ON CYLINDRICAL WINDING FORMS 149

Correction for Insulating Space. The inductance of a winding of wire is, of course, somewhat different from that of the cylindrical current sheet. The difference may in any case be accurately evaluated by a method due to Rosa.[76]

The relation between a single-layer winding of round wire and the equivalent cylindrical current sheet is illustrated in Fig. 40. The current sheet may be regarded as a winding of conducting tape of negligible thickness and with the turns of tape separated by insulation of negligible width. At the center of each turn of tape is located one of the turns of the winding of round wire. The pitch of the winding p is the distance between centers of adjacent wires. Each turn of tape has also a width p. For the 4-turn winding shown in the figure the length of the equivalent current sheet is $b = 4p$.

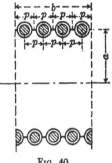

Fig. 40

The inductance of the coil differs from that of the equivalent current sheet for two reasons:

(a) each turn of *wire* has a self-inductance differing from a turn of the *tape;*

(b) each pair of turns of the coil has a mutual inductance differing slightly from that of the corresponding pair of turns of tape.

These differences may be evaluated by formulas (119a) and (119b). Summing them over the coil the total correction is

$$\Delta L = 0.004\pi a N (G + H), \qquad (120)$$

in which the quantity G is a function of the ratio $\dfrac{\delta}{p}$ between the bare diameter of the wire and the pitch of the winding that is given in Table 38 and H is obtained from Table 39 for the given number of turns N on the coil.

The calculated value ΔL is to be subtracted from the inductance of the current sheet given by formula (118) to obtain the inductance of the actual winding of wire.

The correction ΔL is unimportant in many cases met in practice.

Example 46: A single-layer coil wound with enameled wire on a very accurately constructed cylinder of marble yielded the following measurements: [77]

	mean radius,	$a = 27.0862$ cm.
	equivalent length,	$b = 30.5510$
	diameter of bare wire,	$\delta = 0.0634$ cm.
	number of turns,	$N = 440.$

The parameter $\dfrac{b}{2a} = \dfrac{30.5510}{54.1724} = 0.5639587.$

150 CALCULATION OF MUTUAL INDUCTANCE AND SELF-INDUCTANCE

TABLE 39. CORRECTION TERM H IN FORMULAS (120) AND (135)

N	H	N	H	N	H
1	0	31	0.3087	110	0.3278
2	0.1137	32	.3095	120	.3285
3	.1663	33	.3102	130	.3291
4	.1973	34	.3109	140	.3296
5	0.2180	35	0.3115	150	0.3301
6	.2329	36	.3121	160	.3305
7	.2443	37	.3127	170	.3309
8	.2532	38	.3132	180	.3312
9	.2604	39	.3137	190	.3315
10	0.2664	40	0.3142	200	0.3318
11	.2715	41	.3147	220	.3323
12	.2758	42	.3152	240	.3327
13	.2795	43	.3156	260	.3330
14	.2828	44	.3160	280	.3333
15	0.2857	45	0.3164	300	0.3336
16	.2883	46	.3168	350	.3341
17	.2906	47	.3172	400	.3346
18	.2927	48	.3175	450	.3349
19	.2946	49	.3179	500	0.3351
20	0.2964	50	0.3182	550	.3354
21	.2980	55	.3197	600	.3356
22	.2994	60	.3210	650	.3357
23	.3008	65	.3221	700	.3358
24	.3020	70	.3230	750	0.3360
25	0.3032	75	0.3238	800	.3361
26	.3043	80	.3246	850	.3362
27	.3053	85	.3253	900	.3362
28	.3062	90	.3259	950	.3363
29	.3071	95	.3264	1000	0.3364
30	0.3079	100	0.3269	∞	0.3379

Interpolating from Table 36, the value of K for this value of $\frac{b}{2a}$ is found to be

$$K = 0.552988 + 0.001698 + 0.000010$$

$$= 0.554696.$$

Thus the inductance of the equivalent current sheet is, by (118),

$$L = \frac{0.002\pi^2 a N^2 K}{b/2a} = 101810.2 \ \mu\text{h}$$

$$= 0.101802 \text{ henry.}$$

SINGLE-LAYER COILS ON CYLINDRICAL WINDING FORMS 151

The pitch of the winding is
$$p = \frac{30{,}5510}{440},$$
and the ratio
$$\frac{\delta}{p} = 0.9131.$$

Interpolating for this value in Table 38, $G = 0.4659$
and from Table 39 for $N = 440$, $\quad H = 0.3349$
so that the correction is $\quad$ Sum $= 0.8008$,

$$\Delta L = 0.004\pi(440)(0.8008)a$$
$$= 120.0 \ \mu\text{h} = 0.0001200 \text{ henry.}$$

So, finally, the inductance of the coil is

$$L = 0.1018102 - 0.0001200 = 0.1016902 \text{ henry.}$$

The correction amounts to only 1.2 parts in 1000 of the total inductance, but since its value is certain to the number of places given, the value of the inductance of the coil is as accurate as is the value of L_s. The values of K in Table 36 are given correct to enough places of figures to give a calculated value for L_s which is as certain as the measured data on which the calculation is based.

The values of K in Table 36 are carried out to a higher degree of accuracy than will usually be necessary except for calculations of inductance standards of precision. For use in calculations of the mutual inductance of coaxial coils of equal radii, however, accurate values of K are also required (see p. 139). For coils of only a moderate degree of precision the inductance will be more readily calculated by the formulas and tables of the following section on general design.

General Design of Single-layer Coils[78] **on Cylindrical Forms.** In practice the following types of design problem are of frequent occurrence:

- A. Given the dimensions and the number of turns (or the winding density), to calculate the inductance of the coil.
- B. Given the length and winding density, to calculate the diameter that the coil must have to give a desired inductance.
- C. Given the diameter and winding density, to calculate length of coil to give a specified inductance.
- D. Given the coil diameter and coil length, to calculate the winding density necessary to give a desired inductance.
- E. Given that a certain ratio of diameter and length is desired, to calculate the dimensions necessary to give a specified inductance, when a certain winding density is assumed.

These problems will be treated and the formulas given both for metric units and for English units. The inductance will be given in microhenrys, coil dimensions in centimeters or inches, winding densities in turns per cm. or

152 CALCULATION OF MUTUAL INDUCTANCE AND SELF-INDUCTANCE

in turns per inch of axial length. The coil length must be taken as equal to the number of turns multiplied by the winding pitch (distance between centers of adjacent wires). That is, the inductance formulas apply to a coil length which extends half the winding pitch beyond the center of the last turn at each end of the coil. If this rule is followed, the correction for insulating space is usually negligible, except in precision work.

Let d_1, b_1, n_1, δ_1 be, respectively, the coil diameter, equivalent length, winding density, and bare diameter of wire, all in metric units, while d_2, b_2, n_2, δ_2 will denote the corresponding values in English units. The total number of turns will be denoted by N; and the shape ratios r = diameter divided by length, R = length divided by diameter have the same values in either system of units.

TABLE 40. DESIGN DATA. SINGLE-LAYER COILS

$$r = \frac{\text{diameter}}{\text{length}}$$

r	k	F	β	γ	r	k	F	β	γ
0	9.87	0	0	∞	0.25	8.90	2.225	0.556	35.6
0.01	9.83	0.098	0.0010	983	.26	8.86	2.304	.599	34.1
.02	9.79	.196	.0039	489	.27	8.83	2.384	.644	32.7
.03	9.74	.292	.0088	325	.28	8.79	2.462	.689	31.4
.04	9.70	.388	.0155	243	.29	8.76	2.540	.736	30.2
0.05	9.66	0.483	0.0242	193.2	0.30	8.72	2.617	0.785	29.1
.06	9.62	.577	.0346	160.3	.31	8.69	2.693	.835	28.0
.07	9.58	.671	.0470	136.9	.32	8.65	2.769	.886	27.0
.08	9.54	.763	.0611	119.2	.33	8.62	2.844	.939	26.1
.09	9.50	.855	.0770	105.6	.34	8.59	2.919	.993	25.2
0.10	9.46	0.946	0.0946	94.6	0.35	8.55	2.993	1.048	24.4
.11	9.42	1.037	.1140	85.7	.36	8.52	3.067	1.104	23.7
.12	9.38	1.126	.1351	78.2	.37	8.49	3.140	1.162	22.9
.13	9.35	1.215	.1580	71.9	.38	8.45	3.212	1.221	22.2
.14	9.31	1.303	.1825	66.5	.39	8.42	3.284	1.281	21.6
0.15	9.27	1.390	0.2086	61.8	0.40	8.39	3.355	1.342	20.97
.16	9.23	1.477	.2363	57.7	.41	8.36	3.426	1.404	20.38
.17	9.19	1.563	.2656	54.1	.42	8.32	3.496	1.468	19.82
.18	9.16	1.648	.2958	50.9	.43	8.29	3.565	1.533	19.28
.19	9.12	1.732	.3292	48.0	.44	8.26	3.634	1.599	18.77
0.20	9.08	1.816	0.363	45.4	0.45	8.23	3.703	1.666	18.29
.21	9.04	1.899	.399	43.1	.46	8.20	3.771	1.735	17.82
.22	9.01	1.982	.436	40.9	.47	8.17	3.838	1.804	17.37
.23	8.97	2.063	.475	39.0	.48	8.14	3.905	1.874	16.95
.24	8.94	2.144	.515	37.2	.49	8.10	3.972	1.946	16.54
0.25	8.90	2.225	0.556	35.6	0.50	8.07	4.037	2.019	16.15

SINGLE-LAYER COILS ON CYLINDRICAL WINDING FORMS 153

PROBLEM A. Given diameter, length, and winding density; to calculate the inductance.

$$L = 0.001 k d_1{}^2 b_1 n_1{}^2 \text{ (metric)}, \tag{121}$$

$$L = 0.00254 k d_2{}^2 b_2 n_2{}^2 \text{ (English)}, \tag{121a}$$

in which k is to be taken from Tables 40, 41 for the given value of r or R, whichever is less than unity.

If, instead of the winding density, the total number of turns is given,

$$L = 0.001 F d_1 N^2 \text{ (metric)}, \tag{122}$$

$$L = 0.00254 F d_2 N^2 \text{ (English)}. \tag{122a}$$

F is to be taken from Tables 40, 41 for the value of r or R (whichever is less than unity) calculated from the dimensions.

TABLE 40. DESIGN DATA, LONG SINGLE-LAYER COILS (*Concluded*)

$$r = \frac{\text{diameter}}{\text{length}}$$

r	k	F	β	γ	r	k	F	β	γ
0.50	8.07	4.04	2.019	16.15	0.75	7.38	5.54	4.15	9.84
.51	8.04	4.10	2.092	15.77	.76	7.35	5.59	4.25	9.68
.52	8.01	4.17	2.167	15.41	.77	7.33	5.64	4.35	9.52
.53	7.98	4.23	2.243	15.07	.78	7.30	5.70	4.44	9.36
.54	7.96	4.30	2.320	14.73	.79	7.28	5.75	4.54	9.22
0.55	7.93	4.36	2.398	14.41	0.80	7.26	5.80	4.64	9.07
.56	7.90	4.42	2.476	14.10	.81	7.23	5.86	4.74	8.93
.57	7.87	4.48	2.556	13.80	.82	7.21	5.91	4.84	8.79
.58	7.84	4.55	2.637	13.52	.83	7.18	5.96	4.95	8.65
.59	7.81	4.61	2.719	13.24	.84	7.16	6.01	5.05	8.52
0.60	7.78	4.67	2.802	12.97	0.85	7.13	6.06	5.15	8.39
.61	7.75	4.73	2.885	12.71	.86	7.11	6.11	5.26	8.27
.62	7.73	4.79	2.97	12.46	.87	7.09	6.16	5.36	8.15
.63	7.70	4.85	3.06	12.22	.88	7.06	6.22	5.47	8.03
.64	7.67	4.91	3.14	11.99	.89	7.04	6.27	5.58	7.91
0.65	7.64	4.97	3.23	11.76	0.90	7.02	6.32	5.68	7.80
.66	7.62	5.03	3.32	11.54	.91	6.99	6.36	5.79	7.69
.67	7.59	5.08	3.41	11.33	.92	6.97	6.41	5.90	7.58
.68	7.56	5.14	3.50	11.12	.93	6.95	6.46	6.01	7.47
.69	7.54	5.20	3.59	10.92	.94	6.93	6.51	6.13	7.37
0.70	7.51	5.26	3.68	10.73	0.95	6.90	6.56	6.23	7.27
.71	7.48	5.31	3.77	10.54	.96	6.88	6.61	6.34	7.17
.72	7.46	5.37	3.87	10.36	.97	6.86	6.65	6.45	7.07
.73	7.43	5.42	3.96	10.18	.98	6.84	6.70	6.57	6.98
.74	7.41	5.48	4.06	10.01	0.99	6.82	6.75	6.68	6.88
0.75	7.38	5.54	4.15	9.84	1.00	6.79	6.79	6.79	6.79

154 CALCULATION OF MUTUAL INDUCTANCE AND SELF-INDUCTANCE

TABLE 41. DESIGN DATA FOR SHORT SINGLE-LAYER COILS

$$R = \frac{\text{length}}{\text{diameter}} \lessgtr 1$$

R	k	F	β	γ	R	k	F	β	γ
1.00	6.79	6.79	6.79	6.79	0.75	6.15	8.20	10.93	4.61
0.99	6.77	6.84	6.91	6.70	.74	6.13	8.27	11.17	4.53
.98	6.75	6.89	7.03	6.62	.73	6.09	8.34	11.42	4.44
.97	6.73	6.94	7.15	6.53	.72	6.05	8.41	11.68	4.36
.96	6.71	6.98	7.28	6.44	.71	6.02	8.48	11.94	4.28
0.95	6.68	7.03	7.40	6.35	0.70	5.99	8.55	12.22	4.19
.94	6.66	7.08	7.54	6.26	.69	5.95	8.63	12.51	4.11
.93	6.64	7.14	7.67	6.17	.68	5.92	8.70	12.80	4.03
.92	6.61	7.19	7.81	6.08	.67	5.88	8.78	13.11	3.94
.91	6.59	7.24	7.96	6.00	.66	5.85	8.86	13.43	3.86
0.90	6.56	7.29	8.10	5.91	0.65	5.81	8.94	13.76	3.78
.89	6.54	7.35	8.25	5.82	.64	5.78	9.02	14.10	3.70
.88	6.51	7.40	8.41	5.73	.63	5.74	9.11	14.46	3.62
.87	6.49	7.46	8.57	5.64	.62	5.70	9.19	14.83	3.53
.86	6.46	7.51	8.74	5.56	.61	5.66	9.28	15.22	3.45
0.85	6.44	7.57	8.91	5.47	0.60	5.62	9.37	15.62	3.37
.84	6.41	7.63	9.08	5.38	.59	5.58	9.46	16.04	3.29
.83	6.38	7.69	9.26	5.30	.58	5.54	9.55	16.47	3.21
.82	6.35	7.75	9.45	5.21	.57	5.50	9.65	16.93	3.14
.81	6.33	7.81	9.64	5.12	.56	5.46	9.75	17.40	3.06
0.80	6.30	7.87	9.84	5.04	0.55	5.42	9.85	17.90	2.98
.79	6.27	7.94	10.04	4.95	.54	5.37	9.95	18.42	2.90
.78	6.24	8.00	10.26	4.87	.53	5.33	10.05	18.96	2.82
.77	6.21	8.06	10.47	4.78	.52	5.28	10.15	19.53	2.75
.76	6.18	8.13	10.70	4.70	.51	5.23	10.26	20.12	2.67
0.75	6.15	8.20	10.93	4.61	0.50	5.19	10.37	20.74	2.59

Example 47: Wire of bare diameter 0.015 inch is wound 50 turns per inch on a winding form 4 inches in diameter, until a coil 8.5 inches long is obtained. Calculate the inductance.

The shape ratio $r = \frac{4}{8.5} = 0.4706$, and for this value Table 40 gives $k = 8.17$, so that, by (121a),

$$L = 0.00254(8.17)(16)(8.5)(50)^2 = 7055 \; \mu h.$$

This value is checked by formula (122a), using $N = 50(8.5) = 425$. The value of F for $r = 0.4706$ is 3.842. Therefore

$$L = 0.00254(3.842)(4)(425)^2 = 7051 \; \mu h.$$

SINGLE-LAYER COILS ON CYLINDRICAL WINDING FORMS

TABLE 41. DESIGN DATA FOR SHORT SINGLE-LAYER COILS (Concluded)

$$R = \frac{\text{length}}{\text{diameter}} \leqq 1$$

R	k	F	β	γ	R	k	F	β	γ
0.50	5.19	10.37	20.7	2.593	0.25	3.61	14.43	57.7	0.902
.49	5.14	10.48	21.4	2.518	.24	3.52	14.67	61.1	.845
.48	5.09	10.60	22.1	2.443	.23	3.43	14.93	64.9	.790
.47	5.04	10.72	22.8	2.368	.22	3.34	15.20	69.1	.736
.46	4.99	10.84	23.6	2.294	.21	3.25	15.48	73.7	.683
0.45	4.94	10.97	24.4	2.221	0.20	3.16	15.78	78.9	0.631
.44	4.88	11.10	25.2	2.148	.19	3.06	16.10	84.8	.581
.43	4.83	11.23	26.1	2.076	.18	2.96	16.43	91.3	.532
.42	4.77	11.36	27.0	2.004	.17	2.85	16.78	98.7	.485
.41	4.72	11.50	28.0	1.993	.16	2.74	17.15	107.2	.439
0.40	4.66	11.64	29.1	1.863	0.15	2.63	17.55	119.7	0.395
.39	4.60	11.79	30.2	1.793	.14	2.52	17.98	128.4	.352
.38	4.54	11.94	31.4	1.724	.13	2.40	18.44	141.8	.312
.37	4.48	12.10	32.7	1.656	.12	2.27	18.93	157.8	.273
.36	4.41	12.26	34.0	1.588	.11	2.14	19.47	177.0	.236
0.35	4.35	12.42	35.5	1.522	0.10	2.007	20.07	200.7	0.201
.34	4.28	12.59	37.0	1.456	.09	1.865	20.72	230	.168
.33	4.21	12.77	38.7	1.390	.08	1.717	21.46	268	.137
.32	4.14	12.95	40.5	1.326	.07	1.560	22.29	318	.109
.31	4.07	13.14	42.4	1.263	.06	1.396	23.26	388	.0837
0.30	4.00	13.33	44.4	1.200	0.05	1.220	24.40	488	0.0610
.29	3.92	13.53	46.7	1.138	.04	1.032	25.80	645	.0413
.28	3.85	13.74	49.1	1.078	.03	0.828	27.60	920	.0248
.27	3.77	13.96	51.7	1.018	.02	.603	30.15	1508	.0121
.26	3.69	14.19	54.6	0.959	.01	.345	34.50	3450	0.0034
0.25	3.61	14.43	57.7	0.902	0	0	∞	∞	0

These are current sheet values. To estimate the importance of the correction for insulation the quantity

$$\frac{\Delta L}{L} = \frac{6.28(G + H)}{k n_1 d_1} = \frac{6.28(G + H)}{k n_2 d_2} \quad (123)$$

$$= \frac{6.28(G + H)}{FN} \quad (124)$$

may be calculated, remembering that in Table 38

$$\frac{\delta_1}{p_1} = \frac{\delta_2}{p_2} = n_1 \delta_1 = n_2 \delta_2.$$

In this example $n_2\delta_2 = 50(0.015) = 0.75$, so that from Table 38 $G = 0.27$ and for $N = 425$ Table 39 gives $B = 0.335$. Consequently

$$\frac{\Delta L}{L} = \frac{6.28(0.27 + 0.335)}{8.17(50)(4)} = 0.0023,$$

or the correction amounts to 2.3 parts in 1000 of the whole inductance. This value may also be checked by (124), which gives

$$\frac{\Delta L}{L} = \frac{6.28(0.605)}{3.842(425)} = 0.0023.$$

The corrected inductance will therefore be

$$7051(1 - 0.0023) = 7051 - 1.6 = 7049 \ \mu h.$$

Example 48: The problem often occurs that the inductance of a coil already constructed is to be calculated from the measured dimensions. Suppose a coil is wound with wire of bare diameter 0.8 mm. on a cylindrical form having a diameter of 22.2 cm., the total number of turns being 79. Thus

$$d_1 = 22.2 + 0.08 = 22.3 \text{ cm.}$$

The axial distance between the centers of the first and 79th wire was found to be 9.0 cm. Thus the pitch is $p_1 = \frac{1}{n_1} = \frac{9.0}{78} = 0.1154$ cm., and the equivalent length $b_1 = 79(0.1154) = 9.12$ cm.

The shape ratio is, therefore, $R = \frac{9.12}{22.3} = 0.409$, for which we interpolate from Table 41 the value $F = 11.515$. Therefore

$$L = 0.011515(79)^2 22.3 = 1602.6 \ \mu h.$$

The importance of the correction for insulating space is estimated by calculating

$$\frac{\delta_1}{p_1} = \frac{0.08}{0.1154} = 0.693.$$

Using this and the value $N = 79$, Tables 38 and 39 give $G = 0.19$, $H = 0.33$, so that

$$\frac{\Delta L}{L} = \frac{(0.52)(6.28)}{(79)(11.515)} = 0.0036,$$

and the corrected inductance is $1602.6(1 - 0.0036) = 1597 \ \mu h.$

PROBLEM B. Given required inductance in μh, the coil length, and winding density; to find the necessary coil diameter.

In such a design calculation, the correction due to insulating space will be neglected.

Calculate from the given data the quantity

$$\beta = \frac{1000L}{b_1{}^3 n_1{}^2} \text{ (metric)}, \qquad (125)$$

or

$$\beta = \frac{393.7L}{b_2{}^3 n_2{}^2} \text{ (English)}. \qquad (125a)$$

SINGLE-LAYER COILS ON CYLINDRICAL WINDING FORMS

The quantity β, which is also equal to r^2k or $\dfrac{k}{R^2}$, is given in Tables 40, 41. Interpolating in the table for the value of β just calculated, the value of r (or R) corresponding is obtained, and the required coil diameter is obtained from the relations

$$d_1 = b_1 r = \frac{b_1}{R} \text{ (metric)}, \tag{126}$$

$$d_2 = b_2 r = \frac{b_2}{R} \text{ (English)}. \tag{126a}$$

Example 49: Suppose a coil is desired that is to be wound with 40 turns of wire per inch for a length of 5 inches to give an inductance of 3000 μh.

The data are $n_2 = 40$, $b_2 = 5$, and $L = 3000$, to which corresponds in (125a) the value

$$\beta = \frac{393.7(3000)}{125(1600)} = 5.905.$$

To this corresponds the value $r = 0.9205$ in Table 40, and thence by (126a)

$$d_2 = 5(0.9205) = 4.60 \text{ inches}.$$

Usually there will be no point in estimating the correction for insulating space, since a coil constructed with the dimensions just calculated could be readily adjusted to the desired inductance by an additional turn or two of wire. In case, however, that a more accurate design than the foregoing is desired, the correction ΔL may be calculated by (120) using the value of $a = \dfrac{d_2}{2}$ just found as a first approximation. For a second approximation the calculation of β is repeated using for the assumed inductance the value $L + \Delta L$. The resulting value of d_2 will differ very little from that first obtained.

Problem C. Given the coil diameter and the winding density; to calculate the length of coil necessary to give a specified inductance.

This is the problem to be solved when a certain cylindrical winding form is available and it is proposed to use a given sample of wire for the winding. The mean diameter of the coil will be equal to the diameter of the winding form plus the covered diameter of the wire. The winding density may be ascertained by winding enough turns of the given wire to measure the winding pitch that results. In designing a coil with an open winding, the pitch p will be prescribed. The winding density is the reciprocal of the pitch.

To determine the length of the winding necessary to give the desired inductance, calculate from the given data the value of

$$\gamma = \frac{1000L}{d_1{}^3 n_1{}^2} \text{ (metric)}, \tag{127}$$

$$\gamma = \frac{393.7L}{d_2{}^3 n_2{}^2} \text{ (English)}. \tag{127a}$$

This quantity, which is equal to $\dfrac{k}{r}$ or kR, is tabulated in Tables 40, 41 for values of r and R. Entering the table with the calculated value of γ, the corresponding value of r (or R) may be interpolated. From the value thus obtained the equivalent coil length follows from the relation

$$b_1 = \frac{d_1}{r} = d_1 R \text{ (metric)}, \tag{128}$$

$$b_2 = \frac{d_2}{r} = d_2 R \text{ (English)}, \tag{128a}$$

and the total number of turns of the finished winding will be

$$N = n_1 b_1 \text{ (metric)}, \tag{129}$$

$$N = n_2 b_2 \text{ (English)}. \tag{129a}$$

In general, the correction for insulating space need not be considered, since the inductance of the completed winding may be adjusted by adding a turn or two of wire. A more rigorous calculation of the design could, however, be made by calculating the insulation correction ΔL from the given data and the value of N as found above, and then repeating a second approximation using $L + \Delta L$ for L in the formula for γ.

Example 50: Wire of covered diameter 0.067 cm. is to be wound on a cylindrical winding form of 10 cm. diameter to obtain a coil of 1000 μh inductance. Calculate the length of the winding necessary and the number of turns in the completed coil.

Here $n_1 = 15$ turns per centimeter and the mean diameter is $d_1 = 10 + 0.067$. From these data and (127),

$$\gamma = \frac{1000(1000)}{(10.067)^3 (15)^2} = 4.357.$$

From column 10 of Table 41 it is found that to this corresponds $R = 0.720$ so that $b_1 = 10.067(0.720) = 7.25$ cm. The total number of turns is $N = 15 \times 7.25 = 108.7$, so that 109 turns should be wound, and these will cover a length of about 7.25 cm. on the winding form.

Problem D. Given the dimensions of a winding form; to calculate the winding density necessary in order to obtain a specified inductance.

This problem supposes an existing winding form and its solution determines the size of wire that will be necessary for the winding in order to obtain a desired value of inductance. The winding may not always be practicable: the solution may lead to a size of wire of too small a diameter to be practicable, or it may lead to an open winding of inconveniently large pitch. In such cases a winding form of more suitable dimensions must be used.

SINGLE-LAYER COILS ON CYLINDRICAL WINDING FORMS

Since d_1 and b_1 are given, or d_2 and b_2, the value of r (or R) corresponding may be calculated, and the corresponding value of k interpolated from Tables 40, 41.

Then
$$n_1^2 = \frac{1000L}{kd_1^2 b_1} \text{ (metric)}, \tag{130}$$

or
$$n_2^2 = \frac{393.7L}{kd_2^2 b_2} \text{ (English)}. \tag{130a}$$

Example 51: How many turns must be wound upon a form having a diameter of 5.5 inches and axial length of 9.6 inches to give an inductance of 7100 μh?

From the dimensions there follows $r = \frac{5.5}{9.6} = 0.573$, and Table 40 shows that to this value of r corresponds $k = 7.86$. Consequently, in (130a)

$$n_2 = \sqrt{\frac{(393.7)(7100)}{(5.5)^2(9.6)(7.86)}} = 35 \text{ turns per inch,}$$

and the coil would have $N = 35(9.6) = 336$ turns. To the computed winding density corresponds a pitch of winding of 0.0285 inches, which can be obtained with wire of No. 22 A.W.G. The mean diameter of the actual coil would be greater than the 5.5 inches assumed by an amount equal to the diameter of the covered wire.

PROBLEM E. A coil is to have a given shape ratio (relation of length to diameter). When wound to have a chosen winding density, what must the dimensions be in order that the inductance may have a specified value?

From Tables 40, 41 find the value of γ corresponding to the given r (or R). Then the required coil diameter will be calculated from the relation

$$d_1^3 = \frac{1000L}{\gamma n_1^2} \text{ (metric)}, \tag{131}$$

$$d_2^3 = \frac{393.7L}{\gamma n_2^2} \text{ (English)}, \tag{131a}$$

and the length follows from

$$b_1 = \frac{d_1}{r} = d_1 R, \quad \text{or} \quad b_2 = \frac{d_2}{r} = d_2 R.$$

Or we may first find the coil length by obtaining from Tables 40, 41 the value of β corresponding to the given value of r (or R) and with this obtain the length from the equation

$$b_1^3 = \frac{1000L}{\beta n_1^2} \text{ (metric)}, \tag{132}$$

$$b_2^3 = \frac{393.7L}{\beta n_2^2} \text{ (English)}. \tag{132a}$$

160 CALCULATION OF MUTUAL INDUCTANCE AND SELF-INDUCTANCE

The diameter then follows from the relation

$$d_1 = b_1 r = \frac{b_1}{R}, \quad \text{or} \quad d_2 = b_2 r = \frac{b_2}{R}.$$

The total number of turns is $N = b_1 n_1 = b_2 n_2$.

Example 52: A coil is desired for which the shape ratio $\left(\frac{\text{diameter}}{\text{length}}\right)$ is to be 2.6 and wire is to be used that winds 20 turns per inch to give an inductance of 250 μh.
Here

$$R = \frac{1}{2.6} = 0.3846 \quad \text{and} \quad n_2 = 20.$$

From Table 41, corresponding to this value of R, interpolation gives

$$\beta = 30.85 \quad \text{and} \quad \gamma = 1.756.$$

Accordingly

$$d_2{}^3 = \frac{393.7(250)}{1.756(400)} \quad \text{and} \quad d_2 = 5.19 \text{ inches, from (131a)}.$$

$$b_2 = 5.19 \times 0.3846 = 1.998 \text{ inches}.$$

As a check we have also

$$b_2{}^3 = \frac{393.7(250)}{30.85(400)} \quad [\text{see (132a)}],$$

which yields

$$b_2 = 1.998 \text{ inches} \quad \text{and} \quad d_2 = \frac{1.998}{0.3846} = 5.19 \text{ inches},$$

agreeing with the previous calculation.

The total number of turns required is $N = 20(1.998)$, which gives 40 turns total. Strictly, the winding form should have a diameter less than the calculated value by the diameter of the covered wire (about 0.05 inch) but since no account has been taken of insulating space, the actual coil inductance is slightly less than the current sheet value and it would be better to build the coil with a winding form of the diameter d_2 calculated above and to adjust the number of turns slightly to give the desired inductance.

Inductance as a Function of the Number of Turns. When a number of coils are to be wound using a chosen size of wire and winding forms of the same diameter, it is a convenience for design purposes to prepare a graph showing the inductance obtainable with different numbers of turns of the given wire on the standard winding form. Such graphs may be rapidly calculated and drawn as the occasion demands.

The number of turns that may be wound per inch with the given sample may readily be obtained by experiment or taken from a wire table.

Example 53: Wire of No. 22 A.W.G., double cotton covered, winds 30 turns per inch. Using a winding form that gives a mean coil diameter of 3 inches, the formula (122a) becomes $L = 0.00762FN^2$. For a chosen number of turns, say $N = 20$, the equivalent coil length is $b_2 = \frac{20}{30}$ inch, so that $R = \frac{b_2}{d_2} = \frac{1}{3}\left(\frac{2}{3}\right) = \frac{2}{9}$, for which, by Table 41, the value $F = 15.14$ applies. Then $L = 0.00762(15.14)(20)^2 = 46.1$ μh.

SINGLE-LAYER COILS ON CYLINDRICAL WINDING FORMS   161

In a few minutes the following data for plotting a graph of L as a function of N were obtained:

N	5.0	10.0	20.0	30.0	40.0	50.0	60.0	70.0	80.0	90.0	100.0
L	4.5	14.8	46.1	87.2	134.6	186.5	240.0	299.1	358.7	419.1	481.2

From such a graph the approximate number of turns necessary to give a specified coil inductance may be found at once.

Most Economical Coil Shape. The problem may be stated as follows: given a wire of length l which winds n turns per centimeter of axial length, what shape factor will give a coil of the largest inductance possible with the given wire?

Making use of the relations $l = 2\pi a N$, $N = nb$, it is possible to express the inductance formula (118) in terms of the given data l and n and the shape ratio $\dfrac{2a}{b}$.

There is found

$$L = 0.001\sqrt{\pi n} \cdot \sqrt{\frac{2a}{b}} \cdot l^{3/2} K. \tag{133}$$

Making use of Table 36, it is found that the product $K\sqrt{\dfrac{2a}{b}}$ is a maximum for $R = \dfrac{b}{2a} = 0.408$. For this value of the shape ratio of winding, the largest inductance will be obtained from the given length of wire wound to give the stated number of turns per centimeter. The maximum is, however, very flat: a shape ratio of $R = 0.39$ gives an inductance only 17 parts in 100,000 less and the shape ratio $R = 0.42$ only 7 parts in 100,000 less than the maximum.

The design formulas for the most economical coil become, then,

$$\frac{b}{2a} = 0.408, \quad \frac{2a}{b} = 2.451,$$

$$L_{\max} = 0.0013224 n^{1/2} l^{3/2} \;\mu\text{h},$$

$$l = 100 \left(\frac{L^2_{\max}}{1.7489 n}\right)^{1/3} \text{cm.}, \tag{134}$$

$$N^2 = \frac{0.408 nl}{\pi} = 0.12987 nl$$

$$b = \frac{N}{n}, \quad a = 2.451 b.$$

162 CALCULATION OF MUTUAL INDUCTANCE AND SELF-INDUCTANCE

Example 54: What is the maximum inductance possible with 500 cm. of A.W.G. No. 18 wire wound closely to form a single-layer coil?

This size of wire, double cotton covered, winds 20.4 turns per inch, or $n = \dfrac{20.4}{2.54}$ = 8.03 per centimeter and $l = 500$.

The maximum inductance possible is

$$L_{max} = 0.0013224\sqrt{8.03(500)^3} = 41.9 \; \mu\text{h}.$$

The number of turns in this coil is

$$N = \sqrt{0.12987(8.03)500} = 22.8,$$

the coil length,

$$b = \frac{22.8}{8.03} = 2.84 \text{ cm.},$$

and the coil diameter

$$2a = 2.84(2.451) = 6.97 \text{ cm.}$$

Example 55: As a variant of the problem, we may use the design equations to calculate the dimensions of a coil of optimum shape to have a specified inductance. Suppose, for example, that it is desired to use A.W.G. No. 20 wire, double cotton covered, to construct a close wound single-layer coil to have an inductance of 500 μh.

This wire winds with 24.4 turns per inch, so that $n = \dfrac{24.4}{2.54} = 9.61$ turns per centimeter. The required length of wire is, by (134),

$$l = 100 \left[\frac{(500)^2}{(1.7489)(9.61)} \right]^{\frac{1}{5}} = 2460 \text{ cm.},$$

the required number of turns

$$N = \sqrt{0.12987 \times 9.61 \times 2460} = 55.4,$$

and the dimensions of the coil

$$b = \frac{55.4}{9.61} = 5.77 \text{ cm.} \quad \text{and} \quad 2a = 2.451b = 14.13 \text{ cm.}$$

Chapter 17

SPECIAL TYPES OF SINGLE-LAYER COIL

Helices of Conductor of Large Cross Section. These are single-layer coils of bare wire or strip sufficiently stiff so as to require no winding form. Such coils have often found a use in the tank circuits of radio transmitters, the number of turns employed being adjusted by movable clips on the turns of the coil. In such coils the thickness of the wire is necessarily smaller in comparison with the pitch of the winding than in the usual close wound coil of finer wire.

Consequently, although the inductance calculation is still to be based on the current sheet formula, the correction for insulation space cannot be neglected. The Rosa method of correction already outlined (page 149) for round wire has to be extended to cover the case of rectangular strip also.

Helices of Round Wire. If the helix is wound with N turns of conductor of circular cross section, diameter $\delta = 2\rho$, with a pitch of winding p, the mean radius of the turns being a, the inductance is given by

$$L = L_s - 0.004\pi a N(G + H). \qquad (135)$$

L_s is the inductance of the equivalent cylindrical current sheet of length $b = Np$ and diameter $2a$ and may be calculated by the formulas and tables already given. The quantity G in the correction term is given in Table 38 as a function of the ratio $\dfrac{\delta}{p}$. For small values of this ratio interpolation is uncertain and it is better to calculate directly from the relation

$$G = \frac{5}{4} - \log_e \frac{2p}{\delta} = 0.55685 - \log_e \frac{p}{\delta}. \qquad (136)$$

The value of H may be obtained from Table 39 for the given value of the number of turns. It is to be noticed that in formula (135) the mean radius a is to be expressed in centimeters.

Example 56: To find the inductance of a helix of 20 turns of round wire $\frac{1}{4}$ inch in diameter, wound with a pitch of $\frac{3}{4}$ inch and with a mean diameter of a turn of 12 inches. The equivalent cylindrical current sheet has a length of $b = 20$ times $\frac{3}{4}$ inch $= 15$

inches, so that, in Table 40, r is to be taken as 12 divided by 15 = 0.8. The corresponding value of F is 5.80. Thus, by formula (122a)

$$L_s = 0.00254(5.80)12(20)^2 = 70.71 \; \mu h.$$

To evaluate the correction for insulation, $\dfrac{\delta}{p} = \dfrac{1}{3}$, and $N = 20$, so that from Tables 38 and 39,

$$G = -0.5417$$
$$H = +0.2964$$
$$G + H = -0.2453.$$

Consequently the correction in (135) is

$$\Delta L = 0.004\pi(20)(12 \times 2.54)(-0.2453) = -1.88 \; \mu h,$$

and the inductance of the helix is

$$L = 70.71 - (-1.88) = 72.6 \; \mu h.$$

Helices of Rectangular Strip. Usually the strip will be bent so that the longer dimension of the cross section will lie along the radii of the turns. This allows of a greater inductance for a given length of strip than is obtained with the longer dimension axial.

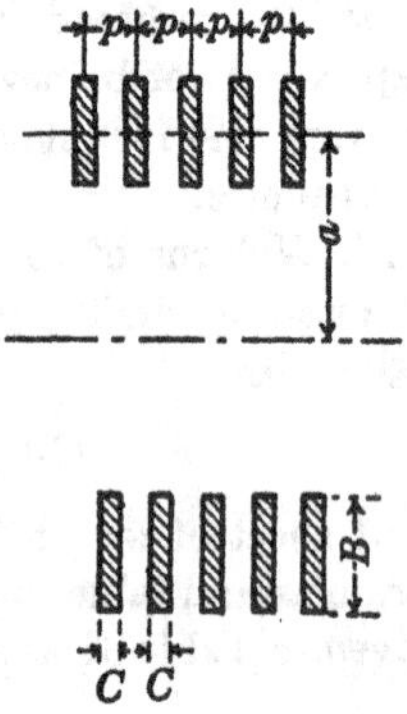

Let the radial dimension be B and the axial C, and let the space ratios be

$$\beta = \frac{B}{p}, \quad \Delta = \frac{C}{B}, \quad \text{and} \quad \gamma = \frac{C}{p} = \beta\Delta.$$

The inductance may then be calculated from the value L_s for the equivalent cylindrical current sheet by the expression

$$L = L_s - 0.004\pi Na(G_1 + H_1). \qquad (137)$$

Fig. 41

The dimensions of the equivalent cylindrical current sheet are diameter $2a$, equal to the mean diameter of the turns of the helix, and the length $b = Np$. The calculation of L_s is made by formulas (121) or (122).

The correction for insulation is based on the tables for the geometric mean differences for rectangles: [79]

$$G_1 = \log_e \frac{B+C}{p} + \log_e e, \qquad (138)$$

$$H_1 = H + 2\left[\frac{N-1}{N}\log_e k + \frac{1}{12}(\beta^2 - \gamma^2)\left(0.6449 - \frac{\log_e \dfrac{N}{2} + 1.270}{N}\right)\right.$$
$$\left. - \frac{1}{60}\left(\beta^4 + \gamma^4 - \frac{5}{2}\beta^2\gamma^2\right)\left(0.0823 - \frac{0.2021}{N}\right)\right]. \quad (139)$$

SPECIAL TYPES OF SINGLE-LAYER COIL

In these equations H is taken from Table 39 for the given number of turns, $\log_e k$ from Tables 1 or 2 for the given shape ratios β $\left(\text{or } \dfrac{1}{\beta}\right)$ and Δ, and $\log_e e$ from Table 3 for the shape ratio $\dfrac{C}{B}$. The natural logarithms are readily calculated from the ordinary logarithms by employing the multiplication table, Auxiliary Table 2. The dimension a in formula (137) is to be expressed in centimeters.

A second method of solution is to base the calculation of the inductance on the value L_u for a circular coil of rectangular cross section having a mean radius a equal to the mean radius of the turns of the helix. The axial dimension of the cross section is taken as $b = Np$, and the radial c is taken equal to the dimension B of the strip conductor.

Then
$$L = L_u + 0.004\pi N a (G_2 + H_2). \tag{140}$$

The calculation of L_u is made by formula (99) and Tables 22, 23. The quantities in the correction term are given by

$$G_2 = \log_e \frac{B+p}{B+C} + (\log_e e_s - \log_e e_w), \tag{141}$$

$$H_2 = 2\left[\frac{N-1}{N}(\log_e k_s - \log_e k_w)\right.$$

$$- \frac{1}{12}(1-\gamma^2)\left(0.6449 - \frac{\log_e \dfrac{N}{2} + 1.270}{N}\right)$$

$$\left.+ \frac{1}{24}\left\{\beta^2(1-\gamma^2) - \frac{2}{5}(1-\gamma^4)\right\}\left\{0.0823 - \frac{0.2021}{N}\right\}\right]. \tag{142}$$

The quantity $\log_e e_s$ is to be taken from Table 3 for the ratio $\dfrac{B}{p}$ $\left(\text{or } \dfrac{p}{B}\right)$ and $\log_e e_w$ from the same table for the ratio $\dfrac{C}{B}$. Table 2 is entered for $\dfrac{1}{\beta} = \dfrac{p}{B}$ and $\Delta = \dfrac{1}{\beta}$ when $B > p$ to find $\log_e k_s$. If $p > B$, $\log_e k_s$ is obtained from Table 1 for $\gamma = 1$ and $\dfrac{1}{\Delta} = \dfrac{B}{p}$. In either case, $\log_e k_w$ is the value in Table 2 for $\dfrac{1}{\beta}$ (or β) and $\Delta = \dfrac{C}{B}$.

In the less frequent cases where the longer dimension of the cross section of the strip lies along the axis of the coil, formula (137) is to be preferred and

166 CALCULATION OF MUTUAL INDUCTANCE AND SELF-INDUCTANCE

$\log_e k$ is taken from Table 1 with $\gamma = \dfrac{C}{p}$ and $\dfrac{1}{\Delta} = \dfrac{B}{C}$ as parameters. It should be noted that in this table all values of $\log_e k$ are negative except only those for $\dfrac{B}{C} = 1$.

As an example of the use of these formulas and tables the inductance of a helix of rectangular strip will be computed by both formulas.

Example 57: Assume a helix of 10 turns of strip of cross sectional dimensions 1 cm. by 0.1 cm., wound with a pitch $p = 1$ cm., with the longer dimension of the strip lying in the radial direction. Furthermore, the mean radius of the coil will be assumed as $a = 15$ cm. The cross sectional dimensions are $B = 1$, $C = 0.1$.

The equivalent cylindrical current sheet will have the dimensions $b = 10p = 10$ cm., $2a = 30$, $N = 10$.

In Table 41 the value F corresponding to $R = \dfrac{b}{2a} = \dfrac{1}{3}$ is 12.71.

Therefore, $L_s = 0.001(12.71)(30)(10)^2 = 38.13$ μh [see (122)].

The shape ratios are $\beta = 1$ and $\Delta = 0.1$, and $\gamma = 0.1$.

From Table 3, for $\dfrac{C}{B} = 0.1$, $\log_e e = 0.0021$. $\log_e \dfrac{B+C}{p} = \log_e 1.1 = 0.0953$, so that in (138) $G_1 = 0.0953 + 0.0021 = 0.0974$.

Table 39 gives for $N = 10$, $H = 0.2664$, and from Table 2 for $\beta = 1$ and $\Delta = 0.1$, $\log_e k = 0.0702$. Substituting in formula (139)

$$H_1 = 0.2664 + 2\left[\dfrac{9}{10}(0.0702) + \dfrac{0.99}{12}\left(0.6449 - \dfrac{\log_e 5 + 1.270}{10}\right)\right.$$
$$\left. - \dfrac{0.975}{60}(0.0823 - 0.0202)\right]$$
$$= 0.2664 + 2[0.0632 + 0.0294 - 0.0010]$$
$$= 0.4496,$$

$G_1 + H_1 = 0.5470,$

and the correction is $0.004\pi(10)(15)(0.5470) = 1.03$ μh. The inductance of the helix is $L = 38.13 - 1.03 = 37.10$ μh.

Making use of the formula (140), we have to find the inductance L_u of a coil of mean radius 15 cm. and a rectangular cross section with dimensions $c = B = 1$ cm. in the radial direction and $b = Np = 10$ cm. in the axial direction. From Table 36, for the ratio $\dfrac{b}{2a} = \dfrac{1}{3}$, we find by interpolation $K = 0.4292$. Table 22, with $\dfrac{c}{2a} = \dfrac{1}{30}$, $\dfrac{c}{b} = 0.1$, yields the value $k = 0.0206$, so that $(K - k) = 0.4086$. Making use of this in formula (99),

$$L_u = 0.019739(3)(10)^2(15)(0.4086) = 36.29 \text{ } \mu h.$$

In the correction term of formula (141), and using Table 3,

$$G_2 = \log_e \dfrac{2}{1.1} + (0.00177 - 0.00210) = 0.5975.$$

SPECIAL TYPES OF SINGLE-LAYER COIL 167

To find $\log_e k_s$ we make use of Table 2 with $\beta = 1$, $\Delta = \beta = 1$, and $\log_e k_s = 0.0065$. For $\log_e k_w$ the parameters are $\beta = 1$ and $\Delta = \dfrac{C}{B} = 0.1$, $\gamma = 0.1$, so that Table 2 gives $\log_e k_w = 0.0702$. Accordingly,

$$H_2 = 2[\tfrac{2}{120}(0.0065 - 0.0702) - \tfrac{1}{12}(1 - 0.01)(0.6449 - 0.2879)$$

$$+ \tfrac{1}{24}\{(1 - 0.01) - \tfrac{2}{3}(0.9999)\}(0.0823 - 0.0202)]$$

$$= 2[-0.05733 - 0.02945 + 0.00153] = -0.1705,$$

and the correction term is $0.004\pi(10)(15)(0.5975 - 0.1705) = 0.805$ μh. The inductance of the helix is $L = 36.29 + 0.80 = 37.095$ μh, which agrees with the value by the other formula. It may be noted that the corrected value lies between the values of L_s and L_u.

Flat Spirals of Strip. Flat spirals of strip conductor have been extensively used in radio work and have the advantage that close coupling of two such coils, arranged coaxially, is readily obtained. Such coils may be regarded as special cases of pancake coils in which the correction for insulating space is not negligible.

The calculation of the inductance is carried out by the same methods as for single-layer coils, two formulas being provided that give a check on the accuracy of the calculations.

a. The inductance may be obtained from the value L_s' calculated for a disc coil of the same mean diameter $2a$ as the spiral, with a radial width $c = Np$ and a zero axial thickness. The mean diameter is taken as the mean of the diameters of the inner and the outer turns. The radial width is taken, as usual, equal to the number of turns N times the distance p between centers of adjacent turns.

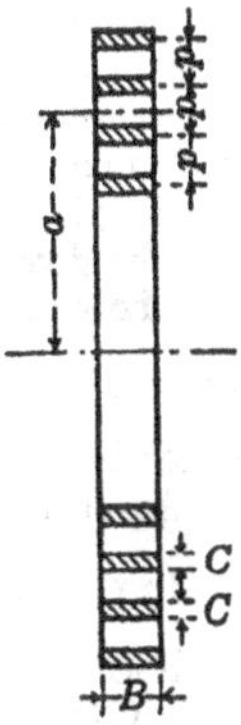

Then for the spiral

$$L = L_s' - 0.004\pi Na(G_1 + H_1) \quad (143)$$

Fig. 42

where G_1 and H_1 are the same corrections, formulas (138) and (139), as occur in the case of single-layer coils already treated and

$$L_s' = 0.001 N^2 aP \text{ (metric)}. \quad (144)$$

The quantity P is a function of the parameter $\dfrac{c}{2a}$ and is given in Table 26. For the region when $\dfrac{c}{2a} < 0.1$, interpolation becomes uncertain and it is better to obtain P by the formula

$$P = 4\pi\left[\left(\log_e \frac{8a}{c} - \frac{1}{2}\right) + \frac{1}{24}\left(\frac{c}{2a}\right)^2\left(\log_e \frac{8a}{c} + 3.583\right)\right]. \quad (145)$$

168 CALCULATION OF MUTUAL INDUCTANCE AND SELF-INDUCTANCE

In calculating H_1 it is to be noticed that here the dimensions B and C are to be measured, respectively, in the axial and the radial directions of the spiral, that is, just the reverse from the case of the solenoid of strip. The convention is in both cases that the dimension B is to be taken perpendicular to the line joining the centers of the cross sections of adjacent turns.

b. The inductance may also be referred to the inductance L_u' of a pancake circular coil of rectangular cross section of the same mean radius as the spiral and with the axial cross sectional dimension b equal to B and the radial dimension c taken equal to Np.

The inductance formula in this case is

$$L = L_u' + 0.004\pi Na(G_2 + H_2). \tag{146}$$

L_u' is to be calculated using formula (100) and Table 24, 26, and G_2 and H_2 are to be calculated by the same formulas (141) and (142) as were used for the solenoid of rectangular strip.

Example 58: The inductance is to be calculated for a flat spiral wound with 10 turns of strip having cross sectional dimensions 1 by 0.2 cm. to have a pitch of 1 cm. The longer dimension of the strip is placed axially. The innermost portion of the winding is 20.5 cm. from the center and the outermost 29.5 cm. Thus the mean radius of the winding is $a = \frac{1}{2}(20.5 + 29.5) = 25$ cm.

The given data yield the values $p = 1$ cm., $B = 1$ cm., $C = 0.2$ cm., $N = 10$.

The equivalent disc coil will, therefore, have $a = 25$, $c = 10p = 10$, so that the value of P is to be found from Table 26 for $\dfrac{c}{2a} = 0.2$. The value is $P = 31.500$, which is to be used in (144).

Accordingly, $L_u' = 0.001(25)(10)^2(31.500) = 78.75$ μh.

From Table 3 we find, for the ratio $\dfrac{C}{B} = 0.2$, that $\log_e e = 0.00249$.

Formula (138) gives $G_1 = \log_e 1.2 + 0.00249 = 0.18231 + 0.00249$, so that $G_1 = 0.1848$.

To calculate H_1 we find $\beta = \dfrac{B}{p} = 1$, $\Delta = \dfrac{C}{B} = 0.2$, and for these parameters Table 2 yields $\log_e k = 0.0685$. The parameter $\gamma = \dfrac{C}{p} = 0.2$, and for $N = 10$, Table 39 gives $H = 0.2664$, so that formula (139) becomes

$H_1 = 0.2664 + 2[0.9(0.0685) + \frac{1}{12}(1 - 0.04)(0.6449 - 0.2879)$

$\qquad\qquad\qquad\qquad\qquad - \frac{1}{60}(1 + 0.0016 - \frac{5}{2}0.04)(0.0823 - 0.0202)]$

$= 0.2664 + 0.1784$

$= 0.4448.$

The correction term in formula (143) is therefore

$$0.004\pi(10)(25)(0.1848 + 0.4448) = 1.98 \ \mu\text{h}.$$

Consequently, $L = 78.75 - 1.98 = 76.77$ μh.

SPECIAL TYPES OF SINGLE-LAYER COIL

By the second method we have to find L_u' for a pancake circular coil of mean diameter $a = 25$ and with a cross section measuring $b = B = 1$ cm. in the axial direction and $c = 10p = 10$ cm. in the radial direction.

In formula (100) we have already found the value $P = 31.500$ and Table 24 gives for $\frac{c}{2a} = 0.2$ and $\frac{b}{c} = 0.1$, the factor $F = 0.9612$.

$$L_u' = 0.001(25)(10)^2(31.500)(0.9612) = 75.69 \,\mu\text{h}.$$

To calculate G_2, Table 3 gives for a rectangle $B \times p$ (here a square), the value $\log_e e_s = 0.00177$ and for the rectangular strip $\frac{C}{B} = 0.2$, $\log_e e_w = 0.00249$. The difference is -0.00072.

Therefore,

$$G_2 = \log_e \frac{2}{1.2} - 0.00072 = 0.5101.$$

To calculate H_2 we have in Table 2 for rectangles $\beta = 1$ and $\Delta = 1$, $\log_e k_s = 0.0065$. For $\log_e k_w$ Table 2 gives for $\beta = 1.0$, $\Delta = \frac{C}{B} = 0.2$, the value 0.0685. The difference is -0.0620.

$$H_2 = 2[0.9(-0.0620) - \tfrac{1}{12}(1 - 0.04)(0.6449 - 0.2879)$$
$$+ \tfrac{1}{24}\{0.96 - \tfrac{2}{3}(0.9984)\}(0.0823 - 0.0202)]$$
$$= -0.1660.$$

Thus the correction term in (146) is

$$0.004\pi(10)(25)(0.5101 - 0.1660) = 1.08 \,\mu\text{h},$$

and the inductance of the spiral is $L = 75.69 + 1.08 = 76.77 \,\mu\text{h}$, which checks the other method.

Toroidal Coils. Frequent use has been made of coils of toroidal form, that is, of coils that are wound in a single layer on a ring-shaped form. Such coils have the advantage of very small external magnetic field.

Formulas for the ideal case of a current sheet, such as would be attained by a winding of very thin tape with negligible insulating space between the turns, are very simple. Close windings of thin round wire possess an inductance sensibly the same as the current sheet of the same number of turns. The small correction for the space occupied by insulation may be calculated with sufficient accuracy by Rosa's method, see page 149.

Closely Wound Single-layer Coil on a Torus.[80] Here the winding form is doughnut-shaped.

Let $R =$ the mean radius of the winding form,

$2a =$ the diameter of the circle enclosed by each turn,

$N =$ the number of turns.

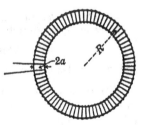

Fig. 43

Then the inductance of the current sheet is

$$L_s = 0.01257N^2 (R - \sqrt{R^2 - a^2}). \tag{147}$$

The correction for insulating space ΔL, which is to be subtracted from L_s, is given by

$$\Delta L = 0.002Nl(G + H)$$
$$= 0.004\pi aN(G + H), \tag{148}$$

in which G is to be obtained from Table 38 for the ratio of the bare diameter of the wire to the mean of the inner and outer pitches of the winding. The term H is to be taken from Table 39 for the given number of turns.

Toroidal Coils of Rectangular Turns.[81] Let r_1 and r_2 be, respectively, the inner and outer radii of the toroidal winding form; and $h =$ the axial thickness of the winding form. Each of the N turns has, therefore, a length $l = 2[h + (r_2 - r_1)]$ cm. The inductance is given by

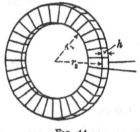

FIG. 44

$$L = 0.002N^2h \log_e \frac{r_2}{r_1} - 0.002Nl(G + H) \tag{149}$$

in which the last turn is the correction for insulating space. The value of G is obtained from Table 38 for the ratio of the bare diameter of the wire to the mean of the inner and outer pitches of the winding, and H is taken from Table 39 for the given value of the number of turns.

Example 59: A winding form of toroidal shape was wound with 2738 turns and the mean radii of the inner and outer portions of each turn were measured as $r_2 = 35.054$ cm. and $r_1 = 24.975$ cm. The axial dimension h taken between centers of wires was 20.085 cm. The diameter of the bare wire is $\delta = 0.02230$ cm.

These data, used in formula (149), give for the current sheet value, first term, 0.102089 henry. The mean pitch of the winding is $p = \dfrac{\pi(r_1 + r_2)}{N} = 0.0689$, so that $\dfrac{\delta}{p}$ in Table 38 is 0.324, for which is found $G = -0.572$. From Table 39 for $N = 2738$, $H = 0.337$. The length of one turn of the coil is $l = 60.327$ cm.

Accordingly, the correction term is

$$0.002(2738)(60.33)(-0.572 + 0.337) = -77.6 \; \mu h,$$

and the corrected inductance of the coil is 0.102167 henry.

Single-layer Polygonal Coils. Single-layer coils wound on polygonal cylindrical forms have the advantages of simplicity of construction and usefulness where it is desired to employ low-loss, small capacitance coils. For the construction of the winding form it is necessary only to prepare polygonal

SPECIAL TYPES OF SINGLE-LAYER COIL 171

end pieces and to mount between them longitudinal strips slightly slotted to hold the wires. The edges of these strips form the vertices of the polygons formed by the turns of the wire. Thus the wire is in contact with no solid insulator except for the short distances at the vertices of the polygon where it is held in place in the slots of the spacing strips.

Exact formulas are known for the inductance of square single-layer coils or solenoids of any length [82] and series expressions [83] for other polygonal coils whose axial length is small compared with the side of the polygon. These formulas show that, as might be expected, the inductance of the polygonal coil is nearly the same as for a circular solenoid whose turns enclose the same area as the area of the polygon. This is more closely the case, the greater the length of the coil as compared with the side of the polygon.

This suggests a simple general method for calculating the inductance. From the dimensions of the given coil the radius of the circular coil of the same number of turns and the same winding pitch is found, which has the same inductance as the polygonal coil. The value of the inductance of the polygonal coil may then be calculated by the formulas and tables given for circular solenoids.

For polygons having an even number of sides the procedure is simple. The coil is calipered over opposite vertices of the polygon. This dimension, minus the diameter of the wire, is equal to the diameter $2r_0$ of the circumscribed circle of the polygon. Sometimes it may be more convenient to measure the side s of the polygon and obtain the diameter of the circumscribed circle from that.

Let N = number of turns of the coil,
$b = pN$ = equivalent length of the coil,
ν = the number of sides of the polygon,
s = the length of the side of the polygon,
$2r_0$ = the diameter of the circumscribed circle,
$2a_0$ = the diameter of the circle whose area is the same as the area of the polygon.

Then there exist between these the relations

$$2r_0 = \frac{s}{\sin\frac{\pi}{\nu}}$$
$$\left(\frac{2a_0}{2r_0}\right)^2 = \frac{\nu}{2\pi}\sin\frac{2\pi}{\nu} \tag{150}$$

The diameter $2a$ of a circular solenoid of the same number of turns N and winding pitch p as the polygonal coil whose inductance is the same as the

172 CALCULATION OF MUTUAL INDUCTANCE AND SELF-INDUCTANCE

inductance of the polygonal coil may be obtained from Table 42, which gives the value of $\frac{2a}{2a_0}$ for different numbers of sides of the polygon and for different ratios of $\frac{b}{2r_0}$. The closeness of the ratio $\frac{2a}{2a_0}$ to unity, except for very short coils, is striking.

TABLE 42. DATA FOR CALCULATIONS OF POLYGONAL SINGLE-LAYER COILS

Triangular Coils			Square Coils			Hexagonal Coils		
$\frac{b}{2r_0}$	$\frac{2a}{2a_0}$	$\frac{2a}{2r_0}$	$\frac{b}{2r_0}$	$\frac{2a}{2a_0}$	$\frac{2a}{2r_0}$	$\frac{b}{2r_0}$	$\frac{2a}{2a_0}$	$\frac{2a}{2r_0}$
0	1.2861	0.8270	0	1.1284	0.9003	0	1.0501	0.9549
0.01	1.1341	.7294	0.01	1.0578	.8440	0.01	1.0203	.9278
.02	1.1168	.7181	.02	1.0500	.8378	.02	1.0172	.9250
.03	1.1052	.7107	.03	1.0449	.8337	.03	1.0151	.9231
.04	1.0964	.7050	.04	1.0410	.8306	.04	1.0136	.9218
0.05	1.0892	0.7004	0.05	1.0378	0.8280	0.05	1.0124	0.9206
.06	1.0831	.6965	.06	1.0351	.8259	.06	1.0113	.9197
.07	1.0779	.6931	.07	1.0328	.8241	.07	1.0104	.9189
.08	1.0732	.6901	.08	1.0308	.8224	.08	1.0097	.9183
.09	1.0691	.6875	.09	1.0290	.8210	.09	1.0090	.9176
0.10	1.0654	.6851	0.10	1.0274	.8198	0.10	1.0085	.9171
.125	1.0576	.6801	.125	1.0241	.8171	.125	1.0073	.9160
.15	1.0512	.6760	.15	1.0214	.8149	.15	1.0064	.9152
.175	1.0460	.6726	.175	1.0191	.8132	.175	1.0056	.9145
.20	1.0416	.6698	.20	1.0173	.8117	.20	1.0050	.9139
0.25	1.0345	0.6652	0.25	1.0143	0.8093	0.25	1.0040	0.9131
.30	1.0292	.6618	.30	1.0121	.8075	.30	1.0034	.9124
.35	1.0251	.6592	.35	1.0104	.8062	.35	1.0029	.9120
.40	1.0219	.6571	.40	1.0090	.8051	.40	1.0025	.9117
.45	1.0191	.6553	.45	1.0079	.8042	.45	1.0022	.9114
0.50	1.0169	0.6539	0.50	1.0070	0.8035	0.50	1.0020	0.9112
.60	1.0139	.6520	.60	1.0056	.8024	.60	1.0017	.9109
.70	1.0118	.6506	.70	1.0046	.8016	.70	1.0014	.9107
.80	1.0103	.6497	.80	1.0039	.8010	.80	1.0012	.9105
0.90	1.0090	.6488	0.90	1.0034	.8006	0.90	1.0011	.9104
1.00	1.0080	0.6482	1.00	1.0030	0.8003	1.00	1.0010	0.9103
$\frac{2r_0}{b}$	$\frac{2a}{2a_0}$	$\frac{2a}{2r_0}$	$\frac{2r_0}{b}$	$\frac{2a}{2a_0}$	$\frac{2a}{2r_0}$	$\frac{2r_0}{b}$	$\frac{2a}{2a_0}$	$\frac{2a}{2r_0}$
0.9	1.0072	0.6477	0.9	1.0026	0.8000	0.9	1.0009	0.9102
.8	1.0064	.6472	.8	1.0022	.7997	.8	1.0008	.9101
.7	1.0056	.6466	.7	1.0019	.7994	.7	1.0007	.9100
.6	1.0048	.6461	.6	1.0016	.7992	.6	1.0006	.9099
0.5	1.0040	.6456	0.5	1.0013	0.7989	0.5	1.0005	0.9098
.4	1.0032	.6451	.4	1.0010	.7987	.4	1.0004	.9098
.3	1.0024	.6446	.3	1.0007	.7984	.3	1.0003	.9097
.2	1.0016	.6440	.2	1.0004	.7982	.2	1.0002	.9096
0.1	1.0008	.6436	0.1	1.0002	.7980	0.1	1.0001	.9095
0	1	0.6430	0	1	0.7979	0	1	0.9094

TABLE 42. DATA FOR THE CALCULATION OF POLYGONAL SINGLE-LAYER COILS (*Concluded*)

| \multicolumn{3}{c}{Octagonal Coils} | \multicolumn{3}{c}{Twelve-Sided Coils} |

$\dfrac{b}{2r_0}$	$\dfrac{2a}{2a_0}$	$\dfrac{2a}{2r_0}$	$\dfrac{b}{2r_0}$	$\dfrac{2a}{2a_0}$	$\dfrac{2a}{2r_0}$
0	1.0270	0.9745	0	1.0117	0.9886
0.01	1.0100	.9583	0.01	1.0039	.9810
.02	1.0082	.9566	.02	1.0034	.9805
.03	1.0071	.9556	.03	1.0029	.9800
.04	1.0063	.9548	.04	1.0025	.9796
0.05	1.0056	0.9542	0.05	1.0022	0.9794
.06	1.0051	.9537	.06	1.0019	.9791
.07	1.0047	.9533	.07	1.0017	.9789
.08	1.0043	.9529	.08	1.0015	.9787
.09	1.0040	.9526	.09	1.0013	.9785
0.10	1.0037	0.9523	0.10	1.0012	0.9784
.125	1.0031	.9518	.125	1.0010	.9782
.15	1.0026	.9514	.15	1.0008	.9780
.175	1.0023	.9510	.175	1.0007	.9779
.20	1.0020	.9508	.20	1.0006	.9778
0.25	1.0016	0.9504	0.25	1.0005	0.9777
.30	1.0013	.9501	.30	1.0004	.9776
.35	1.0011	.9499	.35	1.0004	.9776
.40	1.0009	.9497	.40	1.0003	.9775
0.45	1.0008	.9496	.45	1.0003	.9775
.50	1.0008	0.9496	0.50	1.0002	0.9774
.60	1.0007	.9495	.60	1.0002	.9774
.70	1.0006	.9494	.70	1.0002	.9774
.80	1.0005	.9493	.80	1.0001	.9773
0.90	1.0004	.9492	0.90	1.0001	.9773
1.00	1.0004	0.9492	1.00	1.0001	0.9773
$\dfrac{2r_0}{b}$	$\dfrac{2a}{2a_0}$	$\dfrac{2a}{2r_0}$	$\dfrac{2r_0}{b}$	$\dfrac{2a}{2a_0}$	$\dfrac{2a}{2r_0}$
0.9	1.0004	0.9492	0.9	1.0001	0.9773
.8	1.0003	.9491			
.7	1.0003	.9491			
.6	1.0002	.9491			
0.5	1.0002	0.9490	0.5	1.0000	0.9772
.4	1.0002	.9490			
.3	1.0001	.9489			
.2	1.0001	.9489			
0.1	1.0000	.9489	0.1	1.0000	0.9772
0	1	0.9488	0	1	0.9772

For greater convenience in the calculation of the inductance, Table 42 gives also the ratio $\dfrac{2a}{2r_0}$ of the diameter of the equivalent circular solenoid to the diameter of the circle circumscribed about the polygon.

The inductance calculated from the dimensions of the equivalent circular solenoid is, of course, the value for a polygonal current sheet and the

174 CALCULATION OF MUTUAL INDUCTANCE AND SELF-INDUCTANCE

usual correction for insulating space has to be applied. For this, formula (135) or (137) and Tables 38, 39 or formulas (138)–(142) will be used, depending upon whether the coil is wound with round wire or rectangular strip. The whole procedure is illustrated in the following example.

Example 60: To calculate the inductance of an octagonal coil of 50 turns of round wire, bare diameter 0.2 cm. wound with a pitch of 0.4 cm. The diameter of the covered wire is 0.25 cm. Calipered over the opposite vertices of the octagon, the value 11.29 cm. was found, so that the diameter of the circumscribed circle is $2r_0 = 11.04$ cm.

The length of the equivalent current sheet is $b = 50 \times 0.4 = 20$ cm. so that the shape ratio $\frac{2r_0}{b} = 0.552$ is found. For an octagonal coil with this ratio, Table 42 gives by interpolation $\frac{2a}{2r_0} = 0.9491$, so that the equivalent circular solenoid will have a diameter $2a = 0.9491 \times 11.04 = 10.48$ cm. (It is interesting to note from the table that the diameter of this equivalent circular solenoid bears to the diameter of the circle having the same area as the polygon the ratio 1.00025.)

Calculating the inductance by formula (122) for a circular solenoid, we have to obtain the value of F from Table 40 for the ratio $r = \frac{2a}{b} = \frac{10.48}{20} = 0.524$. The interpolated value is $F = 4.195$, so that the inductance of the equivalent current sheet is

$$L_s = 0.001(4.19)(10.48)(50)^2 = 109.9 \ \mu h.$$

The correction for insulating space is derived from formula (120). Since $\frac{\delta}{p} = \frac{0.2}{0.4} = 0.5$, the value of G derived from Table 38 is -0.1363 and for $N = 50$, Table 39 gives $H = 0.3182$. The correction is, therefore,

$$0.002\pi(10.48)(50)(-0.1363 + 0.3182) = 0.60 \ \mu h.$$

The inductance of the octagonal coil is, therefore, $L = 109.9 - 0.60 = 109.3 \ \mu h$.

Table 42 allows the equivalent radius to be calculated for the most important polygonal coils likely to be found in practice. In case, however, values are required for polygons not there included, values sufficiently accurate may be obtained by plotting values taken from Table 42 for the parameter $\frac{b}{2r_0}$ in question against the reciprocal of the number of sides as abscissas. The desired value can be interpolated for the given value of the number of sides. For example, if it be required to find the ratio $\frac{2a}{2a_0}$ for a pentagon with the parameter $\frac{b}{2r_0} = 0.2$, the values 1.0416, 1.0173, 1.0050, 1.0020 and 1.0006 taken from the table for this parameter are plotted with the abscissas $\frac{1}{3}$, $\frac{1}{4}$, $\frac{1}{6}$, $\frac{1}{8}$, and $\frac{1}{12}$, respectively. The value for zero abscissa is known to be unity exactly. From the plotted curve the value interpolated for the pentagon will have an abscissa $\frac{1}{5}$. This value is $\frac{2a}{2a_0} = 1.0137$. From this, by formula (150) the required value of $\dfrac{2a}{2r_0}$ is $1.0137 \sqrt{\dfrac{5}{2\pi}} \sin \dfrac{2\pi}{5} =$ 0.8819.

For very short polygonal coils the general method for the calculation of the inductance suffers from the difficulty of obtaining accurately interpolated values. In such cases the following series formulas [84] are accurate and simple to use for coils whose lengths b do not exceed a quarter of the diameter $2r_0$ of the circumscribed circle.

Series Formulas for Short Polygonal Coils.

Short Triangular Coil:

$$L_s = 0.006N^2s \left[\log_e \frac{2r_0}{b} - 0.04931 + 0.8503 \frac{b}{2r_0} \right.$$
$$\left. - 0.2037 \left(\frac{b}{2r_0}\right)^2 + 0.0278 \left(\frac{b}{2r_0}\right)^4 + \cdots \right]. \quad (151)$$

Short Square Coil:

$$L_s = 0.008N^2s \left[\log_e \frac{2r_0}{b} + 0.37942 + 0.4714 \frac{b}{2r_0} \right.$$
$$\left. - 0.0143 \left(\frac{b}{2r_0}\right)^2 - 0.0290 \left(\frac{b}{2r_0}\right)^4 + \cdots \right]. \quad (152)$$

Short Hexagonal Coil:

$$L_s = 0.012N^2s \left[\log_e \frac{2r_0}{b} + 0.65533 + 0.2696 \frac{b}{2r_0} \right.$$
$$\left. + 0.0774 \left(\frac{b}{2r_0}\right)^2 - 0.055 \left(\frac{b}{2r_0}\right)^4 - \cdots \right]. \quad (153)$$

Short Octagonal Coil:

$$L_s = 0.016N^2s \left[\log_e \frac{2r_0}{b} + 0.75143 + 0.1869 \left(\frac{b}{2r_0}\right) \right.$$
$$\left. + 0.1197 \left(\frac{b}{2r_0}\right)^2 + 0.082 \left(\frac{b}{2r_0}\right)^4 + \cdots \right]. \quad (154)$$

These are current sheet values and require correction for insulating space. For this purpose there has to be subtracted the value $0.002\nu sN(G + H)$ for a winding of round wire and for rectangular strip G and H have to be replaced by G_1 and H_1. This formula is consistent with that for circular solenoids. In both cases the factor outside the brackets is, except for the decimal point, equal to twice the total length of wire on the coil.

Example 61: To find the inductance of a square coil constructed by winding bare wire of 0.2 cm. diameter with a pitch of 1 cm. to form a square coil of 5 turns, with the side of the square equal to 4 feet.

Here $s = 4$ feet $= 121.92$ cm., $\nu = 4$, so that by (150)

$$2r_0 = \frac{121.92}{\sin\frac{\pi}{4}} = 172.42 \text{ cm.}$$

The equivalent length is $b = 5$, and $\frac{b}{2r_0} = 0.02900$.

In formula (152) $\log_e \frac{2r_0}{b} = 3.5405$, and

$$L_s = 0.008(5)^2(121.92)[3.5405 + 0.3794 + 0.0137] = 95.91 \ \mu\text{h}.$$

Twice total length of wire is $2 \times 5 \times 4 \times 121.92 = 4877$ cm. Since $\frac{\delta}{p} = 0.2$, G from Table 38 is -1.0526, and from Table 39, for $N = 5$, $H = 0.2180$. The correction is therefore $4.877(-1.0526 + 0.2180) = -4.06 \ \mu\text{h}$, which subtracted from L_s gives $L = 100.0 \ \mu\text{h}$.

If we use the general method, the interpolated value from Table 42 for the given ratio $\frac{b}{2r_0}$ comes out $\frac{2a}{2r_0} = 0.8341$, giving $2a = 143.8$ cm. and $\frac{b}{2a} = 0.03477$.

It is, however, difficult to interpolate F accurately from Table 41. The value found is about 26.70, which gives $L_s = 95.99 \ \mu\text{h}$, but using the more accurate Table 36 for K and formula (118) the value found checks the value by the series formula exactly. The series formula is the more convenient for this example.

Flat Spirals with Polygonal Turns. The calculation could in this case be made by finding the diameter of the circumscribed circle of the mean turns of the spiral and thence by Table 42 the approximate mean diameter of an equivalent disc coil. This method would probably give an accuracy as great as is justified by the difficulty of measuring the dimensions of a spiral, but the use of the following series formulas is simple and gives sufficient accuracy in practical cases where the turns of the spiral are not carried down close to the center of the spiral.

Using the nomenclature of the preceding section, with s now the side of the mean turn, the current sheet values [86] are:

Triangular Spiral:
$$L_s = 0.006 N^2 s \left[\log_e \frac{s}{b} + 0.09543 + 0.4132 \frac{b}{s} + 0.3194 \frac{b^2}{s^2} + \cdots \right]. \quad (155)$$

Square Spiral:
$$L_s = 0.008 N^2 s \left[\log_e \frac{s}{b} + 0.72599 + 0.1776 \frac{b}{s} + \frac{1}{8} \frac{b^2}{s^2} + \cdots \right]. \quad (156)$$

Hexagonal Spiral:
$$L_s = 0.012 N^2 s \left[\log_e \frac{s}{b} + 1.34848 + 0.0678 \frac{b}{s} + 0.0491 \frac{b^2}{s^2} \right]. \quad (157)$$

Octagonal Spiral:
$$L_s = 0.016 N^2 s \left[\log_e \frac{s}{b} + 1.71198 + 0.0363 \frac{b}{s} + 0.0277 \frac{b^2}{s^2} + \cdots \right]. \quad (158)$$

These values are to be corrected for insulating space by the same formulas as for the single-layer polygonal coils.

Chapter 18

MUTUAL INDUCTANCE OF CIRCULAR ELEMENTS WITH PARALLEL AXES

Mutual Inductance of Circular Filaments of Equal Radii and with Parallel Axes. *Nomenclature.*

Let a = radius of the circular filaments,
 d = distance between planes of circles,
 ρ = distance between axes,
 θ = angle between axes and radius vector between centers of the circles,
 $\mu = \cos \theta$,
 r = distance between centers.

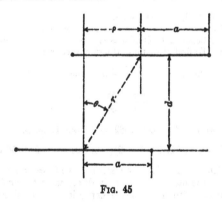

Fig. 45

Usually the given data will be either (a) the radius, distance r, and angle θ; or (b) the radius and the coordinates ρ and d of the center of one circle referred to the center of the other as origin. The following relations are useful:

$$d = r \cos \theta = r\mu,$$

$$\rho = r \sin \theta.$$

177

178 CALCULATION OF MUTUAL INDUCTANCE AND SELF-INDUCTANCE

General Formula. The mutual inductance of the circular filaments is given by the formula

$$M = M_0 F, \qquad (159)$$

in which M_0 = the mutual inductance of the circles when placed in the coaxial position ($\theta = 0$) with a distance r between their centers, and F is a factor depending on the value of μ and the ratio of $\dfrac{r}{2a}$ $\left(\text{or } \dfrac{2a}{r}\right)$. Values of F are given in Table 43 for different values of μ and $\dfrac{r}{2a}$, or $\dfrac{2a}{r}$, depending upon which is less than unity.

The value of $M_0 = fa$, the factor being obtained from Table 16 for δ equal to the given value of $\dfrac{r}{2a}$, from Table 17 for $\Delta = \dfrac{2a}{r}$.

Example 62: Given two circles of radii 15 cm. with the distance between their centers equal to 20 cm. and an angle $\theta = \cos^{-1} 0.8$ between their axes and the line joining centers; to find their mutual inductance.

Arranged in the coaxial position, these circles have a ratio, $\dfrac{\text{distance}}{\text{diameter}} = \dfrac{r}{2a} = \dfrac{20}{30} = \dfrac{2}{3}$. From Table 16 we find by interpolation

$$f = 0.0031239 \text{ and } M_0 = 15 \times 0.0031239 = 0.046858 \; \mu\text{h}.$$

From Table 43 with $\mu = 0.8$ and $\dfrac{r}{2a} = \dfrac{2}{3}$, we find by interpolation $F = 0.9670$ and $M = 0.046858(0.9670) = 0.04531 \; \mu\text{h}$.

Example 63: Two circles of wire, each of diameter equal to 48 inches, are arranged so that the distance between their planes is $d = 15$ and the distance between their axes is $\rho = 47.7$ in. To find their mutual inductance, we have from the given data

$$r = \sqrt{(47.7)^2 + (15)^2} = 50 \text{ inches}, \; \mu = \tfrac{15}{50} = 0.3,$$

and

$$\frac{\text{diameter}}{\text{distance}} = \frac{48}{50} = 0.96 = \frac{2a}{r}.$$

From Table 17 we find $f = 0.0012982$, and from Table 43, $F = -0.3103$, so that

$$M = 24(2.54)(0.0012982)(-0.3103) = -0.02456 \; \mu\text{h}.$$

The negative sign signifies that the electromotive force induced in one circle by a change of current in the other is opposite in direction to the emf resulting from the same change in current with the circles arranged in the coaxial position.

Table 43 has been calculated from several different formulas.[86] It covers the whole range of possible arrangements of equal circles with parallel axes. It suffices for routine calculations, and for single interpolation gives satisfactory accuracy except in regions where the ratio F is changing rapidly. For those cases and for the usual case where a double interpolation is necessary, it will be better to embody the data of Table 43 in curves.

INDUCTANCE OF CIRCULAR ELEMENTS WITH PARALLEL AXES

TABLE 43. VALUES OF F FOR EQUAL CIRCLES WITH PARALLEL AXES, FORMULA (159)

μ	$r/2a = 0$	0.1	0.2	0.3	0.4	0.5	0.6	0.7	0.8	0.9	$r/2a = 1$	μ
1.0	1	1	1	1	1	1	1	1	1	1	1	1.0
0.9	1	1.0267	1.0330	1.0329	1.0265	1.0146	0.9982	0.9790	0.9584	0.9376	0.9176	0.9
0.8	1	1.0552	1.0692	1.0699	1.0568	1.0313	0.9954	0.9527	0.9070	.8613	.8180	0.8
0.7	1	1.0857	1.1087	1.1112	1.0919	1.0509	0.9917	.9200	.8428	.7665	.6959	0.7
0.6	1	1.1155	1.1517	1.1580	1.1328	1.0750	0.9876	.8787	.7619	.6472	.5441	0.6
0.5	1	1.1536	1.1997	1.2111	1.1812	1.1052	0.9842	.8291	.6585	.4938	.3515	0.5
0.4	1	1.1917	1.2524	1.2717	1.2390	1.1440	0.9836	.7668	.5246	.2914	.1014	0.4
0.3	1	1.2330	1.3109	1.3411	1.3085	1.1952	0.9997	.6964	.3489	.0137	−0.2378	0.3
0.2	1	1.2780	1.3760	1.4212	1.3929	1.2641	1.0102	.5850	.1178	−0.3874	−0.7240	0.2
0.1	1	1.3274	1.4489	1.5139	1.4959	1.3577	1.0557	.5505	−0.1681	−1.0231	−1.5087	0.1
0	1	1.3820	1.5311	1.6214	1.6220	1.4851	1.145	.5253	−0.4672	−1.953	−4.053	0

μ	$2a/r = 1$	0.9	0.8	0.7	0.6	0.5	0.4	0.3	0.2	0.1	$2a/r = 0$	μ
1.0	1	1	1	1	1	1	1	1	1	1	1	1.0
0.9	0.9176	0.8968	0.8736	0.8482	0.8231	0.7946	0.7693	0.7471	0.7298	0.7188	0.7150	0.9
0.8	.8180	.7741	.7266	.6771	.6292	.5812	.5398	.5058	.4806	.4652	.4600	0.8
0.7	.6959	.6267	.5551	.4848	.4196	.3626	.3157	.2798	.2546	.2399	.2350	0.7
0.6	.5441	.4477	.3543	.2700	.1992	.1434	.1019	.0729	.0540	.0434	.0400	0.6
0.5	.3515	.2275	.1190	.0336	−0.0287	−0.0697	−0.0956	−0.1109	−0.1195	−0.1237	−0.1250	0.5
0.4	.1014	−0.0470	−.1551	−.2212	−.2551	−.2681	−.2704	−.2678	−.2640	−.2610	−.2600	0.4
0.3	−0.2378	−.3940	−.4670	−.4809	−.4704	−.4410	−.4156	−.3941	−.3780	−.3683	−.3650	0.3
0.2	−0.7240	−0.8337	−.7992	−.7204	−.6357	−.5764	−.5250	−.4867	−.4604	−.4451	−.4400	0.2
0.1	−1.5087	−1.3509	−1.0900	−.8992	−.7586	−.6632	−.5930	−.5434	−.5102	−.4912	−.4850	0.1
0	−4.053	−1.677	−1.2154	−0.9636	−0.8030	−0.6931	−0.6160	−0.5624	−0.5269	−0.5066	−0.5000	0

180 CALCULATION OF MUTUAL INDUCTANCE AND SELF-INDUCTANCE

An inspection of the table brings out some interesting facts. With a given pair of circles, if one of the circles is moved, keeping the axes parallel and maintaining the distance between their centers constant, so that θ is varied from zero to 90°, the mutual inductance varies in a manner depending strongly on the ratio of r to $2a$. For circles near together, $\frac{r}{2a}$ small, the mutual inductance increases continuously from the coaxial to the coplanar position, owing to the decrease in distance between the planes. This is true for all cases up to $\frac{r}{2a} = 0.6$, for which the effect of decreasing distance between planes is compensated by the opposite effect of the increasing distance between the axes. For more distant circles, the latter effect is predominant. The mutual inductance decreases continuously with increasing θ, passes through zero, and becomes negative. The table would be of use in placing two circles or coils of wire, with parallel axes, so as to have zero mutual inductance.

From the tabulated values of F may be derived the following values of θ_0 and $\cos \theta_0$ necessary to give zero mutual inductance for different values of $\frac{2a}{r}$ and $\frac{r}{2a}$. For values of $\frac{r}{2a}$ smaller than about 0.76 the mutual inductance does not become zero for any value of θ.

TABLE 44. ANGULAR POSITION FOR ZERO MUTUAL INDUCTANCE OF PARALLEL EQUAL CIRCLES

$2a/r$	μ	θ_0	$2a/r$	μ	θ_0	$2a/r$	μ	θ_0
0	0.58	54°6	0.5	0.53	58°0	1.0	0.37	68°3
0.1	0.58	54°6	0.6	0.51	59°3	0.9	0.30	72°5
0.2	0.575	54°9	0.7	0.485	61°0	0.8	0.16	80°8
0.3	0.565	55°6	0.8	0.455	62°9			
0.4	0.55	56°6	0.9	0.42	65°2			
0.5	0.53	58°0	1.0	0.37	68°3			

For value of $\frac{r}{2a} \leq 0.76$ the mutual inductance is never equal to zero, whatever the angle.

Mutual Inductance of Coplanar Circular Filaments of Equal Radii. As is evident from Table 43, interpolation would be difficult for this special case. It is better to calculate the mutual inductance directly without recourse to the coaxial case. Available formulas show that for intersecting

INDUCTANCE OF CIRCULAR ELEMENTS WITH PARALLEL AXES 181

coplanar circles $\left(\dfrac{r}{2a}\text{ small}\right)$ the mutual inductance varies logarithmically, while for distant circles $\left(\dfrac{2a}{r}\text{ small}\right)$ it varies with the cube of $\dfrac{2a}{r}$.

For equal intersecting coplanar circles

$$M = Ca, \qquad (160)$$

the value of C being interpolated from Table 45 for the given value of $\dfrac{r}{2a}$.

For the more distant coplanar circles

$$M = -0.001\left(\frac{\pi^2}{8}\right)\left(\frac{2a}{r}\right)^3 Da, \qquad (161)$$

values of D being also obtainable from Table 45.

TABLE 45. CONSTANTS FOR EQUAL COPLANAR CIRCULAR FILAMENTS, FORMULAS (160) AND (161)

$r/2a$	C	Diff.	$\log C$	$2a/r$	C	D	Diff.
0.1	0.029766		$\bar{2}.47372$	1.0	−0.005749	4.6604	
		−9085					
0.2	.020681		$\bar{2}.31557$	0.9	− .001886	2.0969	
		−5608					−4487
0.3	.015073		$\bar{2}.17821$	0.8	− .001041	1.6482	
		−4233					−2350
0.4	.010840		$\bar{2}.03505$	0.7	− .0006160	1.4132	
		−3506					−1463
0.5	0.007334		$\bar{3}.86537$	0.6	− .0003376	1.2669	
		−3062					− 983
0.6	.004272		$\bar{3}.63068$	0.5	− .0001802	1.1686	
		−2767					− 680
0.7	+ .001506		$\bar{3}.17776$	0.4		1.1006	
		−2551					− 468
0.8	− .001045		$n\bar{3}.01915$	0.3		1.0538	
		−2412					− 306
0.9	− .003457		$n\bar{3}.53869$	0.2		1.0232	
		−2292					− 175
1.0	−0.005749		$n\bar{3}.75964$	0.1		1.0057	
							− 57
				0		1	

It is evident that for values of $\dfrac{2a}{r}$ in the neighborhood of unity, interpolation is only rough. Unfortunately for such cases—nonintersecting circles very

182 CALCULATION OF MUTUAL INDUCTANCE AND SELF-INDUCTANCE

close together—no simple formulas exist and accurate values of the mutual inductance can be obtained, in any given case, only with the expenditure of considerable effort. However, in practical cases of this kind, the impossibility of obtaining precision in the given data would render such labor unjustifiable. A practical solution of the difficulty is to obtain a rough value of the mutual inductance from a curve of values plotted from the table and including the value for $\frac{2a}{r} = 1$ (circles tangent to each other).

Example 64: Two coplanar circles of 1 foot diameter are placed with their centers 1.5 feet apart. The corresponding value of $\frac{2a}{r}$ is $\frac{1.0}{1.5} = \frac{2}{3}$. The value of D interpolated from the table is 1.358, so that by (161)

$$M = -\frac{\pi^2}{8}(.001)(1.358)\left(\frac{2}{3}\right)^3 (6 \times 2.54) = -0.007566 \ \mu h.$$

If the circles be moved together until they are tangent, the mutual inductance is, by (160),

$$M = -(6 \times 2.54)(0.005749) = -0.08762 \ \mu h.$$

When they are moved together into positions where they intersect, the mutual inductance decreases, and becomes zero when their centers are 9.1 inches apart. Bringing their centers still nearer together, the mutual inductance reverses sign and increases. With their centers 3 inches apart, $\frac{r}{2a} = \frac{3}{12} = \frac{1}{4}$. From Table 45, by interpolation, log C is found to be $\bar{2}.24762$, corresponding to $C = 0.017686$, so that,

$$M = 6(2.54)(0.017686) = 0.2695 \ \mu h.$$

If the circles could be made concentric, the mutual inductance would, of course, become logarithmically infinite. This is, of course, practically impossible, on account of the finite cross section of the wire.

Mutual Inductance of Circular Filaments Having Parallel Axes and Unequal Radii. This case has less practical importance than that where the radii are equal. Existing formulas due to Snow [87] and Butterworth [88] have poor convergence in many cases and a method employing mechanical quadrature is necessary. Calculations are, therefore, tedious and time consuming. Furthermore, tables to be generally useful would have to involve three parameters and would have to be based on an amount of calculation not justified by the importance of the case. Accordingly, much simpler methods, even though they yield results of rougher accuracy, have to suffice for routine calculations. Three cases, in general, have to be considered.

Case 1. Distant circles, where the distance r between the centers of the circles is large compared with the sum of their radii $(A + a)$.

INDUCTANCE OF CIRCULAR ELEMENTS WITH PARALLEL AXES

Calculate first, the mutual inductance M_0 of these circles placed in the coaxial position with a distance r between their centers.

$$M_0 = f\sqrt{Aa},$$

in which f is obtained from Table 13 (page 79), for the value of

$$k'^2 = \frac{(A-a)^2 + r^2}{(A+a)^2 + r^2}.$$

To obtain the mutual inductance of these circles with their axes parallel, it is necessary to obtain a factor F which depends on $\mu = \cos\theta$. The following empirical method is capable of giving satisfactory accuracy.

Using the Table 43 for equal circles, calculate the ratios $\dfrac{2a}{r}$ and $\dfrac{2A}{r}$ and using these as arguments in the table and the given value of μ, interpolate the corresponding values F_1 and F_2 of the factor. Then the required value F for the unequal circles is approximately the mean of F_1 and F_2, and

$$M = M_0 F \qquad (162)$$

The auxiliary equations

$$d = r\cos\theta = r\mu$$

$$\rho = r\sin\theta$$

apply, the same as for equal circles.

The accuracy of this method is greater, the more distant the circles and the more nearly equal the radii.

Example 65: Assume two circles for which

$$A = 10, \quad a = 8, \quad r = 50, \quad \mu = 0.4.$$

For these circles in the coaxial position, formula (78) (page 77) gives

$$k'^2 = \frac{(2)^2 + (50)^2}{(18)^2 + (50)^2} = \frac{2504}{2824} = 0.88668,$$

and interpolating in Table 13 for this value of k'^2, $f = 1.02896 \times 10^{-4}$, so that $M_0 = \sqrt{80}\cdot(1.02896)10^{-4} = 0.00092033$ μh.

Using $\mu = 0.4$ in Table 43 there is found for the ratios

$$\frac{2a}{r} = \frac{16}{50} = 0.32, \qquad F_1 = -0.2686$$

$$\frac{2A}{r} = \frac{20}{50} = 0.4, \qquad F_2 = -0.2704$$

$$\text{average} = -0.2695,$$

which leads to

$$M = -0.2695\,M_0 = -0.0002480 \text{ μh}.$$

184 CALCULATION OF MUTUAL INDUCTANCE AND SELF-INDUCTANCE

Using Butterworth's formula [88] the true value of F is found to be -0.26970, or a difference of about 1 part in 1000. This calculated value of M by the empirical method is entirely satisfactory for all practical purposes and is obtained with much less labor than by the precision formula.

Even with very unequal radii the accuracy obtainable suffices.

Example 66: For two circles nearer together

$$A = 10, \quad a = 8, \quad r = 20, \quad \mu = 0.6$$

the details of the calculation by the empirical method are as follows:

$$k'^2 = \frac{(10-8)^2 + (20)^2}{(10+8)^2 + (20)^2} = 0.55801,$$

and from Table 13, $f = 0.0010965$,

$$M_0 = \sqrt{80} \cdot (0.0010965) = 0.009807 \ \mu\text{h}.$$

Entering Table 43 with $\mu = 0.6$

$$\frac{2a}{r} = \frac{16}{20} = 0.8, \qquad\qquad F_1 = 0.3543$$

$$\frac{2A}{r} = \frac{20}{20} = 1, \qquad\qquad F_2 = 0.5441$$

$$\text{average} = 0.4492,$$

which leads to

$$M = 0.4492(0.009807) = 0.004405 \ \mu\text{h}.$$

For this case the precision formulas converge so poorly that it is necessary to use a mechanical quadrature method. The value found after a tedious calculation is $M = 0.004465$. The empirical method gives a value in error by less than 1 per cent.

If the smaller radius is taken $a = 3$, the other given data remaining the same, we find $k'^2 = \dfrac{4.49}{5.69} = 0.78912$, and from Table 13, $f = 2.8420 \times 10^{-4}$, so that

$$M_0 = \sqrt{30}(2.8420 \times 10^{-4}) = 0.0015570 \ \mu\text{h}.$$

Using Table 43 with $\mu = 0.6$

$$\frac{2a}{r} = 0.3, \qquad\qquad F_1 = 0.0729$$

$$\frac{2A}{r} = 1, \qquad\qquad F_2 = 0.5441$$

$$\text{average} = 0.3085,$$

so that

$$M = 0.3085 \times 0.0015570 = 0.0004803 \ \mu\text{h}.$$

Calculation by the elaborate Butterworth formula leads to the correct value 0.0004808. The approximate method is entirely satisfactory in this case also.

For greater accuracy the mutual inductance may be calculated by the mechanical quadrature formulas (163) or (166) given below.

Case 2. Circles Close Together. For circles where r is less than about twice the larger radius A, the empirical method just described is unsatisfactory. Series formulas for such cases are known.[87] These are somewhat com-

INDUCTANCE OF CIRCULAR ELEMENTS WITH PARALLEL AXES 185

plicated and unfortunately they are only very slowly convergent in many practical cases. To supplement these expressions the formula (163) below has been derived and is recommended for routine calculations in general. A moderate amount of labor only is necessary to attain an accuracy sufficient for practical requirements.

Using the nomenclature of Fig. 45 and with $\alpha = \dfrac{a}{A}$, $\delta = \dfrac{d}{A}$, the mutual inductance of the circular filaments is given by the general formula

$$M = \sqrt{Aa} \int_0^\pi \frac{f}{\pi} \frac{\left(1 - \dfrac{\rho}{a}\cos\phi\right)}{V^{3/2}} d\phi \ \mu\text{h.} \tag{163}$$

In this formula

$$V = \sqrt{1 - 2\frac{\rho}{a}\cos\phi + \frac{\rho^2}{a^2}}, \tag{164}$$

and f is obtained from Table 13 for values of

$$k'^2 = \frac{(1 - \alpha V)^2 + \delta^2}{(1 + \alpha V)^2 + \delta^2}. \tag{165}$$

The integration indicated in (163) may be performed by mechanical quadrature and for this the following procedure has been found effective. Both Simpson's [89] formula and Weddle's formula are used, one serving to check the other. To apply Simpson's formula the interval of the integration has to be divided into an even number of equal intervals. Weddle's formula assumes the number of intervals to be a multiple of six. If, therefore, in formula (163) twelve values of the integrand, one for every 15°, are calculated the integral may be found by both formulas.

Denoting by $y_0, y_1, y_2 \cdots y_{12}$ the calculated values of the integrand, Simpson's formula gives for the integral

$$S = \frac{1}{3}\left(\frac{\pi}{12}\right)[2(y_0 + y_2 + y_4 + \cdots + y_{12}) + 4(y_1 + y_3 + y_5 + \cdots + y_{11}) \\ - (y_0 + y_{12})], \quad \text{(A)}$$

and the value calculated by the Weddle formula is

$$W = \frac{3}{10}\left(\frac{\pi}{12}\right)[5(y_1 + y_5 + y_7 + y_{11}) + 6(y_3 + y_9) \\ + (y_0 + y_2 + y_4 + \cdots + y_{12}) + y_6]. \quad \text{(B)}$$

The closeness of agreement of the two values S and W is an indication of the accuracy of the evaluation of the integral. In the event of a considerable difference in the two results, a calculation may be made for 24 points ($7\frac{1}{2}°$ interval). This requires, of course, the calculation of 12 ordinates in

186 CALCULATION OF MUTUAL INDUCTANCE AND SELF-INDUCTANCE

addition to those already found. The formulas for mechanical integration in this case become

$$S' = \frac{1}{3}\left(\frac{\pi}{24}\right)[2(y_0 + y_2 + y_4 + \cdots + y_{24})$$
$$+ 4(y_1 + y_3 + y_5 + \cdots + y_{23}) - (y_0 + y_{24})], \quad (C)$$

and

$$W' = \frac{3}{10}\left(\frac{\pi}{24}\right)[5(y_1 + y_5 + y_7 + y_{11} + y_{13} + y_{17} + y_{19} + y_{23})$$
$$+ 6(y_3 + y_9 + y_{15} + y_{21})$$
$$+ (y_0 + y_2 + y_4 + \cdots + y_{22} + y_{24}) + (y_6 + y_{12} + y_{18})]. \quad (D)$$

Unless the curve of the ordinates is very steep or sharply curved, a very accurate value of the integral is then given by $S' + \frac{1}{15}(S' - S)$. In general, the difference $(S' - S)$ is inconsiderable and the agreement of W' and S' is closer than that of W and S.

Example 67: $a = 8$, $A = 10$, $r = 20$, $d = 12$.
Therefore

$$\alpha = 0.8, \quad \delta = 1.2, \quad \mu = \cos\theta = \frac{d}{r} = 0.6$$

and

$$\sin\theta = 0.8, \quad \rho = r\sin\theta = 16, \quad \frac{\rho}{a} = 2.$$

Thus, for this case, $V = \sqrt{5 - 4\cos\phi}$. The following table shows some of the details of the calculation:

	$\phi = 0$	15°	30°	45°	60°	75°	90°
$4\cos\phi$	4.0	3.8637	3.4641	2.8284	2.0	1.0353	0.0
V^2	1.0	1.1363	1.5359	2.1716	3.0	3.9647	5.0
k'^2	1.48	1.4617	1.4401	1.4720	1.5887	1.7915	2.0623
	4.68	4.8729	5.4059	6.1878	7.1314	8.6630	9.2178
	0.31624	0.29996	0.26639	0.23789	0.22277	0.21947	0.22373
$1000f$	3.0172	3.2242	3.7041	4.1799	4.4636	4.5287	4.4449
$1 - \frac{\rho}{a}\cos\phi$	−1.0	−0.93185	−0.73205	−0.4142	0.0	0.48235	1.0
y_n	−3.0172	−2.7298	−1.9654	−0.9678	0.0	0.7775	1.3293

	180°	165°	150°	135°	120°	105°
$4\cos\phi$	−4.0	−3.8637	−3.4641	−2.8284	−2.0	−1.0353
V^2	9.0	8.8637	8.4641	7.8284	7.0	6.0353
aV	2.4	2.3818	2.3274	2.2383	2.1166	1.9654
k'^2	0.26154	0.26011	0.25593	0.24930	0.24089	0.23179
$1000f$	3.7802	3.8029	3.8707	3.9808	4.1263	4.2915
$1 - \frac{\rho}{a}\cos\phi$	3.0	2.9318	2.7320	2.4142	2.0	1.5176
y_n	2.1827	2.1704	2.1311	2.0534	1.0177	1.6915

INDUCTANCE OF CIRCULAR ELEMENTS WITH PARALLEL AXES

The values of f, multiplied by 1000 for convenience, have been interpolated from Table 13 for the calculated values of k'^2. The calculation of the integral in (163) from the calculated ordinates y_n follows:

Even	Odd	Extreme		
−3.0172	−2.7298	−3.0172	2 × 2.5782 =	5.1564
−1.9654	−0.9678	2.1827	4 × 2.9952 =	11.9808
0	0.7775	−0.8345 = Sum		17.1372
1.3293	1.6915			−0.8345
1.9177	2.0534		Diff. =	17.9717
2.1311	2.1704		÷ 36 =	$0.49921 = \dfrac{S}{\pi}$
2.1827	6.6928			
7.5608	−3.6976			
−4.9826	2.9952 = Sum			
2.5782 = Sum				

$y_1 = -2.7298$	$y_3 = -0.9678$	5 × 1.9096 =	9.5480
$y_5 = 0.7775$	$y_9 = 2.0534$	6 × 1.0856 =	6.5136
$y_7 = 1.6915$	Sum = 1.0856	$\Sigma y_{2n} =$	2.5782
$y_{11} = 2.1704$		$y_6 =$	1.3293
4.6394		Sum =	19.9691
−2.7298		÷ 40 =	$0.49923 = \dfrac{W}{\pi}$
Sum = 1.9096			

Therefore, from (163),

$$M = \sqrt{80}(0.49921) = 4.465 \text{ abhenrys}$$
$$= 0.004465 \ \mu\text{h.}$$

Graphical Solution for Circular Filaments with Parallel Axes. In order to save the labor of calculating values of k'^2 and obtaining from them values of f by interpolation in Table 13, the interpolation may be carried out graphically [90] without serious sacrifice of accuracy.

Formula (163) may be written in the form

$$\frac{M}{A} = \frac{1}{\pi} \int_0^\pi \frac{\left(1 - \dfrac{\rho}{a}\cos\phi\right)}{V^2} \left(\frac{m}{A}\right) d\phi \qquad (166)$$

in which $\dfrac{m}{A}$ is the ratio of the mutual inductance in abhenrys of a pair of coaxial circles referred to the radius A of the larger of the given circles. The radius a_1 of the smaller circle is a function of ϕ and is given by $\dfrac{a_1}{A} = \alpha V$ and the spacing of ratio $\dfrac{d_1}{A}$ of the circles is equal to $\dfrac{d}{A} = \delta$.

188 CALCULATION OF MUTUAL INDUCTANCE AND SELF-INDUCTANCE

The curves of Figs. 46 and 47 are the loci of the coordinates of pairs of circles having the indicated values of $\frac{m}{A}$ as parameters. Abscissas are values of the ratio of the radii and ordinates are values of the spacing ratio. Thus the curves are maps of the flux lines due to a current in a circular filament of unit radius, whose axis is the axis of δ and whose radius is the unit of abscissas. That is, the circular filament cuts the plane of the paper in $x = 1$. Fig. 47 exhibits a magnified portion of Fig. 46.

Example 68: To illustrate the use of formula (166) and Figs. 46 and 47, the problem of example 67 will be solved. The method is to calculate, for values of ϕ taken every 15°, the coordinates $x = \alpha V$ and $y = \delta$. For each pair the corresponding value of $\frac{m}{A}$ will be interpolated between the curves of Fig. 46 and the ordinates y_n will be calculated by formula (166). The integration is carried out by mechanical quadrature based on the calculated ordinates y_n just as in example 67.

For the given values $a = 8$, $A = 10$, $d = 12$, we have $\alpha = 0.8$ and $\delta = 1.2$. The value $y = 1.2$ is, therefore, to be used for each interpolation.

	$\phi = 0°$	15°	30°	45°	60°	75°	90°
$x = \alpha V$	0.8	0.8528	0.9914	1.1789	1.3856	1.5929	1.7889
$y = \delta$	1.2	1.2	1.2	1.2	1.2	1.2	1.2
$\frac{m}{A}$	2.66	2.94	3.70	4.57	5.20	5.70	5.90
$1 - \frac{\rho}{a}\cos\phi$	−1.0	−0.9318	−0.7320	−0.4142	0.0	0.4823	1.0
V^2	1.0	1.1363	1.5359	2.1716	3.0	3.9647	5.0
y_n	−2.660	−2.411	−1.763	−0.872	0.0	0.693	1.180

	$\phi = 180°$	165°	150°	135°	120°	105°	
$x = \alpha V$	2.4	2.3818	2.3274	2.2383	2.1166	1.9654	
$\frac{m}{A}$	5.85	5.85	5.90	6.00	6.00	6.00	
$1 - \frac{\rho}{a}\cos\phi$	3.0	2.9318	2.7320	2.4142	2.0	1.5176	
V^2	9.0	8.8637	8.4641	7.8284	7.0	6.0353	
y_n	1.950	1.935	1.904	1.850	1.714	1.509	

INDUCTANCE OF CIRCULAR ELEMENTS WITH PARALLEL AXES 189

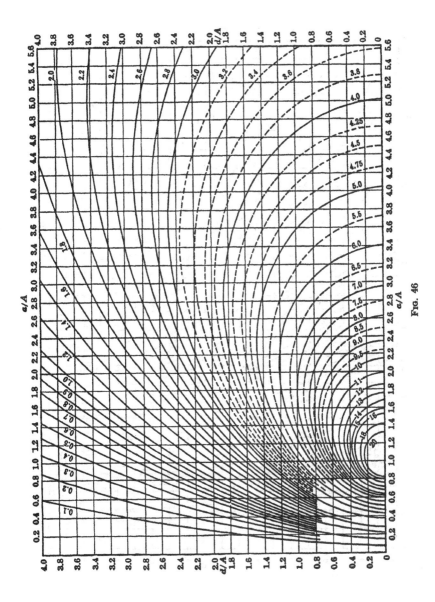

Fig. 46

190 CALCULATION OF MUTUAL INDUCTANCE AND SELF-INDUCTANCE

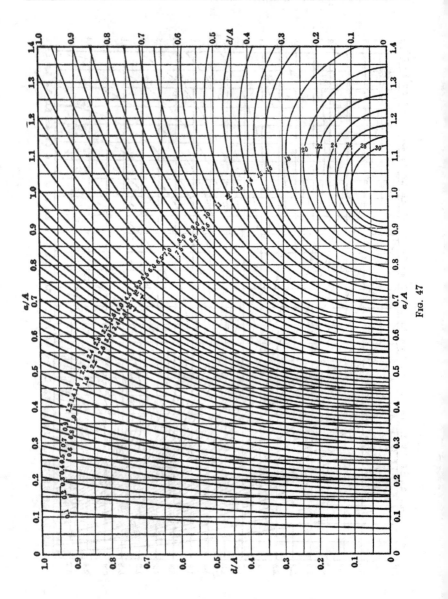

Fig. 47

INDUCTANCE OF CIRCULAR ELEMENTS WITH PARALLEL AXES

The mechanical quadrature yields the following calculation:

Even	Odd	Extremes
−2.660	−2.411	−2.660
−1.763	−0.872	1.950
0	0.693	−0.710
1.180	1.509	
1.714	1.850	
1.904	1.935	
1.950		
6.748	5.987	
−4.423	−3.283	
Sum = 2.325	2.704 = Sum	

$2 \times 2.325 = 4.650$ −2.411
$4 \times 2.704 = 10.816$ 0.693
                      15.466   1.509
                      −0.710   1.935
                                ─────
      Diff. = 16.176   4.137
         ÷ 36 = $0.4493 = \dfrac{S}{\pi}$ −2.411
                                ─────
                                1.726 = Sum

$5 \times 1.726 = 8.630$ −0.872
$6 \times 0.978 = 5.868$ 1.850
                      2.325   ─────
                      1.180   0.978 = Sum
                     ──────
           Sum = 18.003
            ÷ 40 = $0.4501 = \dfrac{W}{\pi}$.

Therefore, by (166)

$$\frac{M}{A} = 0.4493$$

$$M = 4.493 \text{ abhen.}$$

$$= 0.004493\ \mu\text{h.}$$

This differs from the more accurate value calculated from (163) by about 7 parts in 1000.

Case 3. Circles for Which $(A - a) < r < (A + a)$. Series formulas for this case have been derived by Snow.[87] They are, however, complicated and for many cases converge very slowly. It is, therefore, advisable to treat this case also by the formulas and methods of case 2.

Mutual Inductance of Eccentric Circular Coils. If the cross sectional dimensions of the coils are small compared with the distance between the coils, the mutual inductance is found by calculating the mutual inductance M_c of the central turns of the two coils by any suitable formula in the preceding section. This value is to be multiplied by the product of the number of turns in the two coils:

$$M = N_1 N_2 M_c. \tag{167}$$

Although the values of M calculated in the preceding examples are very small, it is only necessary to choose N_1 and N_2 sufficiently large to obtain a desired value of M. If, however, N_1 and N_2 are large and the wire thick, the accuracy of formula (167) decreases and the effect of the considerable cross sectional dimensions must be corrected for by a quadrature formula.

Chapter 19

MUTUAL INDUCTANCE OF CIRCULAR FILAMENTS WHOSE AXES ARE INCLINED TO ONE ANOTHER

Circular Filaments Whose Axes Intersect at the Center of One of the Coils. Distances are measured from the center of the larger circle of radius A. The center of the smaller circle of radius a lies on the axis of the larger circle at a distance D. The axes of the two circles are inclined at an angle θ. For the coaxial position $\theta = 0$.

Let $\dfrac{a}{A} = \alpha$, $\dfrac{D}{A} = \delta$, and $\mu = \cos\theta$. The mutual inductance is given by the formula

$$M = RM_0 \cos\theta, \qquad (168)$$

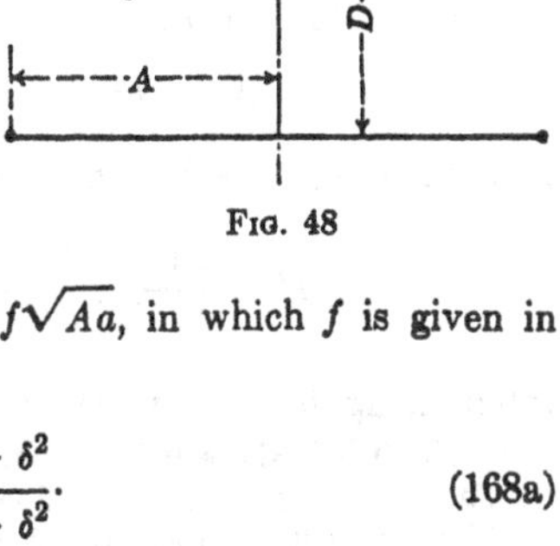

in which M_0 equals the mutual inductance of the circles in the coaxial position and R is a factor that is a function of the parameters μ, α, and δ. For the case of a very small radius a or a very great distance D, this factor would be unity.

FIG. 48

The value of M_0 is calculated as illustrated on page 78 by formula (77); $M_0 = f\sqrt{Aa}$, in which f is given in Table 13 as a function of the parameter

$$k'^2 = \frac{(1-\alpha)^2 + \delta^2}{(1+\alpha)^2 + \delta^2}. \qquad (168\mathrm{a})$$

The factor R is obtained from Table 46 for the given values of α, μ, and $\delta \left(\text{or } \dfrac{1}{\delta}\right)$. The table has been prepared for values of α from 0.5 to 0.9, inclusive, since these are the most difficult cases to calculate, and covers the

194 CALCULATION OF MUTUAL INDUCTANCE AND SELF-INDUCTANCE

usual range of practical coils. For smaller values of α values of R may be calculated from the expression

$$R = \frac{1 - \frac{1}{4}\beta^2 P_3'(\nu)\frac{P_3(\mu)}{\mu} + \frac{1}{8}\beta^4 P_5'(\nu)\frac{P_5(\mu)}{\mu} - \frac{5}{64}\beta^6 P_7'(\nu)\frac{P_7(\mu)}{\mu} + \cdots}{1 - \frac{1}{4}\beta^2 P_3'(\nu) + \frac{1}{8}\beta^4 P_5'(\nu) - \frac{5}{64}\beta^6 P_7'(\nu) + \cdots},$$

(169)

in which

$$\beta^2 = \frac{\alpha^2}{1+\delta^2} = \frac{\alpha^2\left(\frac{1}{\delta}\right)^2}{1+\left(\frac{1}{\delta}\right)^2}, \quad \nu^2 = \frac{\delta^2}{1+\delta^2} = \frac{1}{1+\left(\frac{1}{\delta}\right)^2}.$$

It is to be noted that each of the terms in the numerator of (169) is obtained from the corresponding term of the denominator by multiplying by a single factor. The numerical coefficients are derived from the formula for the general term of the denominator,

$$(-1)^n \frac{1\cdot 3\cdot 5 \cdots (2n-1)}{2\cdot 4\cdot 6 \cdots (2n)(n+1)} \beta^{2n} P'_{2n+1}(\nu).$$

Equations for the zonal harmonics $P_{2n+1}(\mu)$ and the differential coefficients $P'_{2n+1}(\nu)$ are given on page 242, but it will be well to make use of the values tabulated for μ and δ in Auxiliary Tables 3 and 4, obtaining the values of R for the region of the data of the problem and interpolating.

Example 69: Assume two circular filaments for which

$$A = 20 \text{ cm.}, \quad a = 14, \quad D = 0, \quad \text{and} \quad \mu = 0.3.$$

From these data $\alpha = 0.7$ and $\delta = 0$,

$$k'^2 = \left(\frac{1-0.7}{1+0.7}\right)^2 = \left(\frac{0.3}{1.7}\right)^2 \quad \text{and} \quad \log k'^2 = \bar{2}.49334.$$

Interpolating for this value in Table 14 there is found $f = 0.014721$ and $M_0 = \sqrt{280}(0.014721) = 0.24634 \,\mu\text{h}$.
Table 46 gives for $\alpha = 0.7$, $\delta = 0$, and $\mu = 0.3$ the factor $R = 0.64195$.
Thus $M = (0.64195)(0.3)(0.24634) = 0.04744 \,\mu\text{h}$.

Example 70: The case $A = 10$ inches, $a = 3$ inches, $D = 3$, and $\mu = 0.4$ gives $\alpha = 0.3$, $\delta = 0.3$, $\mu = 0.4$, which is not covered in the table for values of the factor R. The calculation of this quantity by means of the formula (169) above is here illustrated for this case. From the data of the problem $\beta^2 = \frac{0.09}{1.09}$. Values of the quantity

MUTUAL INDUCTANCE OF INCLINED CIRCULAR FILAMENTS   195

$\dfrac{P_{2n+1}(\mu)}{\mu}$ are obtained from Auxiliary Table 3 (page 238) and from Auxiliary Table 4 (page 244), values of $P'_{2n+1}(\nu)$ corresponding to $\nu = 0.2873$. These data together with the logarithms of the coefficients and the logarithms of the powers of β^2 are summarized below:

$\log \beta^2 = \bar{2}.91681,$ $P_3(\mu)/\mu = -1.10,$ $\log P_3'(\nu) = n\bar{1}.94454,$
$\log \beta^4 = \bar{3}.83362,$ $P_5(\mu)/\mu = 0.6766,$ $\log P_5'(\nu) = n\bar{2}.37997,$
$\log \beta^6 = \bar{4}.75043,$ $P_7(\mu)/\mu = -0.0365,$ $\log P_7'(\nu) = 0.12006,$
$\log \beta^8 = \bar{5}.66724,$ $P_9(\mu)/\mu = -0.4719,$ $\log P_9'(\nu) = n0.38631.$

The calculation may then be arranged as follows:

$2n + 1$	3	5	7	9
log coeff.	$= n\bar{1}.39794$	$\bar{1}.09691$	$n\bar{2}.89279$	$\bar{2}.73789$
$\log P'_{2n+1}(\nu)$	$= n\bar{1}.94484$	$n\bar{2}.37997$	0.12006	$n0.38631$
$\log \beta^{2n}$	$= \bar{2}.91681$	$\bar{3}.83362$	$\bar{4}.75043$	$\bar{5}.66724$
log (terms in denominator) $=$	$\bar{2}.25959$	$n\bar{5}.31050$	$n\bar{5}.76328$	$n\bar{6}.79144$
$\log \dfrac{P_{2n+1}(\mu)}{\mu}$	$= n0.04139$	$\bar{1}.83033$	$n\bar{2}.56205$	$n\bar{1}.67385$
log (terms in numerator)	$= n\bar{2}.30098$	$n\bar{5}.14083$	$\bar{6}.32533$	$\bar{6}.46529$

Numerator $= 1 - 0.019998 - 0.000014 + 0.000002 + 0.000003$
$= 0.97999.$

Denominator $= 1 + 0.018178 - 0.000020 - 0.000058 - 0.000006$
$= 1.01809.$

$$R = \frac{0.97999}{1.01809} = 0.96263.$$

To finish the calculation of the mutual inductance we find by (168a)

$$k'^2 = \frac{49 + 9}{169 + 9} = \frac{58}{178} = 0.32585,$$

and from Table 13, $f = 0.0029017,$

$$M_0 = \sqrt{30}(2.54)(0.0029017) = 0.04037 \ \mu\text{h}.$$

The mutual inductance of the inclined circles is, by (168)

$$M = 0.4(0.9626)(0.04037) = 0.015543 \ \mu\text{h}.$$

Best Proportions for a Variometer. For small values of α and large values of δ it is easy to approximate a cosine law of variation of the mutual inductance with the angle of inclination. Larger values of M are, however, obtained with more nearly equal radii. Table 46 shows that for values of α from 0.5 to 0.8 a spacing δ of a value about 0.45 gives a fair approximation to the cosine law.

196 CALCULATION OF MUTUAL INDUCTANCE AND SELF-INDUCTANCE

TABLE 46. VALUES OF CONSTANT R FOR INCLINED CIRCLES, FORMULA (168)

Ratio of radii, $\alpha = 0.5$.

μ	$1/\delta = 0$	0.1	0.2	0.3	0.4	0.5	0.6	0.7	0.8	0.9	$1/\delta = 1$	μ
0	1	1.0092	1.0349	1.0719	1.1129	1.1496	1.1793	1.1965	1.2019	1.1988	1.1852	0
0.1	1	1.0091	1.0346	1.0712	1.1116	1.1489	1.1774	1.1944	1.1999	1.1969	1.1837	0.1
0.2	1	1.0088	1.0335	1.0688	1.1080	1.1439	1.1715	1.1882	1.1938	1.1912	1.1790	0.2
0.3	1	1.0084	1.0317	1.0650	1.1018	1.1357	1.1618	1.1778	1.1837	1.1816	1.1710	0.3
0.4	1	1.0077	1.0292	1.0598	1.0934	1.1243	1.1483	1.1634	1.1694	1.1681	1.1596	0.4
0.5	1	1.0069	1.0260	1.0530	1.0827	1.1099	1.1312	1.1450	1.1511	1.1507	1.1442	0.5
0.6	1	1.0059	1.0221	1.0449	1.0698	1.0927	1.1107	1.1227	1.1286	1.1292	1.1247	0.6
0.7	1	1.0047	1.0175	1.0355	1.0550	1.0727	1.0870	1.0968	1.1020	1.1034	1.1007	0.7
0.8	1	1.0033	1.0123	1.0248	1.0382	1.0506	1.0605	1.0674	1.0714	1.0734	1.0719	0.8
0.9	1	1.0017	1.0064	1.0129	1.0199	1.0262	1.0313	1.0350	1.0373	1.0390	1.0382	0.9
1.0	1	1	1	1	1	1	1	1	1	1	1	1.0

μ	$\delta = 1$	0.9	0.8	0.7	0.6	0.5	0.4	0.3	0.2	0.1	$\delta = 0$	μ
0	1.1852	1.1660	1.1376	1.0993	1.0522	0.9966	0.9407	0.8835	0.8361	0.8045	0.7934	0
0.1	1.1837	1.1648	1.1368	1.0989	1.0523	.9972	.9414	.8841	.8373	.8058	.7947	0.1
0.2	1.1790	1.1612	1.1343	1.0979	1.0526	0.9989	.9434	.8873	.8410	.8097	.7987	0.2
0.3	1.1710	1.1548	1.1300	1.0958	1.0529	1.0016	.9468	.8928	.8472	.8164	.8054	0.3
0.4	1.1596	1.1455	1.1234	1.0925	1.0529	1.0050	.9517	.9007	.8563	.8261	.8154	0.4
0.5	1.1442	1.1328	1.1143	1.0874	1.0523	1.0088	0.9582	0.9111	0.8696	0.8394	0.8289	0.5
0.6	1.1247	1.1161	1.1016	1.0797	1.0502	1.0125	.9663	.9242	.8844	.8568	.8468	0.6
0.7	1.1007	1.0950	1.0847	1.0685	1.0457	1.0153	.9757	.9401	.9046	.8794	.8701	0.7
0.8	1.0719	1.0688	1.0627	1.0523	1.0372	1.0159	.9858	.9587	.9299	.9067	.9008	0.8
0.9	1.0382	1.0372	1.0346	1.0300	1.0227	1.0121	0.9944	0.9795	0.9614	0.9474	0.9420	0.9
1.0	1	1	1	1	1	1	1	1	1	1	1	1.0

MUTUAL INDUCTANCE OF INCLINED CIRCULAR FILAMENTS 197

TABLE 46. VALUES OF R FOR INCLINED CIRCLES, FORMULA (168) (*Continued*)

Ratio of radii, $\alpha = 0.6$.

μ	$1/\delta = 0$	0.1	0.2	0.3	0.4	0.5	0.6	0.7	0.8	0.9	$1/\delta = 1$	μ
0	1	1.0133	1.0507	1.1052	1.1665	1.2234	1.2670	1.2925	1.2998	1.2942	1.2722	0
0.1	1	1.0131	1.0502	1.1040	1.1646	1.2208	1.2639	1.2894	1.2968	1.2913	1.2702	0.1
0.2	1	1.0127	1.0486	1.1006	1.1589	1.2131	1.2549	1.2800	1.2880	1.2826	1.2641	0.2
0.3	1	1.0121	1.0459	1.0949	1.1496	1.2005	1.2400	1.2643	1.2731	1.2690	1.2535	0.3
0.4	1	1.0111	1.0422	1.0870	1.1368	1.1830	1.2194	1.2425	1.2521	1.2505	1.2378	0.4
0.5	1	1.0099	1.0375	1.0770	1.1206	1.1651	1.1934	1.2147	1.2247	1.2253	1.2163	0.5
0.6	1	1.0085	1.0319	1.0650	1.1014	1.1351	1.1624	1.1810	1.1910	1.1933	1.1882	0.6
0.7	1	1.0067	1.0252	1.0510	1.0794	1.1055	1.1268	1.1420	1.1510	1.1537	1.1526	0.7
0.8	1	1.0047	1.0177	1.0356	1.0549	1.0727	1.0874	1.0983	1.1053	1.1087	1.1090	0.8
0.9	1	1.0025	1.0092	1.0185	1.0283	1.0373	1.0449	1.0506	1.0546	1.0570	1.0579	0.9
1.0	1	1	1	1	1	1	1	1	1	1	1	1.0

μ	$\delta = 1$	0.9	0.8	0.7	0.6	0.5	0.4	0.3	0.2	0.1	$\delta = 0$	μ
0	1.2722	1.2425	1.1996	1.1431	1.0754	0.9970	0.9166	0.8404	0.7746	0.7306	0.7149	0
0.1	1.2702	1.2411	1.1988	1.1430	1.0758	0.9979	.9179	.8418	.7761	.7321	.7164	0.1
0.2	1.2641	1.2366	1.1963	1.1425	1.0771	1.0006	.9216	.8461	.7806	.7366	.7209	0.2
0.3	1.2535	1.2288	1.1918	1.1413	1.0790	1.0051	.9278	.8533	.7883	.7444	.7287	0.3
0.4	1.2378	1.2169	1.1844	1.1388	1.0810	1.0111	.9366	.8638	.7996	.7560	.7403	0.4
0.5	1.2163	1.2000	1.1732	1.1340	1.0826	1.0183	.9480	.8779	0.8150	0.7718	0.7563	0.5
0.6	1.1882	1.1768	1.1566	1.1253	1.0823	1.0258	.9619	.8960	.8354	.7932	.7779	0.6
0.7	1.1526	1.1460	1.1327	1.1105	1.0779	1.0323	.9777	.9187	.8621	.8218	.8071	0.7
0.8	1.1090	1.1064	1.0995	1.0866	1.0660	1.0344	.9936	.9460	.8972	.8608	.8472	0.8
0.9	1.0579	1.0575	1.0554	1.0505	1.0417	1.0260	1.0048	0.9764	0.9429	0.9160	0.9055	0.9
1.0	1	1	1	1	1	1	1	1	1	1	1	1.0

198 CALCULATION OF MUTUAL INDUCTANCE AND SELF-INDUCTANCE

TABLE 46. VALUES OF R FOR INCLINED CIRCLES, FORMULA (168) (Continued)

Ratio of radii, $\alpha = 0.7$.

μ	$1/\delta = 0$	0.1	0.2	0.3	0.4	0.5	0.6	0.7	0.8	0.9	$1/\delta = 1$	μ
0	1	1.0181	1.0696	1.1459	1.2331	1.3150	1.3778	1.4136	1.4221	1.4080	1.3788	0
0.1	1	1.0180	1.0689	1.1442	1.2303	1.3112	1.3734	1.4091	1.4182	1.4049	1.3765	0.1
0.2	1	1.0174	1.0667	1.1393	1.2220	1.2999	1.3602	1.3958	1.4062	1.3952	1.3693	0.2
0.3	1	1.0165	1.0630	1.1312	1.2084	1.2813	1.3394	1.3735	1.3858	1.3783	1.3565	0.3
0.4	1	1.0152	1.0578	1.1199	1.1899	1.2558	1.3083	1.3422	1.3566	1.3534	1.3370	0.4
0.5	1	1.0136	1.0515	1.1058	1.1666	1.2240	1.2704	1.3021	1.3180	1.3193	1.3091	0.5
0.6	1	1.0115	1.0435	1.0890	1.1393	1.1866	1.2256	1.2536	1.2698	1.2750	1.2708	0.6
0.7	1	1.0092	1.0344	1.0697	1.1084	1.1446	1.1749	1.1976	1.2125	1.2197	1.2205	0.7
0.8	1	1.0064	1.0240	1.0482	1.0744	1.0988	1.1194	1.1354	1.1470	1.1540	1.1574	0.8
0.9	1	1.0034	1.0125	1.0249	1.0381	1.0502	1.0606	1.0689	1.0752	1.0795	1.0827	0.9
1.0	1	1	1	1	1	1	1	1	1	1	1	1.0

μ	$\delta = 1$	0.9	0.8	0.7	0.6	0.5	0.4	0.3	0.2	0.1	$\delta = 0$	μ
0	1.3788	1.3354	1.2746	1.1967	1.1055	1.0028	0.8979	0.7974	0.7104	0.6498	0.6276	0
0.1	1.3765	1.3340	1.2740	1.1970	1.1066	1.0044	.8995	.7990	.7121	.6513	.6291	0.1
0.2	1.3693	1.3294	1.2721	1.1978	1.1097	1.0088	.9044	.8041	.7172	.6562	.6337	0.2
0.3	1.3565	1.3210	1.2684	1.1986	1.1140	1.0159	.9127	.8129	.7259	.6645	.6420	0.3
0.4	1.3370	1.3076	1.2618	1.1986	1.1198	1.0255	.9247	.8258	.7387	.6769	.6542	0.4
0.5	1.3091	1.2872	1.2504	1.1962	1.1254	1.0375	0.9406	0.8434	0.7566	0.6943	0.6714	0.5
0.6	1.2708	1.2572	1.2308	1.1886	1.1293	1.0512	.9607	.8667	.7807	.7182	.6950	0.6
0.7	1.2205	1.2144	1.1992	1.1712	1.1276	1.0642	.9846	.8971	.8135	.7512	.7279	0.7
0.8	1.1574	1.1568	1.1512	1.1375	1.1125	1.0702	1.0097	.9354	.8585	.7984	.7753	0.8
0.9	1.0827	1.0842	1.0840	1.0808	1.0729	1.0557	1.0251	0.9790	0.9217	0.8710	0.8503	0.9
1.0	1	1	1	1	1	1	1	1	1	1	1	1.0

TABLE 46. VALUES OF R FOR INCLINED CIRCLES, FORMULA (168) (*Continued*)

Ratio of radii, $\alpha = 0.8$.

μ	$1/\delta = 0$	0.1	0.2	0.3	0.4	0.5	0.6	0.7	0.8	0.9	$1/\delta = 1$	μ
0	1	1.0238	1.0920	1.1947	1.3140	1.4287	1.5156	1.5633	1.5718	1.5492	1.5060	0
0.1	1	1.0235	1.0910	1.1924	1.3100	1.4232	1.5094	1.5573	1.5668	1.5456	1.5037	0.1
0.2	1	1.0228	1.0880	1.1856	1.2983	1.4070	1.4908	1.5393	1.5517	1.5342	1.4964	0.2
0.3	1	1.0216	1.0830	1.1743	1.2792	1.3805	1.4600	1.5089	1.5254	1.5142	1.4828	0.3
0.4	1	1.0199	1.0761	1.1589	1.2531	1.3444	1.4177	1.4658	1.4868	1.4834	1.4608	0.4
0.5	1	1.0177	1.0674	1.1396	1.2209	1.2997	1.3645	1.4101	1.4346	1.4393	1.4272	0.5
0.6	1	1.0151	1.0570	1.1169	1.1833	1.2478	1.3018	1.3424	1.3682	1.3795	1.3777	0.6
0.7	1	1.0120	1.0449	1.0911	1.1414	1.1901	1.2317	1.2645	1.2881	1.3026	1.3085	0.7
0.8	1	1.0084	1.0314	1.0627	1.0961	1.1285	1.1562	1.1791	1.1970	1.2101	1.2187	0.8
0.9	1	1.0044	1.0162	1.0322	1.0485	1.0646	1.0782	1.0897	1.0992	1.1068	1.1128	0.9
1.0	1	1	1	1	1	1	1	1	1	1	1	1.0

μ	$\delta = 1$	0.9	0.8	0.7	0.6	0.5	0.4	0.3	0.2	0.1	$\delta = 0$	μ
0	1.5060	1.4454	1.3638	1.2627	1.1475	1.0190	0.8911	0.768	0.655	0.566	0.530	0
0.1	1.5037	1.4443	1.3640	1.2636	1.1492	1.0210	.8928	.767	.654	.568	.533	0.1
0.2	1.4964	1.4408	1.3636	1.2664	1.1540	1.0271	.8985	.773	.658	.573	.536	0.2
0.3	1.4828	1.4337	1.3628	1.2706	1.1617	1.0368	.9090	.782	.667	.578	.544	0.3
0.4	1.4608	1.4209	1.3593	1.2752	1.1721	1.0508	.9245	.796	.680	.592	.556	0.4
0.5	1.4272	1.3990	1.3503	1.2782	1.1845	1.0693	.9453	.819	.700	.611	.573	0.5
0.6	1.3777	1.3628	1.3304	1.2754	1.1963	1.0916	.9729	.847	.729	.633	.598	0.6
0.7	1.3085	1.3059	1.2913	1.2583	1.2016	1.1153	1.0077	.886	.765	.668	.627	0.7
0.8	1.2187	1.2235	1.2231	1.2126	1.1855	1.1309	1.0475	.938	.818	.719	.680	0.8
0.9	1.1128	1.1178	1.1222	1.1242	1.1216	1.1069	1.0719	1.008	.905	.807	0.764	0.9
1.0	1	1	1	1	1	1	1	1	1	1	1	1.0

200 CALCULATION OF MUTUAL INDUCTANCE AND SELF-INDUCTANCE

TABLE 46. VALUES OF R FOR INCLINED CIRCLES, FORMULA (168) (*Concluded*)

Ratio of radii, $\alpha = 0.9$.

μ	$1/\delta = 0$	0.1	0.2	0.3	0.4	0.5	0.6	0.7	0.8	0.9	$1/\delta = 1$	μ
0	1	1.0305	1.1178	1.2526	1.4133	1.5682	1.6850	1.7454	1.7477	1.7158	1.6564	0
0.1	1	1.0302	1.1165	1.2495	1.4079	1.5605	1.6765	1.7379	1.7420	1.7120	1.6541	0.1
0.2	1	1.0292	1.1126	1.2403	1.3916	1.5379	1.6514	1.7148	1.7242	1.7003	1.6470	0.2
0.3	1	1.0276	1.1061	1.2251	1.3651	1.5010	1.6091	1.6751	1.6926	1.6787	1.6343	0.3
0.4	1	1.0255	1.0971	1.2045	1.3294	1.4510	1.5508	1.6178	1.6443	1.6437	1.6131	0.4
0.5	1	1.0227	1.0858	1.1789	1.2857	1.3897	1.4777	1.5425	1.5765	1.5899	1.5765	0.5
0.6	1	1.0193	1.0723	1.1490	1.2353	1.3192	1.3922	1.4501	1.4875	1.5118	1.5155	0.6
0.7	1	1.0153	1.0568	1.1154	1.1800	1.2423	1.2975	1.3438	1.3783	1.4066	1.4221	0.7
0.8	1	1.0108	1.0394	1.0789	1.1213	1.1618	1.1977	1.2291	1.2546	1.2781	1.2955	0.8
0.9	1	1.0056	1.0204	1.0402	1.0609	1.0802	1.0972	1.1126	1.1248	1.1374	1.1478	0.9
1.0	1	1	1	1	1	1	1	1	1	1	1	1.0

μ	$\delta = 1$	0.9	0.8	0.7	0.6	0.5	0.4	0.3	0.2	0.1	$\delta = 0$	μ
0	1.6564	1.5738	1.4700	1.3451	1.2034	1.0588	0.9025	0.7511	0.6078	0.4741	0.4099	0
0.1	1.6541	1.5734	1.4705	1.3466	1.2057	1.0580	.9050	.7520	.6095	.4763	.4114	0.1
0.2	1.6470	1.5722	1.4725	1.3517	1.2130	1.0644	.9109	.7596	.6150	.4807	.4161	0.2
0.3	1.6343	1.5686	1.4762	1.3606	1.2260	1.0808	.9273	.7725	.6246	.4890	.4229	0.3
0.4	1.6131	1.5598	1.4793	1.3724	1.2427	1.0973	.9432	.7887	.6389	.5012	.4311	0.4
0.5	1.5765	1.5403	1.4771	1.3845	1.2665	1.1207	.9693	.8121	.6593	.5185	0.4472	0.5
0.6	1.5155	1.5008	1.4620	1.3922	1.2887	1.1564	1.0067	.8473	.6886	.5431	.4673	0.6
0.7	1.4221	1.4276	1.4178	1.3820	1.3091	1.2007	1.0567	.8960	.7308	.5794	.4969	0.7
0.8	1.2955	1.3100	1.3214	1.3226	1.2993	1.2332	1.1190	.9648	.7951	.6383	.5433	0.8
0.9	1.1478	1.1580	1.1698	1.1829	1.1972	1.1918	1.1694	1.0658	0.9064	0.7313	0.6278	0.9
1.0	1	1	1	1	1	1	1	1	1	1	1	1.0

MUTUAL INDUCTANCE OF INCLINED CIRCULAR FILAMENTS 201

An accurately linear variation of the mutual inductance with the angle of inclination is, of course, impossible since the slope of the curve must approach zero as the coaxial position ($\theta = 0$) is approached. The best approximation to the linear law is attained with α about 0.7 and δ about zero.

Calculation in the Most General Case. Cases may readily arise for which interpolation in Table 46 is inaccurate and formula (169) does not converge well. Thus a formula applicable generally is useful.

Such is furnished by the expression [90]

$$M = \sqrt{Aa} \cdot \cos\theta \int_0^\pi \frac{f}{\pi} \cdot \frac{d\phi}{P^{3/2}} \mu h, \qquad (170)$$

in which

$$P = \sqrt{1 - \sin^2\theta \cos^2\phi}, \qquad (171)$$

and f is to be taken from Table 13 for the argument

$$k'^2 = \frac{1 + \alpha^2 + \delta^2 + 2\alpha\delta \cos\phi \sin\theta - 2\alpha P}{1 + \alpha^2 + \delta^2 + 2\alpha\delta \cos\phi \sin\theta + 2\alpha P}. \qquad (171a)$$

Formula (170) may be evaluated by mechanical quadrature. This may be illustrated and checked by the solution of a problem which may also be accurately calculated by the equation (168) and Table 46.

Example 71: Two circles of radii 10 and 20 cm. will be considered with the center of one on the axis of the other and a distance of 20 cm. between centers. The axes will be assumed to be inclined at an angle of 30°.

For this case $a = 10$, $A = 20$, $D = 20$, $\mu = \cos\theta = 0.8660$ and $\sin\theta = 0.5$. The parameters are $\alpha = \frac{1}{2}$ and $\delta = 1$. The details of the calculation of the mutual inductance by the formula (170) follow:

ϕ	k'^2	$1000f$	$\dfrac{1000f}{P^{3/2}}$
0	0.52101	1.2947	1.6065
15°	0.51469	1.3311	1.6245
30	0.49704	1.4368	1.6790
45	0.47137	1.6023	1.7710
60	0.44167	1.8134	1.9033
75	0.41170	2.0501	2.0762
90°	0.38462	2.2884	2.2884
105	0.36275	2.4997	2.5317
120	0.34762	2.6571	2.7889
135	0.33937	2.7472	3.0365
150	0.33683	2.7775	3.2432
165	0.33730	2.7703	3.3809
180°	0.33791	2.7635	3.4288

CALCULATION OF MUTUAL INDUCTANCE AND SELF-INDUCTANCE

The calculation by mechanical quadrature gives (see formulas (A) and (B), p. 185).

Even	Odd	Extreme
1.6065	1.6245	1.6065
1.6790	1.7710	3.4288
1.9033	2.0762	-------
2.2884	2.5317	5.0353
2.7889	3.0365	
3.2432	3.3809	
3.4288	-------	
-------	14.4208 = Sum	
16.9381 = Sum.		

$y_3 = 1.7710$ $2 \times 16.9381 = 33.8762$
$y_9 = 3.0365$ $4 \times 14.4208 = 57.6832$

$ 4.8075$ Sum = 91.5594
                      $\div 36 = 2.4034 = \dfrac{S}{\pi}$

$y_1 = 1.6245$ $5 \times 9.6133 = 48.0665$
$y_5 = 2.0762$ $6 \times 4.8075 = 28.8450$
$y_7 = 2.5317$ $\Sigma y_{2n} = 16.9381$
$y_{11} = 3.3809$ $y_6 = 2.2884$

$ 9.6133$ Sum = 96.1380
                      $\div 40 = 2.4034 = \dfrac{W}{\pi}$.

Therefore, by (170),

$$M = \sqrt{Aa} \cdot \cos\theta\,(2.4034) = 29.436 \text{ abhen.}$$

$$\frac{M}{A} = \sqrt{0.5} \cdot \cos\theta\,(2.4034) = 1.6994 \cos\theta.$$

To check this result by formula (168) and Table 46 we interpolate in Table 46 for $\alpha = 0.5$, $\delta = 1$, and $\mu = 0.8660$, with the result $R = 1.0502$.

To find M_0 Table 13 is used with $k'^2 = \dfrac{(1-0.5)^2 + 1}{(1+0.5)^2 + 1} = 0.38462$ to find $f = 0.0022884$

$$M = \sqrt{Aa}\,(2.2884)(1.0502) \cos\theta \text{ abhenrys,}$$

$$\frac{M}{A} = \sqrt{0.5}\,(2.2884)(1.0502) \cos\theta = 1.6994 \cos\theta,$$

which checks the calculation by (170).

For cases where a moderate accuracy suffices, the calculation may be simplified by the use of the curves of Figs. 46 and 47. The general formula (170) becomes [90]

$$\frac{M}{A} = \frac{\cos\theta}{\pi} \int_0^\pi \frac{1}{P^2}\left(\frac{m}{A}\right) d\phi \text{ abhenrys,} \qquad (172)$$

in which P has the same value as in (171) and $\dfrac{m}{A}$ is to be obtained from Fig. 46 or 47 for circles with coordinates

$$x = \alpha P, \quad y = \delta + \alpha \sin\theta \cos\phi. \qquad (173)$$

MUTUAL INDUCTANCE OF INCLINED CIRCULAR FILAMENTS

Example 72: The use of formula (172) and the curves may be illustrated by solving the problem of the preceding example. The calculation is summarized in the following table, in which

$$x = \alpha P, \quad y = 1 + \frac{1}{4}\cos\phi, \quad y_n = \frac{1}{P^2}\left(\frac{m}{A}\right).$$

ϕ	P^2	x	y	$\dfrac{m}{A}$	y_n
0	0.7500	0.4330	1.2500	0.89	1.187
15°	0.7667	0.4378	1.2415	0.90	1.174
30	0.8125	0.4507	1.2165	0.95	1.169
45	0.8750	0.4677	1.1768	1.10	1.257
60	0.9375	0.4841	1.1250	1.26	1.344
75	0.9833	0.4958	1.0647	1.45	1.475
90°	1.0	0.5	1.0	1.60	1.600
105	0.9833	0.4958	0.9353	1.77	1.800
120	0.9375	0.4841	0.8750	1.86	1.984
135	0.8750	0.4677	0.8232	1.89	2.160
150	0.8125	0.4507	0.7835	1.88	2.314
165	0.7667	0.4378	0.7585	1.85	2.413
180°	0.7500	0.4330	0.7500	1.83	2.440

Integrating these values of y_n by mechanical quadrature

Even	Odd	Extreme		
1.187	1.174	1.187	1.174	1.257
1.169	1.257	2.440	1.475	2.160
1.344	1.475	———	1.600	———
1.600	1.800	3.627	2.413	3 417
1.984	2.160		———	
2.314	2.413		6.862	
2.440	———			
———	10.279			
12.038				

$2 \times 12.038 = 24.076$ $5 \times 6.862 = 34.310$

$4 \times 10.279 = 41.116$ $6 \times 3.417 = 20.502$

Sum $= 65.192$ $\Sigma y_{2n} = 12.038$

          3.627 $y_6 = 1.600$

Diff. $= 61.565$ Sum $= 68.450$

$\div\ 36 = 1.710 = \dfrac{S}{\pi}$ $\div\ 40 = 1.711 = \dfrac{W}{\pi}$

Accordingly, $\dfrac{M}{A} = 1.710 \cos\theta$, a result about 6 parts in 1000 higher than by the more accurate methods.

Mutual Inductance of Inclined Circular Filaments Whose Axes Intersect, but Not at the Center of Either.

Let a_1, a_2, be the radii and 0 be the point of intersection of the axes. The centers of the circles are at distances x_1 and x_2 from 0.

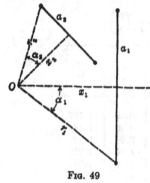

Fig. 49

$$r_1^2 = a_1^2 + x_1^2, \qquad r_2^2 = a_2^2 + x_2^2,$$

$$\cos \alpha_1 = \nu_1 = \frac{x_1}{r_1}, \qquad \cos \alpha_2 = \nu_2 = \frac{x_2}{r_2}.$$

θ is the angle of inclination of the axes. $\mu = \cos \theta$. If the subscripts be chosen so that $r_2 < r_1$, then

$$M = 0.002\pi^2 \frac{a_1^2 a_2^2 \mu}{r_1^3} \left[1 + \frac{1}{3} \frac{r_2}{r_1} \frac{P_2(\mu)}{\mu} P_2'(\nu_1) P_2'(\nu_2) \right.$$

$$+ \frac{1}{6}\left(\frac{r_2}{r_1}\right)^2 \frac{P_3(\mu)}{\mu} P_3'(\nu_1) P_3'(\nu_2) + \frac{1}{10}\left(\frac{r_2}{r_1}\right)^3 \frac{P_4(\mu)}{\mu} P_4'(\nu_1) P_4'(\nu_2) + \cdots$$

$$+ \frac{2}{(n+1)(n+2)} \left(\frac{r_2}{r_1}\right)^n \frac{P_{n+1}(\mu)}{\mu} P'_{n+1}(\nu_1) P'_{n+1}(\nu_2) + \cdots \left. \right]. \qquad (174)$$

The values of the zonal harmonics $P_{n+1}(\mu)$ and the differential coefficients $P'_{n+1}(\nu)$ are to be taken from the auxiliary tables and equations of pages 238–247.

The equation (174) is more convergent, the smaller the ratio $\frac{r_2}{r_1}$. It converges very poorly when r_2 and r_1 are nearly equal.

Example 73: To apply formula (174) to the circular filaments for which the given data are

$$a_1 = 16, \quad a_2 = 10, \quad \mu = \cos \theta = 0.5.$$

$$x_1 = 20, \quad x_2 = 5,$$

then

$$r_1^2 = 656, \quad r_2^2 = 125,$$

$$\cos^2 \alpha_1 = \nu_1^2 = \frac{400}{656} = \frac{1}{1.64} = \frac{1}{1 + \left(\frac{1}{\delta_1}\right)^2},$$

$$\cos^2 \alpha_2 = \nu_2^2 = \frac{25}{125} = \frac{1}{5} = \frac{\frac{1}{4}}{1 + \frac{1}{4}},$$

$$\frac{a_1}{x_1} = \frac{1}{\delta_1} = 0.8,$$

$$\frac{a_2}{x_2} = \delta_2 = 0.5.$$

From the tables were taken the data

$P_2(\mu) = -0.125$,	$\log P_3'(\nu_1) = 0.36970$,	$\log P_3'(\nu_2) = 0.12763$.
$P_3(\mu) = -0.4375$,	$\log P_3'(\nu_1) = 0.48759$,	$P_3'(\nu_2) = 0$
$P_4(\mu) = -0.2891$,	$\log P_4'(\nu_1) = 0.39379$,	$\log P_4'(\nu_2) = n0.25257$.
$P_5(\mu) = 0.0898$,	$\log P_5'(\nu_1) = \bar{1}.70640$,	$\log P_5'(\nu_2) = n0.25527$.
$P_6(\mu) = 0.3232$,	$\log P_6'(\nu_1) = n0.32159$,	$\log P_6'(\nu_2) = \bar{1}.57479$.
$P_7(\mu) = 0.2231$,	$\log P_7'(\nu_1) = n0.61720$,	$\log P_7'(\nu_2) = 0.39164$.

The terms in the series in (174) are

$$1 - 0.11432 + 0 + 0.02130 - 0.00040 - 0.00038 - 0.00112 + 0.00013 = 0.90521.$$

The mutual inductance is then readily obtained from (174) using logarithms:

$$\log 0.002\pi^2 = \bar{2}.29533$$
$$\log (a_1 a_2)^2 = 4.40824$$
$$\log \mu = \bar{1}.69897$$
$$\log (0.90521) = \bar{1}.95674$$
$$\text{Sum} = 2.35928$$
$$\log r_1^3 = 4.22535$$
$$\log M = \bar{2}.13393$$

$$M = 0.013612 \ \mu\text{h}.$$

General Method of Treatment. In cases where formula (174) converges slowly the mutual inductance may be obtained by the following formulas.[90]

Taking the origin in Fig. 49 at the center of the larger circle, radius A, and the Z axis along its axis, the coordinates of the center of the circle of radius a are $y = \rho$, $z = d$. The inclination of the axes is θ. Then

$$M = \frac{\sqrt{Aa}}{\pi} \int_0^\pi \frac{\left(\cos\theta - \frac{\rho}{a}\cos\phi\right)}{Q^{3/2}} f \, d\phi \text{ abhenrys}, \qquad (175)$$

in which f is to be taken from Table 13 for the argument

$$k'^2 = \frac{(1-\alpha Q)^2 + \zeta^2}{(1+\alpha Q)^2 + \zeta^2}, \qquad (175\text{a})$$

with $Q^2 = 1 - \cos^2\phi \sin^2\theta - 2\dfrac{\rho}{a}\cos\phi\cos\theta + \dfrac{\rho^2}{a^2}$,

$$\alpha = \frac{a}{A}, \quad \delta = \frac{d}{A}, \qquad d = x_1 - x_2 \cos\theta$$

$$\zeta = \delta - \alpha \sin\theta\cos\phi. \qquad \rho = x_2 \sin\theta$$

The integration of (175) may be performed by mechanical quadrature as in the preceding cases, formulas (163) and (170).

206 CALCULATION OF MUTUAL INDUCTANCE AND SELF-INDUCTANCE

For moderate accuracy the solution may be made to depend on the curves of Figs. 46 or 47. The formula (175) becomes for this purpose

$$\frac{M}{A} = \frac{1}{\pi} \int_0^\pi \frac{\left(\cos\theta - \frac{\rho}{a}\cos\phi\right)}{Q^2} \left(\frac{m}{A}\right) d\phi \text{ abhenrys,} \qquad (176)$$

in which values of $\frac{m}{A}$ for the chosen values of ϕ are to be interpolated from the curves for abscissas $x = \alpha Q$ and ordinates $y = \zeta$.

The point of intersection of the axes is on the axis of the larger circle at a distance $x_1 = d + \frac{\rho}{\tan\theta}$. The relation $\rho = x_2 \sin\theta$ holds. These give the relations between the nomenclatures of formulas (174) and (175).

Most General Case. Inclined Circular Filaments Placed in Any Desired Position. Taking the origin at the center of the larger circle, radius A, Fig. 50, the axis of Z will be taken along the axis of this circle.

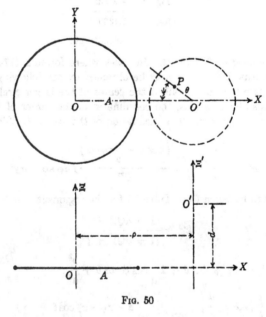

Fig. 50

The XZ plane will be taken to pass through the center O' of the smaller circle, radius a. The coordinates of O' are $x = \rho$, $y = 0$, $z = d$. To orient the axis of the circle of radius a, imagine a sphere taken with O' as center. The point P where this axis cuts the sphere is located by the longitude ψ, reckoned clockwise from the XZ plane and the colatitude θ, reckoned from the line $O'Z'$, taken parallel to OZ.

The general formula based on the use of the curves of Fig. 46 or 47 is found to be [90]

$$\frac{M}{A} = \frac{1}{2\pi} \int_0^{2\pi} \frac{\left[\cos\theta - \dfrac{\rho}{a}(\cos\psi\cos\phi - \sin\psi\sin\phi\cos\theta)\right]}{R^2} \left(\frac{m}{A}\right) d\phi, \quad (177)$$

in which

$$R^2 = (1 - \cos^2\phi\sin^2\theta) + 2\frac{\rho}{a}(\sin\psi\sin\phi - \cos\psi\cos\phi\cos\theta) + \frac{\rho^2}{a^2}, \quad (178)$$

and the coordinates of the circle pairs to be used with Fig. 46 or 47 are $x = \alpha R$, $y = \delta - \alpha \sin\theta\cos\phi = \zeta$, with $\alpha = \dfrac{a}{A}$, $\delta = \dfrac{d}{A}$.

The formula (177) is solved by mechanical quadrature as illustrated in preceding cases [formulas (170) and (172)]. More accurate values are obtained if Table 13 is used to find f in the formula

$$M = \frac{\sqrt{Aa}}{2\pi} \int_0^{2\pi} \frac{\left[\cos\theta - \dfrac{\rho}{a}(\cos\psi\cos\phi - \sin\psi\sin\phi\cos\theta)\right]}{R^{3/2}} f d\phi \; \mu\text{h}, \quad (179)$$

the argument of f being

$$k'^2 = \frac{(1 - \alpha R)^2 + \zeta^2}{(1 + \alpha R)^2 + \zeta^2}, \quad (179a)$$

but the accuracy of (177) is sufficient to render the extra labor of using (179) unnecessary in most cases.

The general formula (177) includes two special cases not previously considered, namely,

$\psi = 90°$ (inclined axes but with the axes in parallel planes, separated by a distance ρ),

$\theta = 90°$ (axis of one circle in a plane perpendicular to the axis of the other).

Mutual Inductance of Circular Coils of Small Cross Section with Inclined Axes. The formulas of the preceding sections for the mutual inductance of circular filaments with parallel axes and with inclined axes apply, strictly, only to filaments. However, they give a sufficient degree of accuracy when applied to circular coils whose cross sectional dimensions are small compared with the other dimensions of the problem. A solution is made for the circular filaments that pass through the centers of the cross sections and this value, multiplied by the product of the numbers of turns on the coils, gives the simplest approximation to the mutual inductance of the coils. To make more accurate allowance for the finite cross sectional dimensions would require the use of a method of quadratures such as that of Rayleigh, formula (C), page 11, with an increase in the labor of calculation that the difficulty of measuring the dimensions would hardly justify.

Chapter 20

MUTUAL INDUCTANCE OF SOLENOIDS WITH INCLINED AXES, AND SOLENOIDS AND CIRCULAR COILS WITH INCLINED AXES

Inclined Solenoids with Center of One on the End Face of the Other. The basic case is that where the center of one solenoid lies at the center of the end face of the other. The inner solenoid has a radius a, an axial length $2m$, and N_2 turns; the radius of the outer solenoid is A, its length x, and its winding density n, so that the number of its turns is $N_1 = nx$. The axes are inclined at an angle θ and $\mu = \cos \theta$. Placing $d = \sqrt{x^2 + A^2}$, $\delta = \sqrt{m^2 + a^2}$, the mutual inductance is given by a modification of Snow's formula:[91]

$$M = \frac{0.002\pi^2 a^2 (nx) N_2 \mu}{d} \left[1 - \frac{1}{2}\frac{A^2}{d^2} \cdot \frac{\delta^2}{d^2} \left\{ \lambda_2 \frac{P_3(\mu)}{\mu} + \lambda_4 \xi_2 \frac{P_5(\mu)}{\mu} \cdot \frac{\delta^2}{d^2} \right. \right.$$
$$\left. \left. + \lambda_6 \xi_4 \frac{P_7(\mu)}{\mu} \cdot \frac{\delta^4}{d^4} + \cdots \right\} \right]. \quad (180)$$

The quantities λ_2, λ_4, etc., are functions of $\frac{a^2}{\delta^2}$ given in Tables 31–34, and ξ_2, ξ_4, etc., are the same functions of $\frac{A^2}{d^2}$ and may be obtained from the same tables. The zonal harmonic functions are given in Auxiliary Table 3 (page 238).

In general, the coil of smaller radius will be of short axial length, since, if this arrangement of coils is to be used for a variometer, δ must not exceed A in size, or the rotation of the coil will be limited in range. The convergence of formula (180) is better, the smaller the ratios $\frac{\delta^2}{d^2}$ and $\frac{A^2}{d^2}$, that is, the greater the length

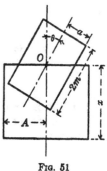

Fig. 51

of the outer coil, and the smaller the radius of the inner coil and the smaller its length.

From the basic formula (180) may be obtained the mutual inductance for any two solenoids with inclined axes when the center of one lies on the axis of the other. This is shown in the cases that follow.

Concentric Solenoids with Inclined Axes. Evidently the mutual inductance of either half of the outer solenoid on the inner is given by formula (180) so that the total is twice the value given by formula (180). That is, formula (180) applies to this case if nx is replaced by $2nx$, which is the total number of turns on the outer coil in this case. The expression for the diagonal d is the same as in the preceding case, that is, $d = \sqrt{x^2 + A^2}$, which is the half diagonal of the outer coil.

Unsymmetrical Cases. (a) The center of the smaller coil lies on the axis of the larger at a point that divides the length of the coil into two segments x_1 and x_2, such that $x = x_1 + x_2$.

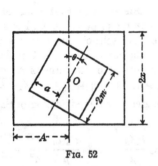
Fig. 52

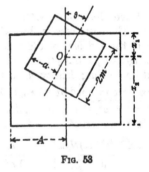
Fig. 53

Then the mutual inductance is $M = M_1 + M_2$. The two terms of the right-hand member are obtained from formula (180) according to the following scheme:

	M_1	M_2
No. of turns	nx_1	nx_2
Diagonal d	$\sqrt{A^2 + x_1^2}$	$\sqrt{A^2 + x_2^2}$

(b) The center of the coil of smaller radius a lies on the axis of the larger coil at a distance x_1 from its end face. As before, the larger coil has a length x and a winding density n. The mutual inductance is given by

$$M = M_1 - M_2,$$

MUTUAL INDUCTANCE OF INCLINED SOLENOIDS

in which each term is calculated from (180) according to the scheme:

	M_1	M_2
Length	$(x + x_1)$	x_1
No. of turns	$n(x + x_1)$	nx_1
Diagonal d	$\sqrt{A^2 + (x + x_1)^2}$	$\sqrt{A^2 + x_1^2}$

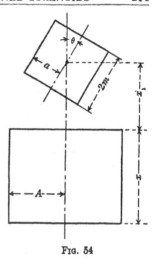

Fig. 54

Note: It may, in some cases, help the convergence if the principle of inter-change of lengths is applied. In every case, however, A is supposed to be the larger radius, and the diagonal d is formed on the radius A and the diagonal δ on the smaller radius. It is necessary for convergence that δ shall be smaller than d.

Mutual Inductance of Solenoid and Circular Filament with Inclined Axes. The basic case is shown in Fig. 55, the center of the circle of radius a lying at the center of the end face of the solenoid of radius A, length x, and winding density n. The number of turns on the solenoid is nx; the axis of the circle forms an angle θ with the axis of the solenoid.

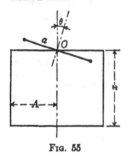

Fig. 55

Placing

$$d = \sqrt{A^2 + x^2}, \quad \rho^2 = \frac{A^2}{d^2}, \quad \text{and}$$

$$\alpha = \frac{a}{A}, \quad \mu = \cos\theta,$$

the mutual inductance is

$$M = \frac{0.002\pi^2 a^2 (nx)\mu}{d}\left[1 + \frac{3}{8}\alpha^2\rho^4\left\{\frac{P_3(\mu)}{\mu} - \frac{5}{6}\xi_2\alpha^2\rho^2\frac{P_5(\mu)}{\mu}\right.\right.$$
$$\left.\left. + \frac{35}{48}\xi_4\alpha^4\rho^4\frac{P_7(\mu)}{\mu} - \cdots\right\}\right]. \quad (181)$$

From this basic case the general case may be treated by the same methods as in the case of solenoids with inclined axes.

(a) If the center of the circle is in the midplane of a solenoid of length $2x$, then formula (181) is to be used, with $2nx$ replacing nx, the other quantities being the same.

212 CALCULATION OF MUTUAL INDUCTANCE AND SELF-INDUCTANCE

(b) If the center of the circle divides the length of the solenoid into two segments x_1 and x_2,
$$M = M_1 + M_2,$$
where

	M_1	M_2
No. of turns	nx_1	nx_2
Diagonal d	$\sqrt{A^2 + x_1^2}$	$\sqrt{A^2 + x_2^2}$

(c) If the center of the circle lies at a distance x_1 on the axis outside the end face of the solenoid of length x,
$$M = M_1 - M_2,$$
where

	M_1	M_2
No. of turns	$n(x + x_1)$	nx_1
Diagonal d	$\sqrt{A^2 + (x + x_1)^2}$	$\sqrt{A^2 + x_1^2}$

If the radius of the circle is larger than the radius of the solenoid, apply the principle of interchange of lengths, so that the solenoid of length x, radius a, and the circle of radius A, are replaced by a solenoid of length x, radius A, and a circle of radius a.

Example 74: Given a solenoid of radius 6 cm., length 12 cm., wound with 10 turns per centimeter and a concentric solenoid of radius 5 cm., length 4 cm., wound with 15 turns per centimeter, to calculate the mutual inductance when their axes are inclined at an angle whose cosine is 0.6.

Here $a = 5$, $A = 6$, $2m = 4$, $2x = 12$, $n = 10$, $2nx = 120$, and the total number of turns on the smaller coil is 4 times 15 = 60 = N_2.

$$\delta^2 = (5)^2 + (2)^2 = 29, \quad d^2 = (6)^2 + (6)^2 = 72,$$

and the parameters are

$$\frac{A^2}{d^2} = \frac{36}{72} = \frac{1}{2}, \quad \frac{\delta^2}{d^2} = \frac{29}{72}, \quad \frac{a^2}{\delta^2} = \frac{25}{29} = 0.8620, \quad \text{and} \quad \mu = 0.6.$$

From Auxiliary Table 3 (page 238) $\frac{P_3(\mu)}{\mu} = -0.600$, $\frac{P_5(\mu)}{\mu} = -0.2543$, $\frac{P_7(\mu)}{\mu} = 0.5377$. From Tables 31–33, with $\gamma^2 = 0.8620$, we find

$$\lambda_2 = -0.5086, \quad \lambda_4 = 0.1862, \quad \lambda_6 = 0.0147,$$

and with $\gamma^2 = \frac{1}{2}$,

$$\xi_2 = 0.1250, \quad \xi_4 = -0.2188.$$

In formula (180), the factor in the curved brackets is
$$(-0.5086)(-0.600) + (0.1862)(0.1250)(-0.2543)(\tfrac{29}{72})$$
$$+ (0.0147)(-0.2188)(0.5377)(\tfrac{29}{72})^2$$
$$= 0.3052 - 0.0024 - 0.0003 = 0.3025.$$

MUTUAL INDUCTANCE OF INCLINED SOLENOIDS

Multiplying this by $\dfrac{1}{2}\dfrac{A^2}{d^2}\cdot\dfrac{\delta^2}{d^2} = 0.1007$, the product $= 0.03046$. Therefore, (180) gives

$$M = \frac{0.002\pi^2(25)(60)(120)(0.6)}{\sqrt{72}}[1 - 0.03046] = 243.6\ \mu\text{h}.$$

Similar calculations for other values of μ lead to the following table, which gives the ratio of the mutual inductance for a given value of μ as compared with that which holds for the coaxial case, $\mu = 1$.

μ	Ratio	$\cos\theta$	θ	Ratio $\div \cos\theta$	Ratio $\div (90 - \theta)$
1.0	1.	1.0	0	1.	0.01111
0.9	0.8799	0.9	25°8	0.9754	0.01187
0.8	0.7659	0.8	36°8	0.9574	0.01441
0.6	0.5538	0.6	53°1	0.9230	0.01501
0.4	0.3591	0.4	66°4	0.8978	0.01522
0.2	0.1765	0.2	78°4	0.8825	0.01525
0.0	0	0			

From these data it is apparent that if these coils were used in a variometer, the mutual inductance would not depart from the simple cosine law of variation by more than 10 per cent. For angles θ greater than 30°, M is nearly proportional to the angle.

Example 75: As an example of the use of the formulas for the mutual inductance of solenoid and inclined circular filament, let us consider the case of a solenoid of radius 6 cm., length 12 cm., and winding density 10 turns per cm., so that the total number of turns is 120. A circular filament of 5 cm. radius is to be centered at different points on the axis of the solenoid and a study made of the mutual inductance with different angles of inclination. Three cases will be investigated:

(a) Center of the circle in the middle plane of the solenoid,
(b) Center of the circle in the end plane of the solenoid,
(c) Center of the circle on the axis of the solenoid 6 cm. outside the end plane.

Case a. The mutual inductance is twice the value for half the solenoid with the circle in its end place.

$$A = 6,\quad a = 5,\quad x = 6,\quad d^2 = (6)^2 + (6)^2 = 72,$$

$$N = 10 \times 6 = 60,\quad \rho^2 = \frac{A^2}{d^2} = \frac{1}{2},\quad \alpha = \frac{5}{6}.$$

From Tables 31-33, for this value of $\dfrac{A^2}{d^2}$,

$$\xi_2 = 0.1250,\quad \xi_4 = -0.2188,\quad \text{and}\quad \xi_6 = -0.0527.$$

With these data, the mutual inductance for $\mu = 1$ is given by formula (181) as twice $3.704 = 7.408\ \mu\text{h}$. Using Auxiliary Table 3, the value of mutual inductance for different values of μ may be studied. The values found are given below.

Case b. We have now $N = 10 \times 12 = 120$,

$$d^2 = (6)^2 + (12)^2 = 180,\quad \text{so that}\quad \rho^2 = \tfrac{36}{180} = 0.2.$$

214 CALCULATION OF MUTUAL INDUCTANCE AND SELF-INDUCTANCE

From Tables 31, 32, $\xi_2 = 0.6500$, $\xi_4 = 0.2650$, and $\alpha^2\rho^2 = \frac{2\,4}{3\,6}$ (0.2). Formula (181) gives for $\mu = 1$, the mutual inductance equal to 4.456 μh.

Case c. The mutual inductance in this case is the difference of the values for two values of x, viz: $x_1 = 6 + 12 = 18$ and $x_2 = 6$, with $N = 180$ and 60, respectively. The latter case has already been treated in case a. For the first,

$$d^2 = (6)^2 + (18)^2 = 360, \text{ so that } \rho^2 = \tfrac{3\,6}{3\,6\,0} = 0.1.$$

Tables 31, 32 give, for this value of $\dfrac{A^2}{d^2}$, the data $\xi_2 = 0.8250$ and $\xi_4 = 0.5912$. There results for this case and $\mu = 1$, the value $M = 4.6999$ μh. Accordingly for the solenoid and circle case c and $\mu = 1$, $M = 4.6999 - 3.7040 = 0.9959$ μh.

The table below summarizes the results of the calculations for these three cases with different assumed angles of inclination. The table includes for each of the three cases the values of μ, the ratio of the mutual inductance to that corresponding to $\mu = 1$, and the ratio of this relation to the values of μ. That is, this last column shows how closely the mutual inductance is proportional to the cosine of the angle of inclination.

Viewing these values it is apparent that, for the design of a variometer, case c is not suitable. Not only is the mutual inductance small, but the law of variation of M with the angle is unfavorable. The curve is very flat for an angular range of 30° to 40°.

The arrangement with the circle in the midplane of the solenoid possesses the advantage of large mutual inductance and the mutual inductance for small inclination is nearly proportional to the cosine of the angle of inclination. However, the arrangement with the circle in the end plane of the solenoid has about 60 per cent as much mutual inductance and the value is proportional to the cosine of the angle with an error not greater than 1 or 2 per cent over most of the range. A slight shift of the center of the circle beyond the end plane of the solenoid would give variation almost in perfect agreement with the cosine law.

SUMMARY OF RESULTS OF EXAMPLE ON SOLENOID AND INCLINED CIRCLE

	Center at Mid Plane			Center in End Plane			Center 6 cm. Beyond End Plane	
μ	$\dfrac{M}{M_0}$	$\dfrac{M}{M_0} \div \mu$	μ	$\dfrac{M}{M_0}$	$\dfrac{M}{M_0} \div \mu$	μ	$\dfrac{M}{M_0}$	$\dfrac{M}{M_0} \div \mu$
1	1	1	1	1	1	1	1	1
0.9	0.8774	0.9749	0.9	0.8963	0.9959	0.9	0.9768	1.085
.8	.7598	.9497	.8	.7935	.9919	.8	.9369	1.171
.6	.5433	.9055	.6	.5907	.9845	.6	.7930	1.322
.4	.3493	.8732	.4	.3914	.9785	.4	.5725	1.431
0.2	0.1709	0.8545	0.2	0.1950	0.9750	0.2	0.2988	1.494
0	0		0	0		0	0	

Chapter 21

CIRCUIT ELEMENTS OF LARGER CROSS SECTIONS WITH PARALLEL AXES

Solenoid and Circular Filament.

Let n_1 = the winding density of the solenoid,
    A = radius of solenoid,
    b = axial length of solenoid,
    a = radius of the circular filament,
    ρ = distance between axes,
    u = distance between centers measured along the axes (see Fig. 56).

Placing

$$d_1 = u + \frac{b}{2}, \qquad d_2 = u - \frac{b}{2},$$

$$r_1 = \sqrt{\rho^2 + d_1^2}, \quad r_2 = \sqrt{\rho^2 + d_2^2},$$

$$\mu_1 = \frac{d_1}{r_1}, \qquad \mu_2 = \frac{d_2}{r_2},$$

it is found that

$$M = 0.001\pi^2 a^2 A^2 n_1 \left[\frac{V_2}{r_2^2} - \frac{V_1}{r_1^2} \right], \quad (182)$$

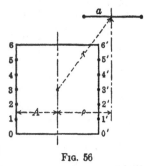

Fig. 56

in which V_2 and V_1 are found by substituting, respectively, μ_2, r_2 and μ_1, r_1, for μ, r in the expression

$$V = \mu \left[1 - \frac{3}{4} K_1 \frac{A^2}{r^2} \frac{P_3(\mu)}{\mu} + \frac{5}{8} K_2 \frac{A^4}{r^4} \frac{P_5(\mu)}{\mu} - \frac{35}{64} K_3 \frac{A^6}{r^6} \frac{P_7(\mu)}{\mu} + \cdots \right]. \quad (183)$$

The general term of this series is

$$(-1)^n \frac{3 \cdot 5 \cdot 7 \cdots (2n+1)}{4 \cdot 6 \cdot 8 \cdots (2n+2)} K_n \left(\frac{A}{r}\right)^{2n} \frac{P_{2n+1}(\mu)}{\mu}.$$

216 CALCULATION OF MUTUAL INDUCTANCE AND SELF-INDUCTANCE

The factors K_1, K_2, and K_3 may be interpolated from Table 47 as a function of the ratio $\alpha^2 = \dfrac{a^2}{A^2}$. The zonal harmonic functions $\dfrac{P_{2n+1}(\mu)}{\mu}$ may be obtained from Auxiliary Table 3 (page 238).

TABLE 47. VALUES OF K_n IN FORMULAS (183) AND (185)

Interpolation may be avoided by using the following formulas directly:

$K_1 = 1 + \alpha^2$, $\qquad K_3 = 1 + 6\alpha^2 + 6\alpha^4 + \alpha^6$.

$K_2 = 1 + 3\alpha^2 + \alpha^4$, $\qquad K_4 = 1 + 10\alpha^2 + 20\alpha^4 + 10\alpha^6 + \alpha^8$.

In general $K_n = F(-n - 1, -n, 2, \alpha^2)$, where F is the hypergeometric series.

α^2	K_1	K_2	Δ_1	K_3	Δ_1	Δ_2	K_4	Δ_1	Δ_2	Δ_3
0	1.0	1.00		1.000			1.000			
			31		661			1210		
0.1	1.1	1.31		1.661		126	2.210		462	
			33		787			1672		62
.2	1.2	1.64		2.448		132	3.882		524	
			35		919			2196		67
.3	1.3	1.99		3.367		138	6.078		591	
			37		1057			2787		69
.4	1.4	2.36		4.424		144	8.866		660	
			39		1201			3447		70
0.5	1.5	2.75		5.625		150	12.312		730	
			41		1351			4177		73
.6	1.6	3.16		6.976		156	16.490		803	
			43		1507			4980		76
.7	1.7	3.59		8.483		162	21.470		879	
			45		1669			5859		78
.8	1.8	4.04		10.152		168	27.330		957	
			47		1837			6816		81
0.9	1.9	4.51		11.989		174	34.146		1038	
			49		2011			7854		
1.0	2.0	5.00		14.000			42.000			
$\Delta_1 = 0.1$		$\Delta_2 = 2$		$\Delta_3 = 6$			$\Delta_4 = 2.7$			

By applying the principle of interchange of lengths, it is evident that the mutual inductance of a solenoid of radius a, length b, and a circular filament of radius A is the same as the mutual inductance of the solenoid of radius A, length b, and the circular filament of radius a in Fig. 56, provided ρ and u are the same in both cases. Therefore, the general formula (182) may be used, whichever element has the larger radius A.

CIRCUIT ELEMENTS OF LARGER CROSS SECTIONS 217

Example 76:

$$a = A = 10 \text{ cm.}, \quad u = 20 \text{ cm.},$$
$$b = 12, \quad \rho = 20.$$

From Table 47, for $\alpha^2 = \dfrac{a^2}{A^2} = 1$, there are found $K_1 = 2$, $K_2 = 5$, $K_3 = 14$. From the given data,

$$d_2 = 14, \quad d_1 = 26,$$
$$r_2{}^2 = 596, \quad r_1{}^2 = 1076,$$
$$\mu_2 = 0.57348, \quad \mu_1 = 0.79263,$$

and from Auxiliary Table 3

$$\frac{P_3(\mu_2)}{\mu_2} = -0.6780, \quad \frac{P_3(\mu_1)}{\mu_1} = 0.0706,$$
$$\frac{P_5(\mu_2)}{\mu_2} = -0.1523, \quad \frac{P_5(\mu_1)}{\mu_1} = -0.5140,$$
$$\frac{P_7(\mu_2)}{\mu_2} = 0.5561, \quad \frac{P_7(\mu_1)}{\mu_1} = -0.2656.$$

Using these values we find in (183)

$$V_2 = 0.57348(1 + 0.17064 - 0.01340 - 0.02011)$$
$$V_2/r_2{}^2 = 0.6521 \div 596 = 0.0010942,$$
$$V_1 = 0.79263(1 - 0.00984 - 0.01387 + 0.00163)$$
$$V_1/r_1{}^2 = 0.7751 \div 1076 = 0.0007204,$$

so that

$$M = 0.001\pi^2(100)^2 n_1(0.0003738)$$
$$= 0.03689 n_1 \; \mu\text{h}.$$

The general formula (183) converges well only if $\dfrac{A^2}{r_n{}^2}$ is small, and furthermore, the mutual inductance is given by the difference of two terms each of which has to be calculated with a greater degree of precision than is required in the result. Unless the distances r_1 and r_2 are considerably larger than the sum of the radii, the accuracy may not be sufficient.

In such cases the Rayleigh quadrature formula (page 11), or some other method of averaging may have to be applied. For equal radii this is not difficult, as may be illustrated for the case treated in example 76.

Example 77: The degree of convergence in the preceding example leaves something to be desired, especially with respect to the calculation of V_2. To test the accuracy of the result, suppose circular filaments to be selected at equal intervals along the length of the solenoid (Fig. 56), in the positions $00'$, $11' \cdots 66'$. The mutual inductance of each of these filaments and the given circular filament will be calculated.

218 CALCULATION OF MUTUAL INDUCTANCE AND SELF-INDUCTANCE

Since the radii are all the same, Table 43 for equal circles with parallel axes may be employed. The main results of the calculation follow:

Circle	00'	11'	22'	33'	44'	55'	66'
d_n	26.0	24.0	22.0	20.0	18.0	16.0	14.0
r_n^2	1076.0	976.0	884.0	800.0	724.0	656.0	596.0
$\dfrac{2A}{r_n}$	0.60999	0.64018	0.67266	0.70710	0.74330	0.78088	0.81922
$\mu_n = \dfrac{d_n}{r_n}$	0.79300	0.76823	0.73995	0.70710	0.66896	0.62470	0.57346

The values of F are interpolated from Table 43 for these values of $\dfrac{2A}{r_n}$ and μ_n and the values of f from Table 17 for coaxial circles using $\dfrac{\text{diameter}}{\text{distance}} = \dfrac{2A}{r_n}$.

	00'	11'	22'	33'	44'	55'	66'
F	0.6197	0.5852	0.5471	0.5040	0.4517	0.3885	0.3127
$1000f$	0.4386	0.4961	0.5621	0.6362	0.7187	0.8091	0.9060
$1000Ff$	0.2718	0.3075	0.3075	0.3206	0.3248	0.3143	0.2833

The mutual inductance m of each circle with the given circular filament is $AFfn_1\,dx$ and the total desired mutual inductance is found by integrating this over the length of the solenoid. The integration may be obtained by Simpson's rule, using these calculated values of the integrand. The interval of integration is $\frac{1}{6}$ of 12, or 2 cm.

```
Even         Odd          Extremes
0.2718       0.2903       0.2718            2 × 1.1874 =   3.3748
0.3075       0.3206       0.2833            4 × 0.9252 =   3.7008
0.3248       0.3143       ------                           --------
0.2833       ------       0.5551 = Sum                     6.0756
------       0.9252 = Sum                                 −0.5551
1.1874 = Sum                                               --------
                                                           5.5205
                                        times ⅓ of 2 cm. = 3.6804
```

$$M = 0.001(10)(3.6804)n_1 = 0.036804 n_1 \;\mu\text{h}.$$

This value is more accurate than that calculated by formula (182). This example is a favorable case for this method, since the calculated points differ only slowly. The increase of m, due to decreasing distance between circles, is offset by the decrease in m, due to decreasing μ.

For this case also the Rayleigh formula is favorable. For this problem this becomes

$$M = \tfrac{1}{6}(4m_{33} + m_{00} + m_{66})(\text{number of turns})$$

$$= 0.3061(0.001)(10)(12n_1)$$

$$= 0.036732 n_1 \ \mu\text{h}.$$

This is nearly as accurate as the preceding value and requires the calculation of only three circles, whereas seven were necessary for the other.

If the radii of the solenoid and circular filament are not equal, the matter is complicated. For values not very different, instead of the parameter $\dfrac{2A}{r_n}$, the ratio of the mean diameter of coil and circle to r_n may be used with a moderate accuracy. Otherwise, the required mutual inductances of the unequal circles with parallel axes that enter in the calculation should be obtained by the graphical method described on page 187.

Solenoids with Parallel Axes. Expressions for the mutual inductance of single-layer coils with parallel axes have been given by Dwight [92] and Purssell and by Clem.[93] It is easy to show that the following formula may be derived from these and it is in an especially convenient form for numerical calculations.

The two coils of radii a and A are shown in Fig. 57. Their lengths x and l are taken as equal to the number of turns times the pitch of the windings. Accordingly, the winding densities n_1 and n_2 are, respectively, $n_1 = \dfrac{N_1}{x}$ and $n_2 = \dfrac{N_2}{l}$.

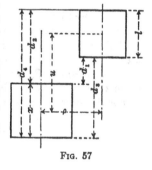

Fig. 57

Let ρ = distance between the axes, and calculate the four distances d_n between the ends of the coils shown in Fig. 57.

$$d_1 = u - \left(\frac{x+l}{2}\right), \quad d_3 = u + \left(\frac{x-l}{2}\right),$$

$$d_2 = u + \left(\frac{l-x}{2}\right), \quad d_4 = u + \left(\frac{x+l}{2}\right),$$

in which u = axial distance between the centers of the coils (see Fig. 57).

From these distances are to be calculated the four radii vectors $r_n = \sqrt{\rho^2 + d_n^2}$ and the four cosines $\mu_n = \dfrac{d_n}{r_n}$.

Then,
$$M = 0.001\pi^2 a^2 A^2 n_1 n_2 \left[\frac{X_1}{r_1} - \frac{X_2}{r_2} - \frac{X_3}{r_3} + \frac{X_4}{r_4}\right] \mu h, \quad (184)$$

in which

$$X_n = \left[1 - \frac{1}{4}K_1 \frac{A^2}{r_n^2} P_2(\mu_n) + \frac{1}{8}K_2 \frac{A^4}{r_n^4} P_4(\mu_n)\right.$$

$$\left. - \frac{5}{64} K_3 \frac{A^6}{r_n^6} P_6(\mu_n) + \frac{7}{128} K_4 \frac{A^8}{r_n^8} P_8(\mu_n) - \cdots \right]. \quad (185)$$

The constants K_1, K_2, K_3, and K_4 are functions of $\alpha^2 = \dfrac{a^2}{A^2}$ and may be calculated from the formulas

$$K_1 = 1 + \alpha^2, \qquad K_2 = 1 + 3\alpha^2 + \alpha^4,$$

$$K_3 = 1 + 6\alpha^2 + 6\alpha^4 + \alpha^6, \quad K_4 = 1 + 10\alpha^2 + 20\alpha^4 + 10\alpha^6 + \alpha^8,$$

or interpolated from Table 47. The zonal harmonics $P_{2n}(\mu_n)$ may be interpolated from Auxiliary Table 3 (page 238).

If the coils are overlapping, some of the distances d_n may be regarded as negative. This does not, however, affect the values of r_n and the zonal harmonics $P_{2n}(-\mu) = P_{2n}(\mu)$, so that the signs of the d_n are immaterial in the formula for the mutual inductance.

For the special case that the coils have their bases in the same plane

$$d_1 = -x \qquad d_3 = 0,$$
$$d_2 = \pm (x - l), \quad d_4 = l;$$

and for equal coils resting on the same plane

$$x = l, \qquad d_2 = 0,$$
$$d_1 = d_4 = x, \quad d_3 = 0;$$

and formula (184) becomes

$$M = 0.002\pi^2 a^2 A^2 n_1 n_2 \left[\frac{X_1}{r_1} - \frac{X_2}{r_2}\right] \mu h. \quad (186)$$

The convergence of (185) is better the smaller the values of $\left(\dfrac{A}{r_n}\right)^2$, and each r_n must be greater than $(A + a)$. The principle of interchange of the lengths holds, but since the d_n, r_n, and μ_n are thereby unchanged, no improvement of the convergence is thereby obtained.

CIRCUIT ELEMENTS OF LARGER CROSS SECTIONS

Formula (184) is of such a form that the individual terms have to be calculated to a higher degree of accuracy than is attainable in the result. This disadvantage is especially acute for distant coils, but for such cases the convergence of formula (185) is good.

The writer finds that for the special case of loosely coupled coils of equal radii A and equal length B the following series formula may be used:

$$M = 0.002\pi^2 \frac{A^4 n_1 n_2 \beta^2}{R} [P_2(\mu) + \beta^2 P_4(\mu) + \beta^4 P_6(\mu) + \beta^6 P_8(\mu) + \cdots$$

$$- \frac{1}{2} \frac{A^2}{R^2} \{6 P_4(\mu) + 15\beta^2 P_6(\mu) + 28\beta^4 P_8(\mu) + \cdots\}$$

$$+ \frac{25}{8} \frac{A^4}{R^4} \{3 P_6(\mu) + 14\beta^2 P_8(\mu) + \cdots\}$$

$$- \frac{245}{8} \frac{A^6}{R^6} \{P_8(\mu) + \cdots\} + \cdots]\ \mu\text{h}, \qquad (187)$$

in which the distance between the axes is assumed to be ρ and the distance between centers $R = \sqrt{\rho^2 + u^2}$. The argument of the zonal harmonics is $\mu = \dfrac{u}{R}$, and $\beta = \dfrac{B}{R}$. The convergence is better the smaller the space ratios $\dfrac{A}{R}$ and β.

Example 78: The mutual inductance of two equal single-layer coils of radii $A = 5$, length $x = 10$, and winding density 20 turns per cm. will be calculated by formula (184). The distance between the axes will be taken as $\rho = 10$ and the axial distance between their nearer ends as 5 cm. That is, $u = 15$.

Then

$$d_1 = 5, \qquad d_2 = d_3 = 15, \qquad d_4 = 25,$$

$$r_1^2 = 125, \qquad r_2^2 = r_3^2 = 325, \qquad r_4^2 = 725.$$

Since the radii are equal, Table 47 gives for $\alpha^2 = 1$,

$$K_1 = 2, \quad K_2 = 5, \quad K_3 = 14, \quad K_4 = 42.$$

We find also

$$\mu_1 = 0.44721, \quad \mu_2 = \mu_3 = 0.83205, \quad \mu_4 = 0.92848,$$

and for these values Auxiliary Table 3 gives

$P_2(\mu_1) = -0.1999$,	$P_2(\mu_2) = 0.5385$,	$P_2(\mu_4) = 0.7931$,
$P_4(\mu_1) = -0.2000$,	$P_4(\mu_2) = -0.1243$,	$P_4(\mu_4) = 0.3936$,
$P_6(\mu_1) = 0.3280$,	$P_6(\mu_2) = -0.4147$,	$P_6(\mu_4) = -0.0366$.
$P_8(\mu_1) = -0.2000$,		

The series for the X_n are

X_1	$X_2 = X_3$	X_4
1.01999	1	1
−0.00500	−0.02071	−0.01367
−0.00287	−0.00046	+0.00029
−0.00072	+0.00021	−0.00002
1.01139	0.97904	0.98662

$$\frac{X_1}{r_1} = 0.090461 \quad \frac{X_2}{r_2} = 0.054307$$

$$\frac{X_4}{r_4} = 0.036642$$

$$\overline{0.127103} - 0.108614 = 0.018489$$

$$M = 0.001\pi^2(25)^2(20)^2(0.018489) = 45.62 \ \mu h.$$

In order to obtain a three figure accuracy in the result, it is necessary that the separate terms shall be accurate to the fourth significant figure. The value of X_1 should be calculated to include one more term to assure this accuracy.

Example 79: To illustrate the use of formula (187) the solution will be found for the case of two equal loosely coupled coils for which the given constants are

$$A = 5, \quad \rho = 15, \quad n_1 = 20,$$
$$B = 5, \quad u = 10, \quad n_2 = 20.$$

The distance between centers is $R = \sqrt{15^2 + 10^2} = \sqrt{325}$. The space ratios are $\beta = \frac{B}{R} = \frac{5}{\sqrt{325}}$ and $\frac{A}{R} = \frac{5}{\sqrt{325}}$. The zonal harmonics for $\mu = \frac{u}{R} = \frac{10}{\sqrt{325}} = 0.55470$, taken from Auxiliary Table 3, are

$$P_2(\mu) = -0.0385, \quad P_6(\mu) = 0.2634,$$
$$P_4(\mu) = -0.3640, \quad P_8(\mu) = 0.0871.$$

It will be necessary to obtain a more precise value of $P_2(\mu)$ by the defining relation $P_2(\mu) = \frac{1}{2}(3\mu^2 - 1)$, which leads to the value -0.0384615.

The four terms in the brackets of formula (187) yield for this case $-0.06486 + 0.07176 + 0.01635 - 0.00121 = 0.02205$.

Substituting in formula (187), $M = 0.4643 \ \mu h$.

For this case, formula (184) requires each of the quantities X_n to be calculated accurately to seven significant figures to give a four figure accuracy in the result. The series given for the X_n are not sufficiently convergent to allow this without further terms. The calculated M comes out $0.472 \ \mu h$.

A further method of attack, applicable to loosely coupled solenoids of unequal radii, where formula (187) cannot be used, is to integrate formula (182) for solenoid and eccentric circle over one of the solenoids. Making the calculation for a number of equally spaced turns, the integration may be accomplished mechanically.

To illustrate the process the solution may be found for the problem just considered. Seven circles, $a, b, c, \cdots g$ equally spaced axially along one

CIRCUIT ELEMENTS OF LARGER CROSS SECTIONS 223

of the coils are taken. The mutual inductance m of each circle and the other solenoid is calculated by formula (182) for solenoid and circle with parallel axes. The calculated mutual inductances in abhenrys are -38.26, -18.23, -3.29, 9.10, 18.07, 24.85, and 30.79. The distance between consecutive circles is $\frac{5}{8}$ cm. Summing by Simpson's rule and multiplying by the winding density $n_2 = 20$, the value $M = 0.4721$ μh is found, which agrees with the value by (187) to about 1 per cent.

Example 80: The last described method is more accurate for still more loosely coupled coils. Assume two equal coils with

$$A = 2.5, \quad u = 0, \quad n_1 = 25,$$
$$B = 5, \quad \rho = 25, \quad n_2 = 25.$$

Here $R = 25$, $\dfrac{A^2}{R^2} = 0.01$, $\beta^2 = 0.04$, and $\mu = 0$. The zonal harmonics for this value, taken from Auxiliary Table 3, are

$$P_2(\mu) = -0.5, \quad P_4(\mu) = 0.375, \quad P_6(\mu) = -0.3125, \quad P_8(\mu) = 0.2734.$$

By (187)

$$M = 0.002\pi^2 \frac{(25)^2(2.5)^4(0.04)}{25}[-0.48448 - 0.01037 - 0.00002]$$
$$= -0.38159 \ \mu\text{h}.$$

The convergence of (187) is excellent.

To apply the Rayleigh quadrature method, calculate by formula (182) the mutual inductance of one solenoid on the three circles taken at the ends of the other solenoid and at its midsection. From the symmetry of this arrangement, the value will be the same for each end section:

$$u = 2.5, \quad d_2 = 0, \quad d_1 = 5.$$
$$\mu_2 = 0, \quad \mu_1 = 0.19611,$$

The zonal harmonics $P_{2n+1}(\mu)$ are all zero for $\mu = 0$, so that the value of $\dfrac{V_2}{r_2^3}$ in (182) is zero. The calculated value of m_e for the end circles comes out -2.968 abhenrys. For the circle at the midsection

$$d_2 = -2.5, \quad d_1 = 2.5, \quad \mu_2 = -\mu_1.$$

Since $P_{2n+1}(-\mu_1) = -P_{2n+1}(\mu_1)$, the two terms of (182) give $-2\dfrac{V_1}{r_1^3}$, and the mutual inductance m_c is -3.107 abhenrys.

The Rayleigh formula for this case gives

$$M = \frac{N_2}{6}(4m_c + 2m_e)10^{-3} \ \mu\text{h}$$

$$= \frac{125}{1000}\left[\frac{4(-3.107) + 2(-2.968)}{6}\right] = -0.3826 \ \mu\text{h}.$$

The value obtained by multiplying the value of the mutual for the midcircle by N_2 is -0.386, which is an approximation that may often be sufficient.

Solenoids with Parallel Axes Having Zero Mutual Inductance. The preceding examples have included cases where the mutual inductance may have either sign, which suggests the possibility of placing the coils so as to have zero mutual inductance. Such an arrangement is used in the familiar case of the coils in a neutrodyne circuit.

To determine how the coils should be placed we may employ the series formula (187) for equal solenoids. Imposing the condition that $M = 0$, the corresponding value of $\mu = \dfrac{u}{R}$ may be found by successive approximations. This is readily accomplished for given numerical data making use of Auxiliary Table 3 for values of the zonal harmonics.

Example 81: Assume coil dimensions and spacing such that $\beta = \dfrac{1}{4}$, $\dfrac{A}{R} = \dfrac{1}{4}$, that is, the coil length is one half of the diameter, and the distance between centers is four times the length of the coil.

Substituting these values in the series of formula (187), the necessary condition for zero mutual inductance is

$$P_2(\mu) - \tfrac{1}{8}P_4(\mu) + \tfrac{33}{2048}P_6(\mu) + \tfrac{1}{32768}P_8(\mu) = 0.$$

Using the first term and Auxiliary Table 3 it is evident that μ must be about 0.55. Calculating the above sum for several values of μ near 0.55 the sum is found to be as follows:

μ	Sum
0.53	−0.0337
.54	−0.0161
.55	+0.0017
0.56	0.0196

The value of $\mu = 0.549$ is, closely, the solution. Denoting by θ the angle between the line joining the coil centers and the direction of the axes, this value of μ corresponds to $\theta = 56°\ 42'$.

If the coil is very short, axially, $\beta = 0$, and the corresponding value of μ is 0.533 or $\theta = 57°\ 48'$. This is checked by the use of Table 43 for eccentric circles, from which the value $\mu = 0.536$ is found.

For longer coils, with $\beta = \tfrac{1}{2}$, with the same value of $\dfrac{A}{R}$ as before, it works out that $\mu = 0.561$, $\theta = 55°\ 53'$.

For coils far apart so that $\beta \approx 0$ and $\dfrac{A}{R} \approx 0$, the limiting value is $\mu = 0.577$, or $\theta = 54°\ 44'$.

It is evident, therefore, that except for coils close together, the condition for zero mutual inductance is not very critical.

Solenoid and Coil of Rectangular Cross Section with Parallel Axes. A general formula, derived from one by Dwight and Purssell,[92] is the following, in which the nomenclature is that of Fig. 58, and n_1 is the winding density of the solenoid and N_2 the total number of turns on the coil.

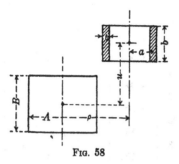

Fig. 58

$$M = 0.001\pi^2 A^2 a^2 n_1 \frac{N_2}{b}\left[\frac{Y_1}{r_1} - \frac{Y_2}{r_2} - \frac{Y_3}{r_3} + \frac{Y_4}{r_4}\right] \mu h, \qquad (188)$$

in which

$$Y_m = \left\{ t_2 - \frac{1}{4}\frac{A^2}{r_m^2}\left(t_2 + \frac{a^2}{A^2}t_4\right)P_2(\mu_m) \right.$$

$$+ \frac{1}{8}\frac{A^4}{r_m^4}\left(t_2 + 3\frac{a^2}{A^2}t_4 + \frac{a^4}{A^4}t_6\right)P_4(\mu_m)$$

$$- \frac{5}{64}\frac{A^6}{r_m^6}\left(t_2 + 6\frac{a^2}{A^2}t_4 + 6\frac{a^4}{A^4}t_6 + \frac{a^6}{A^6}t_8\right)P_6(\mu_m)$$

$$+ \frac{7}{128}\frac{A^8}{r_m^8}\left(t_2 + 10\frac{a^2}{A^2}t_4 + 20\frac{a^4}{A^4}t_6 + 10\frac{a^6}{A^6}t_8 + \frac{a^8}{A^8}t_{10}\right)P_8(\mu_m)$$

$$\left. - \cdots \right\}, \qquad (189)$$

with

$$r_m^2 = d_m^2 + \rho^2, \qquad \mu_m = \frac{d_m}{r_m},$$

$$d_1 = u - \left(\frac{b+B}{2}\right), \quad d_3 = u + \left(\frac{B-b}{2}\right),$$

$$d_2 = u + \left(\frac{b-B}{2}\right), \quad d_4 = u + \left(\frac{B+b}{2}\right).$$

The coefficients t_2, t_4, t_6, etc., are functions of the ratio τ of the thickness of the coil and its mean radius and may be obtained from Table 48, where

CALCULATION OF MUTUAL INDUCTANCE AND SELF-INDUCTANCE

TABLE 48. CORRECTIONS FOR COIL THICKNESS. COILS WITH PARALLEL AXES, FORMULAS (188), (190), AND (192)

τ^2	t_2 or T_2	Δ	t_4 or T_4	Δ	t_6 or T_6	Δ	t_8 or T_8	Δ	τ^2
0	1.0000		1.000		1.00		1.00		0
		83		50		13		24	
0.1	1.0083		1.050		1.13		1.24		0.1
		83		50		13		26	
.2	1.0167		1.100		1.26		1.50		.2
		83		51		13		28	
.3	1.0250		1.151		1.39		1.78		.3
		83		51		14		30	
.4	1.0333		1.202		1.53		2.08		.4
		83		51		14		32	
0.5	1.0417		1.253		1.67		2.39		0.5
		83		51		15		34	
.6	1.0500		1.304		1.82		2.73		.6
		83		52		15		36	
.7	1.0583		1.356		1.97		3.08		.7
		83		52		15		38	
.8	1.0667		1.408		2.12		3.46		.8
		83		52		16		40	
0.9	1.0750		1.460		2.28		3.85		0.9
		83		52		16		42	
1.0	1.0833		1.512		2.44		4.27		1.0
		83		53		16		44	
1.1	1.0917		1.565		2.60		4.71		1.1
		83		53		17		46	
1.2	1.1000		1.618		2.77		5.17		1.2
		83		53		17		48	
1.3	1.1083		1.671		2.95		5.65		1.3
		83		53		17		50	
1.4	1.1167		1.724		3.12		6.15		1.4
		83		54		18		53	
1.5	1.1250		1.778		3.30		6.68		1.5
		83		54		18		55	
1.6	1.1333		1.832		3.49		7.23		1.6
		83		54		19		58	
1.7	1.1417		1.886		3.68		7.81		1.7
		83		54		19		60	
1.8	1.1500		1.940		3.87		8.40		1.8
		83		55		20		62	
1.9	1.1583		1.995		4.07		9.03		1.9
		83		55		20		65	
2.0	1.1667		2.050		4.27		9.67		2.0

they are given for different values of the argument $\tau^2 = \dfrac{t^2}{a^2}$ or from the general formula

$$t_n = 1 + \frac{n(n-1)}{3!}\frac{\tau^2}{2^2} + \frac{n(n-1)(n-2)(n-3)}{5!}\frac{\tau^4}{2^4}$$

$$+ \cdots + \frac{n(n-1)(n-2)\cdots[n-(n-1)]}{(n+1)!}\frac{\tau^n}{2^n}$$

The zonal harmonic functions $P_{2n}(\mu_m)$ are to be interpolated from Auxiliary Table 3 (page 238).

In Fig. 58 the coil of rectangular cross section is shown with the smaller radius a. If, however, it is the solenoid that has the smaller radius, the general formula (188) still applies except that now the functions t_2, t_4, t_6, etc., are obtained from Table 48 for the argument $\tau^2 = \dfrac{t^2}{A^2}$.

Example 82: Take the same case as in Example 76, except that instead of a circular filament or coil of negligible cross section a square cross section 2 cm. on a side will now be assumed.

The data of the problem are

$$a = A = 10, \qquad \rho = 20, \qquad u = 20,$$
$$B = 12, \qquad b = 2, \qquad t = 2.$$

From these are found

$$d_1 = 13, \qquad d_3 = 25, \qquad \tau^2 = 0.04.$$
$$d_2 = 15, \qquad d_4 = 27,$$

Table 48 gives for this value of τ^2

$$t_2 = 1.00333, \qquad t_6 = 1.0504, \qquad t_{10} = 1.154.$$
$$t_4 = 1.0204, \qquad t_8 = 1.095,$$

The calculation with the four values of d is based on formula (188) with the further quantities that are given below:

$d_1 = 13,$	$d_2 = 15,$	$d_3 = 25,$	$d_4 = 27,$
$r_1^2 = 569,$	$r_2^2 = 625,$	$r_3^2 = 1025,$	$r_4^2 = 1129,$
$\mu_1 = 0.54499,$	$\mu_2 = 0.6,$	$\mu_3 = 0.78088,$	$\mu_4 = 0.80357,$
$P_2(\mu_1) = -0.0552,$	$P_2(\mu_2) = 0.0400,$	$P_2(\mu_3) = 0.4148,$	$P_2(\mu_4) = 0.4086,$
$P_4 = -0.3522,$	$P_4 = -0.4080,$	$P_4 = -0.2849,$	$P_4 = -0.2223,$
$P_6 = 0.2785,$	$P_6 = 0.1721,$	$P_6 = -0.3578,$	$P_6 = -0.3955,$
$P_8 = 0.0567,$	$P_8 = 0.2133,$	$P_8 = 0.0752,$	$P_8 = -0.0348.$

1.00333	1.00333	1.00333	1.00333
491	34	35	29
13		2	1
	1.00367		
1.00837		1.00370	1.00363
	−0.00324		
−0.00695	668	−0.02047	−0.02100
159	74	173	111
−0.00854	−0.01066	−0.02220	−0.02211

$Y_1 = 0.99983,\quad Y_2 = 0.99301,\quad Y_3 = 0.98150,\quad Y_4 = 0.98152.$

$\dfrac{Y_1}{r_1} = 0.041915,\quad \dfrac{Y_2}{r_2} = 0.039720,\quad \dfrac{Y_3}{r_3} = 0.030657,\quad \dfrac{Y_4}{r_4} = 0.029211.$

Summing these quantities according to formula (188) there results 0.000749, so that

$$M = 0.001\pi^2(100)(100)(0.000749)\frac{N_2 n_1}{2}$$
$$= 0.03696 n_1 N_2 \ \mu h.$$

If the cross section was of negligible dimensions, the calculation in Example 76 gives the value $0.03689 n_1 N_2$, which indicates an effect of cross section of about ½ per cent. However, the terms in formula (188) have to be very accurately calculated because of the near cancellation of the four terms. In many practical cases, where the cross section is small, it will suffice to calculate the mutual inductance of the solenoid and the center filament of the coil and to multiply by N_2.

Two Coils of Rectangular Cross Sections with Parallel Axes. Fig. 59 will make clear the nomenclature. The coils have mean radii a, A (A the

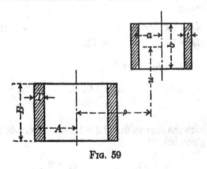

Fig. 59

larger), lengths b, B, and axial thicknesses t, T, respectively. The axes are a distance ρ apart and the centers of the coils displaced an axial distance u. The total number of turns on the coils are N_1 and N_2, respectively.

The general formula of Dwight and Purssell [92] arranged as a series involving zonal harmonics is

$$M = 0.001\pi^2 A^2 a^2 \frac{N_1}{b} \cdot \frac{N_2}{B}\left[\frac{Z_1}{r_1} - \frac{Z_2}{r_2} - \frac{Z_3}{r_3} + \frac{Z_4}{r_4}\right]\ \mu h, \qquad (190)$$

CIRCUIT ELEMENTS OF LARGER CROSS SECTIONS

in which the quantities Z_m are given by

$$Z_m = \left[T_2 t_2 - \frac{1}{4} \frac{A^2}{r_m^2} \left(T_4 t_2 + \frac{a^2}{A^2} T_2 t_4 \right) P_2(\mu_m) \right.$$
$$+ \frac{1}{8} \frac{A^4}{r_m^4} \left(T_6 t_2 + 3 \frac{a^2}{A^2} T_4 t_4 + \frac{a^4}{A^4} T_2 t_6 \right) P_4(\mu_m)$$
$$- \frac{5}{64} \frac{A^6}{r_m^6} \left(T_8 t_2 + 6 \frac{a^2}{A^2} T_6 t_4 + 6 \frac{a^4}{A^4} T_4 t_6 + \frac{a^6}{A^6} T_2 t_8 \right) P_6(\mu_m)$$
$$+ \frac{7}{128} \frac{A^8}{r_m^8} \left(T_{10} t_2 + 10 \frac{a^2}{A^2} T_8 t_4 + 20 \frac{a^4}{A^4} T_6 t_6 \right.$$
$$\left. \left. + 10 \frac{a^6}{A^6} T_4 t_8 + \frac{a^8}{A^8} T_2 t_{10} \right) P_8(\mu_m) + \cdots \right]. \qquad (191)$$

The four radii vectors $r_m = \sqrt{d_m^2 + \rho^2}$ depend upon the four distances d_m which are given by

$$d_1 = u - \left(\frac{b+B}{2} \right), \quad d_3 = u + \left(\frac{B-b}{2} \right),$$
$$d_2 = u + \left(\frac{b-B}{2} \right), \quad d_4 = u + \left(\frac{B+b}{2} \right),$$

and the zonal harmonics $P_{2n}(\mu_m)$ are for argument $\mu_m = \dfrac{d_m}{r_m}$.

The factors t_2 and T_2 are the same functions of $\dfrac{t^2}{a^2}$ and $\dfrac{T^2}{A^2}$, respectively, and the same is true of t_4 and T_4, t_6 and T_6, etc. Values of all of these may be interpolated from Table 48, for the arguments r^2 equal to $\dfrac{t^2}{a^2}$ or $\dfrac{T^2}{A^2}$, as the case may be.

The convergence of the series (191) is sufficient for most purposes as long as all the distances r_m are greater than $(A + a)$. Since the general term of the series is known and that of the series defining the t_{2s} and T_{2s} functions, it should be possible to use (191) over the whole range. However, the calculation of higher power terms becomes very tedious and time consuming. Those here included are covered by the tables.

Example 83: To calculate the mutual inductance of two coils of appreciable thickness to illustrate the effect of the cross sectional dimensions on the result. Given

$$a = 3, \qquad b = 5, \qquad t = 1, \qquad \frac{N_1}{b} = 10,$$
$$A = 5, \qquad B = 10, \qquad T = 2, \qquad \frac{N_2}{B} = 15,$$

suppose the coils are spaced so that $\rho = 10$ and $u = 10.5$ cm.

230 CALCULATION OF MUTUAL INDUCTANCE AND SELF-INDUCTANCE

Then
$$\frac{t^2}{a^2} = \frac{1}{9}, \quad \frac{T^2}{A^2} = 0.16,$$

and, interpolating from Table 48,

$t_2 = 1.0092$,	$T_2 = 1.0133$,	$T_2 t_2 = 1.02275$.
$t_4 = 1.055$,	$T_4 = 1.080$,	
$t_6 = 1.14$,	$T_6 = 1.21$,	
$t_8 = 1.27$,	$T_8 = 1.39$,	
$t_{10} = 1.45$,	$T_{10} = 1.67$,	

The salient points of the calculation follow:

$d_1 = 3$,	$d_2 = 8$,	$d_3 = 13$,	$d_4 = 18$,
$r_1^2 = 109$,	$r_2^2 = 164$,	$r_3^2 = 269$,	$r_4^2 = 424$,
$\mu_1 = 0.28734$,	$\mu_2 = 0.62470$,	$\mu_3 = 0.79262$,	$\mu_4 = 0.87416$,
$P_2(\mu_1) = -0.3761$,	$P_2(\mu_2) = 0.0854$,	$P_2(\mu_3) = 0.4424$,	$P_2(\mu_4) = 0.6463$,
$P_4 = 0.0952$,	$P_4 = -0.4220$,	$P_4 = -0.2539$,	$P_4 = 0.0646$,
$P_6 = 0.1032$,	$P_6 = 0.1081$,	$P_6 = -0.3798$,	$P_6 = -0.3505$,
$P_8 = -0.2207$,	$P_8 = 0.2661$,	$P_8 = 0.0195$,	$P_8 = -0.3658$.

```
    1.02275           1.02275           1.02275           1.02275
    3180                    9                12                 7
     163                                                         3
   ───────           ───────           ───────           ───────
   1.05618           1.02284           1.02287           1.02285
   ───────                             
  -0.00050          -0.00480          -1.01516          -0.01405
      37               319                71
                        15           ───────
   ───────           ───────          -0.01587
  -0.00087          -0.00814
```

$Z_1 = 1.05531$, $Z_2 = 1.01470$, $Z_3 = 1.00700$, $Z_4 = 1.00880$.

$\dfrac{Z_1}{r_1} = 0.101080$, $\dfrac{Z_2}{r_2} = 0.079235$, $\dfrac{Z_3}{r_3} = 0.061398$, $\dfrac{Z_4}{r_4} = 0.048992$.

$$0.101080 - 0.079235 - 0.061398 + 0.048992 = 0.009439,$$

$$M = 0.001\pi^2(25)(9)(10)(15)(0.009439) = 3.144 \; \mu\text{h}.$$

To see what the magnitude of the effect of the finite cross section is, formula (184) may be used to calculate the value of the mutual inductance, supposing the turns to be wound on solenoids having mean radii equal to those of the actual coils. That is, the mutual inductance will be calculated for $t = 0$ and $T = 0$. The data are the same except that the coefficients involving the t's and T's go over into the factors K_1, K_2, K_3, and K_4. From Table 47, for $a^2 = 0.36$, $K_1 = 1.360$, $K_2 = 2.210$, $K_3 = 3.984$, and $K_4 = 7.69$.

The terms $\dfrac{Z_m}{r_m}$ are

$$0.098662 - 0.077524 - 0.060087 + 0.047939 = 0.008990,$$

so that

$$M = 0.001\pi^2(25)(9)(10)(15)(0.008990)$$

$$= 2.994 \; \mu\text{h}.$$

CIRCUIT ELEMENTS OF LARGER CROSS SECTIONS

The mutual inductance, taking into account the thickness of the coils, is about 5 per cent greater than that of the solenoids in the median planes. That is, the influence of the nearer portions of the coils in increasing the mutual inductance is greater than the reduction of mutual inductance resulting from the greater distance between the more widely separated portions.

Example 84: Attention has already been directed to the limitation of the convergence of the formula (191) to cases where the distance between the centers of the coils is greater than the sum of their radii. This is a serious limitation in that it excludes many practical cases where the coils are close together and their cross sections are relatively large. Unfortunately, no formula for this special case is as yet available. It should be observed, however, that, in practice, the difficulty of accurately determining the dimensions of such systems renders illusory the importance of any very accurate calculation of the mutual inductance. Some method of approximation should suffice for purposes of orientation.

This may be illustrated for the case of two equal Brooks coils each having 400 turns wound in a square cross section 2 cm. on a side. The coils are arranged as shown in Fig. 60. The distance between the axes is assumed to be $\rho = 6$ and the displacement of the centers along the axes $u = 4$. The mean radii are $a = A = 3$ cm. Thus, in the Dwight and Purssell formula,[12] the distances are $d_1 = 2$, $d_2 = d_3 = 4$, and $d_4 = 6$,

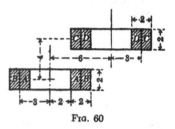

Fig. 60

with the corresponding radii vectors $r_1 = \sqrt{40}$, $r_2 = r_3 = \sqrt{52}$, and $r_4 = \sqrt{72}$. The thickness ratio is $\frac{t}{a} = \frac{T}{A} = \frac{2}{3}$ and it is found that the value $\tau^2 = \frac{4}{9}$ in Table 48 leads to $T_2 = t_2 = 1.0370$, $T_4 = t_4 = 1.225$, $T_6 = t_6 = 1.59$, and $T_8 = t_8 = 2.22$. These values lead to the coefficients in formula (191) of 1.0754, 2.541, 7.800, 28.00, and 111.6. Accordingly the Dwight and Purssell formula does not converge sufficiently well to give any accuracy.

This trouble may be avoided as far as the coefficients are concerned, by supposing the coils to be divided into sections A, B and C, D as shown in Fig. 60. The mutual inductance will, therefore, be given by

$$M = M_{AC} + M_{AD} + M_{BC} + M_{BD}$$
$$= M_{AD} + 2M_{AC} + M_{BC}.$$

For each of these terms the thickness ratios are smaller than for the undivided coils and the convergence is better. However, even for the limiting case of very thin coils, the convergence of the formula is poor and a very large number of terms would have to be calculated and especially in the case of $d = 2$. Furthermore, the three terms $\frac{Z_m}{r_m}$ are individually much larger than their combination.

Thus for M_{AD} the values are $\frac{Z_1}{r_1} = 0.168590$, $\frac{Z_2}{r_2} = \frac{Z_3}{r_3} = 0.142129$, and $\frac{Z_4}{r_4} = 0.119369$, giving the combined value in (191) of 0.003700 and

$$M_{AD} = 0.001\pi^2(2.5)^2(2.5)^2(\tfrac{200}{4})(\tfrac{200}{4})(0.003700) = 14.27\ \mu h.$$

However, each of the $\frac{Z_m}{r_m}$ terms is not certain to the fourth place. Carrying through the calculation, there are found $M_{BC} = 90.30$ μh and $M_{AC} = M_{BD} = 35.94$ μh.

The total is $M = 176.4$ μh. This value is probably 20 per cent too large and the matter is not improved by sectioning each coil into three parts and carrying out the increased amount of calculation. The uncertainty results, not from the size of the Tt coefficients but is inherent in the largeness of the values of $\frac{A^2}{r^2}$ and the near cancellation of the $\frac{Z_m}{r_m}$ terms.

A more fruitful method of solution is to make use of the formula (159) and Table 43 for equal circles with parallel axes (see page 178).

As a first approximation the calculation is made for the central circular filaments OO and $O'O'$, Fig. 61.

With $\rho = 6$, $u = 4$, the distance between centers is $r = \sqrt{6^2 + 4^2} = \sqrt{52}$. The mean radius being $a = 3$ cm., the parameters to be used in Table 43 are $\frac{2a}{r} = \frac{6}{\sqrt{52}} = 0.83205$ and $\mu = \frac{u}{r} = \frac{4}{\sqrt{52}} = 0.55470$.

Interpolating in Table 43 with these parameters there is found $F = 0.2824$. Making use of Table 17 (page 84) with the parameter $\frac{\text{diameter}}{\text{distance}} = \frac{2a}{r} = 0.83205$, for the factor f for the same circular filaments placed at the same distance but in the coaxial position, there is interpolated $f = 0.00093950$. Accordingly, for the central filaments, formulas (159) and (80) give

$$m = 3(0.00093950)(0.2824) = 0.0007959 \text{ μh},$$

and, multiplying by the product of the turn numbers of the two coils, for the coils

$$M = (400)(400)(0.0007959) = 127.34 \text{ μh}.$$

A second approximation will next be obtained by dividing each coil into four sections, $abcd$ and $a'b'c'd'$ in Fig. 61. The mutual inductance will then be calculated for

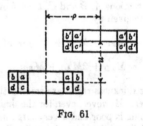

Fig. 61

each pair of circular filaments, one at the center of a section of one coil and the other at the center of a section of the other, assuming one fourth of the turns to be concentrated with each filament.

The second approximation for the mutual inductance will then be

$$M = 2M_{aa'} + 2M_{bb'} + 4M_{ab'} + 2M_{ad'} + 2M_{cb'} + M_{ac'} + M_{bd'} + M_{ca'} + M_{db'}.$$

For some of these terms the use of Table 43 is, strictly speaking, inapplicable since the filaments are not of equal radii. However, since the difference is not great, a good

CIRCUIT ELEMENTS OF LARGER CROSS SECTIONS 233

approximation is to use in this the mean of the radii. Since in formula (77) the geometric mean of the radii appears, that will be used in the present problem. The parameters are then

	d	r	$2a$	$2a/r$	μ
$M_{aa'}$	4	$\sqrt{52}$	5	0.69338	0.55470
$M_{bb'}$	4	$\sqrt{52}$	7	0.97073	0.55470
$M_{ac'}$	3	$\sqrt{45}$	5	0.74538	0.44723
$M_{ca'}$	5	$\sqrt{61}$	5	0.64019	0.64019
$M_{bd'}$	3	$\sqrt{45}$	7	$\dfrac{1}{0.95830}$	0.44723
$M_{db'}$	5	$\sqrt{61}$	7	0.89627	0.64019
$M_{ab'}$	4	$\sqrt{52}$	$\sqrt{35}$	0.82042	0.55470
$M_{ad'}$	3	$\sqrt{45}$	$\sqrt{35}$	0.88192	0.44723
$M_{cb'}$	5	$\sqrt{61}$	$\sqrt{35}$	0.75748	0.64019

The calculation follows the lines already illustrated. Naturally, with so many terms, the interpolations are tedious. The results found are

$$2M_{aa'}] = 2 \times 0.2409 = 0.4818$$
$$2M_{bb'}] = 2 \times 1.9943 = 3.9886$$
$$4M_{ab'}] = 4 \times 0.7291 = 2.9164$$
$$2M_{ad'}] = 2 \times 0.2062 = 0.4124$$
$$2M_{cb'}] = 2 \times 0.8971 = 1.7942$$
$$M_{ac'}] = -0.1261$$
$$M_{bd'}] = 1.5838$$
$$M_{ca'}] = 0.3907$$
$$M_{db'}] = 2.0273$$

$$\text{Sum} = 13.4691.$$

Multiplying by $(100)^2$ and dividing by 1000 to reduce to microhenrys, the result is $M = 134.69 \mu\text{h}$.

It is noticeable how widely the contributions of the sections differ. Those with the larger radii are much the larger. A third approximation is to divide each of the sections b, d, b', d' into two equal sections of different radii.

These sections are separately treated and the radii in those cases where they differ are now more nearly equal than before and the assumption of a mean value is more nearly correct. The summed values for these sections replace the contributions $2M_{bb'} + M_{bd'} + M_{db'}$ of the summation for the second approximations.

They lead to a value of 79.40 in place of the value 76.00 above. Consequently for the whole coils the third approximation is 138.09 μh.

Summarizing

$$\text{1st approximation} = 127.3 \; \mu\text{h}$$
$$\text{2nd approximation} = 134.7$$
$$\text{3rd approximation} = 138.1$$

Further sectioning would lead to a more accurate value, but the labor would be very great. This special problem is a rather unfavorable one, but by no means unusual. Such close spacing of coils of large cross section is not favorable for practical inductance standards.

Mutual Inductance of Disc Coils with Parallel Axes. This case is concerned with coils whose cross section in the axial direction is negligible but which have an appreciable thickness in the radial direction.

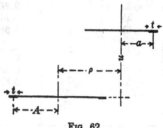

Fig. 62

The two coils of turns N_1 and N_2 have mean radii a and A with radial thicknesses t and T, respectively. Their axes are separated by a distance ρ, and u is the distance between their planes.

The formula for the mutual inductance, derived from the Dwight and Purssell formula [92] is

$$M = \frac{0.002\pi^2 A^2 a^2 N_1 N_2}{r^3} \left[T_2 t_2 P_2(\mu) - \frac{3}{2} \frac{A^2}{r^2} \left(T_4 t_2 + \frac{a^2}{A^2} T_2 t_4 \right) P_4(\mu) \right.$$

$$+ \frac{15}{8} \frac{A^4}{r^4} \left(T_6 t_2 + 3 \frac{a^2}{A^2} T_4 t_4 + \frac{a^4}{A^4} T_2 t_6 \right) P_6(\mu)$$

$$\left. - \frac{35}{16} \frac{A^6}{r^6} \left(T_8 t_2 + 6 \frac{a^2}{A^2} T_6 t_4 + 6 \frac{a^4}{A^4} T_4 t_6 + \frac{a^6}{A^6} T_2 t_8 \right) P_8(\mu) + \cdots \right] \mu\text{h.} \quad (192)$$

In this formula $r = \sqrt{\rho^2 + u^2}$ is the distance between the centers of the coils and the zonal harmonics $P_{2n}(\mu)$ have for argument $\mu = \dfrac{u}{r}$. The functions T_{2s} and t_{2s} are given by the same expressions as for formula (190). Their values may be obtained from Table 48 for the argument $\tau^2 = \dfrac{T^2}{A^2}$ or $\dfrac{t^2}{a^2}$, respectively.

Formula (192) is subject to the same limitations of convergence as (190), from which it is derived, and to a greater degree.

CIRCUIT ELEMENTS OF LARGER CROSS SECTIONS 235

Example 85: As an example take two equal coils having $a = A = 10$, $\rho = 20$, $u = 20$, $t = T = 2.5$ and turns N_1 and N_2.

From these data $r = \sqrt{20^2 + 20^2} = \sqrt{800}$, $\tau^2 = \dfrac{t^2}{a^2} = \dfrac{T^2}{A^2} = 0.0625$, and, from Table 48,

$$t_2 = T_2 = 1.0052, \qquad t_6 = T_6 = 1.0789,$$
$$t_4 = T_4 = 1.0312, \qquad t_8 = T_8 = 1.1492.$$

Also $\mu = \dfrac{u}{r} = \dfrac{20}{\sqrt{800}} = \dfrac{1}{\sqrt{2}} = 0.70711$.

For this value of μ Auxiliary Table 3 gives

$$P_2(\mu) = 0.2500, \qquad P_6(\mu) = -0.1413,$$
$$P_4(\mu) = -0.4062, \qquad P_8(\mu) = 0.2979,$$

and from (192)

$$M = \frac{0.002\pi^2(100)(100)N_1N_2}{800\sqrt{800}}[0.2526 + 0.1579 - 0.0222 - 0.0114]$$

$$= 0.00329 N_1 N_2 \ \mu\text{h}.$$

This is not a favorable case and the convergence is poor. As a check the calculation may be made for the central filaments of the coils using formula (159) and Table 43 for equal circles with parallel axes.

For $\dfrac{2a}{r} = \dfrac{20}{\sqrt{800}} = 0.70711$ and $\mu = 0.70711$ Table 43 gives $F = 0.5040$, and for coaxial circles with $\dfrac{\text{diameter}}{\text{distance}} = 0.70711$, Table 17 yields $f = 0.0006362$, so that for the central filaments $m = 10(0.0006362)(0.5040) = 0.003207$ and for the two coils $M = 0.003207 N_1 N_2 \ \mu\text{h}$.

Chapter 22

AUXILIARY TABLES OF FUNCTIONS WHICH APPEAR FREQUENTLY IN INDUCTANCE FORMULAS

AUXILIARY TABLE 1. NATURAL LOGARITHMS OF NUMBERS

This table is intended to give directly the natural logarithms of the numbers included and of numbers which may be simply factored. For example

$$\log_e 1525 = \log_e 25 + \log_e 61 = 7.3298,$$

$$\log_e 98.4 = \log_e 12 + \log_e 82 - \log_e 10 = 4.5890.$$

N	$\log_e N$	N	$\log_e N$	N	$\log_e N$	N	$\log_e N$	N	$\log_e N$	N	$\log_e N$
0	$-\infty$	25	3.2189	50	3.9120	75	4.3175	100	4.6052	125	4.8283
1	0.0000	26	3.2581	51	3.9318	76	4.3307	101	4.6151	126	4.8363
2	0.6931	27	3.2958	52	3.9512	77	4.3438	102	4.6250	127	4.8442
3	1.0986	28	3.3322	53	3.9703	78	4.3567	103	4.6347	128	4.8520
4	1.3863	29	3.3673	54	3.9890	79	4.3694	104	4.6444	129	4.8598
5	1.6094	30	3.4012	55	4.0073	80	4.3820	105	4.6540	130	4.8675
6	1.7918	31	3.4340	56	4.0254	81	4.3944	106	4.6634	131	4.8752
7	1.9459	32	3.4657	57	4.0431	82	4.4067	107	4.6728	132	4.8828
8	2.0794	33	3.4965	58	4.0604	83	4.4188	108	4.6821	133	4.8903
9	2.1972	34	3.5264	59	4.0775	84	4.4308	109	4.6913	134	4.8978
10	2.3026	35	3.5553	60	4.0943	85	4.4427	110	4.7005	135	4.9053
11	2.3979	36	3.5835	61	4.1109	86	4.4543	111	4.7095	136	4.9127
12	2.4849	37	3.6109	62	4.1271	87	4.4659	112	4.7185	137	4.9200
13	2.5649	38	3.6376	63	4.1431	88	4.4773	113	4.7274	138	4.9273
14	2.6391	39	3.6636	64	4.1589	89	4.4886	114	4.7362	139	4.9345
15	2.7081	40	3.6889	65	4.1744	90	4.4998	115	4.7449	140	4.9416
16	2.7726	41	3.7136	66	4.1897	91	4.5109	116	4.7536	141	4.9488
17	2.8332	42	3.7377	67	4.2047	92	4.5218	117	4.7622	142	4.9558
18	2.8904	43	3.7612	68	4.2195	93	4.5326	118	4.7707	143	4.9628
19	2.9444	44	3.7842	69	4.2341	94	4.5433	119	4.7791	144	4.9698
20	2.9957	45	3.8067	70	4.2485	95	4.5559	120	4.7875	145	4.9767
21	3.0445	46	3.8286	71	4.2627	96	4.5643	121	4.7958	146	4.9836
22	3.0910	47	3.8501	72	4.2767	97	4.5747	122	4.8040	147	4.9904
23	3.1355	48	3.8712	73	4.2905	98	4.5850	123	4.8122	148	4.9972
24	3.1781	49	3.8918	74	4.3041	99	4.5951	124	4.8203	149	5.0039
25	3.2189	50	3.9120	75	4.3175	100	4.6052	125	4.8283	150	5.0106

AUXILIARY TABLES OF FUNCTIONS

AUXILIARY TABLE 2. FOR CONVERTING COMMON LOGARITHMS INTO NATURAL LOGARITHMS

Common	Natural	Common	Natural	Common	Natural	Common	Natural
0	0	25	57.565	50	115.129	75	172.694
1	2.3026	26	59.867	51	117.432	76	174.996
2	4.6052	27	62.170	52	119.734	77	177.299
3	6.9078	28	64.472	53	122.037	78	179.602
4	9.2103	29	66.775	54	124.340	79	181.904
5	11.513	30	69.078	55	126.642	80	184.207
6	13.816	31	71.380	56	128.945	81	186.509
7	16.118	32	73.683	57	131.247	82	188.812
8	18.421	33	75.985	58	133.550	83	191.115
9	20.723	34	78.288	59	135.853	84	193.417
10	23.026	35	80.590	60	138.155	85	195.720
11	25.328	36	82.893	61	140.458	86	198.022
12	27.631	37	85.196	62	142.760	87	200.325
13	29.934	38	87.498	63	145.063	88	202.627
14	32.236	39	89.801	64	147.365	89	204.930
15	34.539	40	92.103	65	149.668	90	207.233
16	36.841	41	94.406	66	151.971	91	209.535
17	39.144	42	96.709	67	154.273	92	211.838
18	41.447	43	99.011	68	156.576	93	214.140
19	43.749	44	101.314	69	158.878	94	216.443
20	46.052	45	103.616	70	161.181	95	218.746
21	48.354	46	105.919	71	163.484	96	221.048
22	50.657	47	108.221	72	165.786	97	223.351
23	52.959	48	110.524	73	168.069	98	225.653
24	55.262	49	112.827	74	170.391	99	227.956
25	57.565	50	115.129	75	172.694	100	230.259

Examples of the use of the table.

To find the natural logarithm of 37.48. The common logarithm is 1.57380. The natural logarithm is M times 1.57380, where $M = 2.30259$. From the table

$$1.5M = 3.4539$$
$$0.073M = 0.1681$$
$$0.00080M = 0.0018$$
$$\text{Sum} = 3.6238 = \log_e 37.48.$$

To find the natural logarithm of 0.00748. The common logarithm is $\overline{3}.87390 = 0.87390 - 3$. From the table

$$0.87M = 2.00325$$
$$0.0039M = 0.00898$$
$$\text{Sum} = 2.01223$$
$$-3M = -6.9078$$
$$\text{Sum} = -4.8956 = \log_e 0.00748.$$

(Auxiliary Table 2 is a reproduction of the table in *Bureau of Standards Circular* 74, 241.)

238 CALCULATION OF MUTUAL INDUCTANCE AND SELF-INDUCTANCE

AUXILIARY TABLE 3. VALUES OF ZONAL HARMONIC FUNCTIONS

μ	$P_2(\mu)$	$P_3(\mu)$	$P_3(\mu)/\mu$	$P_4(\mu)$	$P_5(\mu)$	$P_5(\mu)/\mu$	$P_6(\mu)$	$P_7(\mu)$	$P_7(\mu)/\mu$	$P_8(\mu)$	μ
0	−0.5000	0	−1.500	0.3750	0	1.875	−0.3125	0	−2.187	0.2734	0
0.01	−.4998	−0.0150	−1.500	.3746	0.0187	1.874	−.3118	−0.0219	−2.185	.2724	0.01
.02	−.4994	−.0300	−1.499	.3735	.0374	1.871	−.3099	−.0436	−2.180	.2695	.02
.03	−.4986	−.0449	−1.498	.3716	.0560	1.867	−.3066	−.0651	−2.170	.2646	.03
.04	−.4976	−.0598	−1.496	.3690	.0744	1.861	−.3021	−.0862	−2.156	.2579	.04
0.05	−0.4972	−0.0747	−1.494	0.3657	0.0927	1.854	−0.2962	−0.1069	−2.138	0.2491	0.05
.06	−.4946	−.0895	−1.491	.3616	.1106	1.844	−.2891	−.1270	−2.118	.2387	.06
.07	−.4926	−.1041	−1.488	.3567	.1283	1.832	−.2808	−.1464	−2.091	.2265	.07
.08	−.4904	−.1187	−1.484	.3512	.1455	1.819	−.2713	−.1651	−2.062	.2126	.08
.09	−.4878	−.1332	−1.480	.3449	.1624	1.804	−.2606	−.1828	−2.028	.1972	.09
0.10	−0.4850	−0.1475	−1.475	0.3379	0.1788	1.788	−0.2488	−0.1995	−1.995	0.1803	0.10
.11	−.4818	−.1617	−1.470	.3303	.1947	1.770	−.2360	−.2151	−1.955	.1621	.11
.12	−.4784	−.1757	−1.464	.3219	.2101	1.751	−.2220	−.2295	−1.912	.1426	.12
.13	−.4746	−.1895	−1.458	.3129	.2248	1.729	−.2071	−.2427	−1.867	.1220	.13
.14	−.4706	−.2031	−1.451	.3032	.2389	1.706	−.1913	−.2545	−1.818	.1006	.14
0.15	−0.4662	−0.2166	−1.444	0.2928	0.2523	1.682	−0.1746	−0.2649	−1.766	0.0783	0.15
.16	−.4616	−.2298	−1.438	.2819	.2650	1.656	−.1572	−.2738	−1.711	.0554	.16
.17	−.4566	−.2427	−1.428	.2703	.2769	1.629	−.1389	−.2812	−1.654	.0319	.17
.18	−.4514	−.2554	−1.419	.2581	.2880	1.600	−.1201	−.2870	−1.594	+.0062	.18
.19	−.4458	−.2679	−1.410	.2453	.2982	1.570	−.1006	−.2911	−1.532	−.0157	.19
0.20	−0.4400	−0.2800	−1.400	0.2320	0.3075	1.538	−0.0806	−0.2935	−1.468	−0.0396	0.20
.21	−.4338	−.2918	−1.389	.2181	.3159	1.504	−.0601	−.2943	−1.401	−.0633	.21
.22	−.4274	−.3034	−1.379	.2037	.3234	1.470	−.0394	−.2933	−1.333	−.0865	.22
.23	−.4206	−.3146	−1.368	.1889	.3299	1.434	−.0183	−.2906	−1.263	−.1093	.23
.24	−.4136	−.3254	−1.356	.1735	.3353	1.397	+.0029	−.2861	−1.192	−.1313	.24
0.25	−0.4062	−0.3359	−1.344	0.1577	0.3397	1.359	+0.0243	−0.2799	−1.120	−0.1525	0.25

AUXILIARY TABLE 3. VALUES OF ZONAL HARMONIC FUNCTIONS (Continued)

μ	$P_2(\mu)$	$P_3(\mu)$	$P_3(\mu)/\mu$	$P_4(\mu)$	$P_5(\mu)$	$P_5(\mu)/\mu$	$P_6(\mu)$	$P_7(\mu)$	$P_7(\mu)/\mu$	$P_8(\mu)$	μ
0.25	−0.4062	−0.3359	−1.344	0.1577	0.3397	1.359	0.0243	−0.2799	−1.120	−0.1525	0.25
.26	−.3986	−.3461	−1.331	.1415	.3431	1.320	.0456	−.2720	−1.046	−.1725	.26
.27	−.3906	−.3558	−1.318	.1249	.3453	1.279	.0669	−.2625	−.972	−.1914	.27
.28	−.3824	−.3651	−1.304	.1079	.3465	1.237	.0879	−.2512	−.897	−.2088	.28
.29	−.3738	−.3740	−1.290	.0906	.3465	1.195	.1087	−.2384	−.822	−.2247	.29
0.30	−0.3650	−0.3825	−1.275	.0729	.3454	1.151	0.1292	−0.2241	−.747	−0.2391	0.30
.31	−.3558	−.3905	−1.260	.0550	.3431	1.107	.1492	−.2082	−.672	−.2516	.31
.32	−.3464	−.3981	−1.244	.0369	.3397	1.062	.1686	−.1910	−.597	−.2621	.32
.33	−.3366	−.4052	−1.228	.0185	.3351	1.015	.1873	−.1724	−.522	−.2706	.33
.34	−.3266	−.4117	−1.211	+.0000	.3294	.969	.2053	−.1527	−.449	−.2770	.34
0.35	−0.3162	−0.4178	−1.194	−.0187	.3225	.921	0.2225	−0.1318	−.377	−0.2812	0.35
.36	−.3056	−.4234	−1.176	−.0375	.3144	.873	.2388	−.1098	−.305	−.2831	.36
.37	−.2946	−.4284	−1.158	−.0564	.3051	.825	.2540	−.0870	−.235	−.2826	.37
.38	−.2834	−.4328	−1.139	−.0753	.2948	.776	.2681	−.0635	−.167	−.2798	.38
.39	−.2718	−.4367	−1.120	−.0942	.2833	.726	.2810	−.0393	−.101	−.2746	.39
0.40	−0.2600	−0.4400	−1.100	−.1130	.2706	.6766	0.2926	−0.0146	−.036	−0.2670	0.40
.41	−.2478	−.4427	−1.080	−.1317	.2569	.6266	.3029	+.0104	+0.0253	−.2570	.41
.42	−.2354	−.4448	−1.059	−.1504	.2421	.5764	.3118	.0356	.0848	−.2448	.42
.43	−.2226	−.4462	−1.038	−.1688	.2263	.5263	.3191	.0608	.1414	−.2302	.43
.44	−.2096	−.4470	−1.016	−.1870	.2095	.4761	.3249	.0859	.1952	−.2134	.44
0.45	−0.1962	−0.4472	−.994	−.2050	.1917	.4260	0.3290	0.1106	0.2458	−0.1945	0.45
.46	−.1826	−.4467	−.971	−.2226	.1730	.3761	.3314	.1348	.2930	−.1737	.46
.47	−.1686	−.4454	−.948	−.2399	.1534	.3264	.3321	.1584	.3370	−.1510	.47
.48	−.1544	−.4435	−.924	−.2568	.1330	.2771	.3310	.1811	.3773	−.1266	.48
.49	−.1398	−.4409	−.900	−.2732	.1118	.2281	.3280	.2027	.4137	−.1008	.49
0.50	−0.1250	−0.4375	−.875	−.2891	.0898	.1796	0.3232	0.2231	.4462	−0.0736	0.50

240 CALCULATION OF MUTUAL INDUCTANCE AND SELF-INDUCTANCE

AUXILIARY TABLE 3. VALUES OF ZONAL HARMONIC FUNCTIONS (Continued)

μ	$P_2(\mu)$	$P_3(\mu)$	$P_3(\mu)/\mu$	$P_4(\mu)$	$P_5(\mu)$	$P_5(\mu)/\mu$	$P_6(\mu)$	$P_7(\mu)$	$P_7(\mu)/\mu$	$P_8(\mu)$	μ
0.50	−0.1250	−0.4375	−0.875	−0.2891	0.0898	0.1796	0.3232	0.2231	0.4462	−0.0736	0.50
.51	−.1098	−.4334	−.850	−.3044	.0673	.1320	.3166	.2422	.4749	−.0454	.51
.52	−.0944	−.4285	−.824	−.3191	.0441	.0848	.3080	.2596	.4992	−.0164	.52
.53	−.0786	−.4228	−.798	−.3332	+.0204	+.0345	.2975	.2753	.5194	+.0133	.53
.54	−.0626	−.4163	−.771	−.3465	−.0037	−.0068	.2851	.2891	.5354	.0432	.54
.55	−.0462	−.4091	−.744	−.3590	−.0282	−.0513	.2708	.3007	.5467	.0731	.55
.56	−.0296	−.4010	−.716	−.3707	−.0529	−.0945	.2546	.3102	.5540	.1029	.56
.57	−.0126	−.3920	−.688	−.3815	−.0779	−.1366	.2366	.3172	.5565	.1320	.57
.58	+.0046	−.3822	−.659	−.3914	−.1028	−.1773	.2168	.3217	.5547	.1601	.58
.59	.0222	−.3716	−.630	−.4002	−.1278	−.2166	.1953	.3235	.5483	.1869	.59
.60	.0400	−.3600	−.600	−.4080	−.1526	−.2543	.1721	.3226	.5377	.2123	.60
.61	.0582	−.3475	−.570	−.4146	−.1772	−.2905	.1473	.3188	.5226	.2357	.61
.62	.0766	−.3332	−.539	−.4200	−.2014	−.3249	.1211	.3121	.5034	.2569	.62
.63	.0954	−.3199	−.508	−.4242	−.2251	−.3573	.0935	.3023	.4798	.2753	.63
.64	.1144	−.3046	−.476	−.4270	−.2482	−.3878	.0646	.2895	.4524	.2909	.64
0.65	0.1338	−0.2884	−0.4437	−0.4284	−0.2705	−0.4162	0.0347	0.2737	0.4211	0.3032	0.65
.66	.1534	−.2713	−.4111	−.4284	−.2919	−.4423	+.0038	.2548	.3861	.3120	.66
.67	.1734	−.2531	−.3778	−.4268	−.3122	−.4660	−.0278	.2329	.3476	.3169	.67
.68	.1936	−.2339	−.3440	−.4236	−.3313	−.4872	−.0601	.2081	.3060	.3179	.68
.69	.2142	−.2137	−.3097	−.4187	−.3490	−.5058	−.0926	.1805	.2616	.3145	.69
0.70	0.2350	−0.1925	−0.2750	−0.4121	−0.3652	−0.5216	−0.1253	0.1502	0.2146	0.3067	0.70
.71	.2562	−.1702	−.2397	−.4036	−.3796	−.5346	−.1578	.1173	.1652	.2942	.71
.72	.2776	−.1469	−.2040	−.3933	−.3922	−.5447	−.1899	.0822	.1142	.2771	.72
.73	.2994	−.1225	−.1678	−.3810	−.4026	−.5515	−.2214	.0450	.0617	.2553	.73
.74	.3214	−.0969	−.1310	−.3666	−.4107	−.5550	−.2518	+.0061	+.0082	.2288	.74
0.75	0.3438	−0.0703	−0.0937	−0.3501	−0.4164	−0.5552	−0.2808	−0.0342	−0.0446	0.1976	0.75

AUXILIARY TABLE 3. VALUES OF ZONAL HARMONIC FUNCTIONS (Concluded)

μ	$P_2(\mu)$	$P_3(\mu)$	$P_3(\mu)/\mu$	$P_4(\mu)$	$P_5(\mu)$	$P_5(\mu)/\mu$	$P_6(\mu)$	$P_7(\mu)$	$P_7(\mu)/\mu$	$P_8(\mu)$	μ
0.75	0.3438	−0.0703	−0.0937	−0.3501	−0.4164	−0.5552	−0.2808	−0.0342	−0.0446	0.1976	0.75
.76	.3664	−.0426	−.0661	−.3314	−.4193	−.5517	−.3081	.0754	.0992	.1621	.76
.77	.3894	−.0137	−.0178	−.3104	−.4193	−.5451	−.3333	.1171	.1521	.1226	.77
.78	.4126	+.0164	+.0210	−.2871	−.4162	−.5336	−.3559	.1588	.2036	.0792	.78
.79	.4362	.0476	.0602	−.2613	−.4097	−.5186	−.3756	.1999	.2530	+.0325	.79
0.80	0.4600	0.0800	0.1000	−0.2330	−0.3995	−0.4994	−0.3918	−0.2397	−0.2996	−0.0167	0.80
.81	.4842	.1136	.1402	−.2021	−.3855	−.4759	−.4041	−.2774	−.3425	−.0677	.81
.82	.5086	.1484	.1810	−.1685	−.3674	−.4480	−.4119	−.3124	−.3810	−.1199	.82
.83	.5334	.1845	.2223	−.1321	−.3449	−.4155	−.4147	−.3437	−.4141	−.1720	.83
.84	.5584	.2218	.2640	−.0928	−.3177	−.3782	−.4120	−.3703	−.4408	−.2227	.84
0.85	0.5838	0.2603	0.3062	−0.0506	−0.2857	−0.3361	−0.4030	−0.3913	−0.4604	−0.2710	0.85
.86	.6094	.3001	.3490	−.0053	−.2484	−.2888	−.3872	−.4055	−.4715	−.3151	.86
.87	.6354	.3413	.3923	+.0431	−.2056	−.2363	−.3638	−.4116	−.4731	−.3531	.87
.88	.6616	.3837	.4360	.0947	−.1570	−.1784	−.3322	−.4083	−.4640	−.3830	.88
.89	.6882	.4274	.4802	.1496	−.1023	−.1150	−.2916	−.3942	−.4429	−.4027	.89
0.90	0.7150	0.4725	0.5250	0.2079	−0.0411	−0.0457	−0.2412	−0.3678	−0.4087	−0.4097	0.90
.91	.7422	.5189	.5702	.2698	+.0268	+.0294	−.1802	−.3274	−.3597	−.4009	.91
.92	.7696	.5667	.6160	.3352	.1017	.1105	−.1077	−.2713	−.2949	−.3738	.92
.93	.7974	.6159	.6623	.4044	.1842	.1981	−.0229	−.1975	−.2124	−.3243	.93
.94	.8254	.6665	.7090	.4773	.2744	.2919	+.0751	−.1040	−.1106	−.2490	.94
0.95	0.8538	0.7184	0.7562	0.5541	0.3727	0.3923	0.1875	+0.0112	+0.0118	−0.1441	0.95
.96	.8824	.7718	.8040	.6349	.4796	.4996	.3151	.1506	.1569	−.0046	.96
.97	.9114	.8267	.8523	.7198	.5954	.6138	.4590	.3165	.3263	+.1740	.97
.98	.9406	.8830	.9010	.8089	.7204	.7351	.6204	.5115	.5219	.3970	.98
0.99	0.9702	0.9407	0.9502	0.9022	0.8552	0.8638	0.8003	0.7384	0.7458	0.6704	0.99
1.00	1.0000	1.0000	1.0000	1.0000	1.0000	1.0000	1.0000	1.0000	1.0000	1.0000	1.00

Series for Zonal Harmonics. With $\mu = \cos\theta$, the zonal harmonic functions are

$$P_0(\mu) = 1, \quad P_1(\mu) = \mu,$$

$$P_2(\mu) = \frac{1}{2}(3\mu^2 - 1),$$

$$P_3(\mu) = \frac{\mu}{2}(5\mu^2 - 3),$$

$$P_4(\mu) = \frac{1}{8}(35\mu^4 - 30\mu^2 + 3),$$

$$P_5(\mu) = \frac{\mu}{8}(63\mu^4 - 70\mu^2 + 15),$$

$$P_6(\mu) = \frac{1}{16}(231\mu^6 - 315\mu^4 + 105\mu^2 - 5),$$

$$P_7(\mu) = \frac{\mu}{16}(429\mu^6 - 693\mu^4 + 315\mu^2 - 35),$$

$$P_8(\mu) = \frac{1}{128}(6435\mu^8 - 12012\mu^6 + 6930\mu^4 - 1260\mu^2 + 35),$$

and, in general,

$$P_m(\mu) = \frac{(2m-1)(2m-3)\cdots 1}{m!}\left[\mu^m - \frac{m(m-1)}{2(2m-1)}\mu^{m-2} \right.$$
$$\left. + \frac{m(m-1)(m-2)(m-3)}{2.4(2m-1)(2m-3)}\mu^{m-4} - \cdots\right].$$

The series terminates with the term in μ, if m is odd, and with the term independent of μ, if m is even.

For large values of m, $P_m(\cos\theta) \cong \left(\frac{2}{m\pi\sin\theta}\right)^{1/2} \sin\left\{\left(m+\frac{1}{2}\right)\theta + \frac{\pi}{4}\right\}$.

The recursion formula $(m+1)P_{m+1}(\mu) = (2m+1)\mu P_m(\mu) - mP_{m-1}(\mu)$ may be useful.

Differential Coefficients, $\frac{d}{d\mu}[P_m(\mu)] = P_m'(\mu)$.

$$P_0'(\mu) = 0, \quad P_1'(\mu) = 1, \quad P_2'(\mu) = 3\mu,$$

$$P_3'(\mu) = \frac{1}{2}(3.5\mu^2 - 1\cdot 3),$$

AUXILIARY TABLES OF FUNCTIONS

$$P_4'(\mu) = \frac{1}{2}(5.7\mu^3 - 3\cdot 5\mu),$$

$$P_5'(\mu) = \frac{1}{2\cdot 4}(5\cdot 7\cdot 9\mu^4 - 2\cdot 3\cdot 5\cdot 7\mu^2 + 1\cdot 3\cdot 5),$$

$$P_6'(\mu) = \frac{1}{2\cdot 4}(7\cdot 9\cdot 11\mu^5 - 2\cdot 5\cdot 7\cdot 9\mu^3 + 3\cdot 5\cdot 7\mu),$$

$$P_7'(\mu) = \frac{1}{2\cdot 4\cdot 6}(7\cdot 9\cdot 11\cdot 13\mu^6 - 3\cdot 5\cdot 7\cdot 9\cdot 11\mu^4 + 3\cdot 3\cdot 5\cdot 7\cdot 9\mu^2 - 1\cdot 3\cdot 5\cdot 7),$$

with the general relation,

$$(\mu^2 - 1)P_m'(\mu) = m\mu P_m(\mu) - mP_{m-1}(\mu).$$

244 CALCULATION OF MUTUAL INDUCTANCE AND SELF-INDUCTANCE

AUXILIARY TABLE 4. VALUES OF DIFFERENTIAL COEFFICIENTS OF ZONAL HARMONICS

ν	$P_5'(\nu)$	Δ'	$P_6'(\nu)$	Δ'	Δ''	$P_7'(\nu)$	Δ'	Δ''	ν
0	−1.5000		1.8750			−2.1873			0
		+ 8		− 25	−55		57	120	
0.01	1.4992		1.8725			2.1816			0.01
		+ 22		− 80	−51		177	117	
.02	1.4970		1.8645			2.1639			.02
		+ 38		− 131	−52		294	115	
.03	1.4932		1.8514			2.1345			.03
		+ 52		− 183	−52		409	115	
.04	1.4880		1.8331			2.0936			.04
		+ 68		− 235	−51		524	111	
0.05	−1.4812		1.8096			−2.0412			0.05
		+ 82		− 286	−51		635	110	
.06	1.4730		1.7810			1.9777			.06
		+ 98		− 337	−50		745	104	
.07	1.4632		1.7473			1.9032			.07
		+112		− 387	−50		849	102	
.08	1.4520		1.7086			1.8183			.08
		+128		− 437	−48		951	98	
.09	1.4392		1.6649			1.7232			.09
		+142		− 485	−48		1049	92	
0.10	−1.4250		1.6164			−1.6183			0.10
		+158		− 533	−47		1141	88	
.11	1.4092		1.5631			1.5042			.11
		+172		− 580	−45		1229	81	
.12	1.3920		1.5051			1.3813			.12
		+188		− 625	−45		1310	76	
.13	1.3732		1.4426			1.2503			.13
		+202		− 670	−43		1386	70	
.14	1.3530		1.3756			1.1117			.14
		+218		− 713	−42		1456	62	
0.15	−1.3312		1.3043			−0.9661			0.15
		+232		− 755	−40		1518	55	
.16	1.3080		1.2288			.8143			.16
		+248		− 795	−40		1573	48	
.17	1.2832		1.1493			.6570			.17
		+262		− 835	−36		1621	41	
.18	1.2570		1.0658			.4949			.18
		+278		− 871	−36		1662	30	
.19	1.2292		0.9787			.3287			.19
		+292		− 907	−34		1692	25	
0.20	−1.2000		0.8880			−0.1595			0.20
		+308		− 941	−31		1717	14	
.21	1.1692		.7939			+0.0122			.21
		+322		− 972	−29		1731	+ 5	
.22	1.1370		.6967			.1853			.22
		+338		−1001	−29		1736	− 4	
.23	1.1032		.5966			.3589			.23
		+352		−1030	−24		1732	−15	
.24	1.0680		.4936			.5321			.24
		+368		−1054			1717		
0.25	−1.0312		0.3882			0.7038			0.25

AUXILIARY TABLES OF FUNCTIONS

AUXILIARY TABLE 4. DIFFERENTIAL COEFFICIENTS OF ZONAL HARMONICS (*Continued*)

ν	$P_3'(\nu)$	Δ'	$P_5'(\nu)$	Δ'	Δ''	$P_7'(\nu)$	Δ'	Δ''	ν
0.25	−1.0312		0.3882		−23	0.7038		− 20	0.25
		382		−1077			1697		
.26	0.9930		.2805		−21	0.8735		− 32	.26
		398		−1098			1665		
.27	.9532		.1707		−18	1.0400		− 41	.27
		412		−1116			1624		
.28	.9120		+0.0591		−16	1.2024		− 51	.28
		428		−1132			1573		
.29	.8692		−0.0541		−13	1.3597		− 60	.29
		442		−1145			1513		
0.30	−0.8250		.1686		− 9	1.5110		− 71	0.30
		458		−1154			1442		
.31	.7792		.2840		− 7	1.6552		− 79	.31
		472		−1161			1363		
.32	.7320		.4001		− 4	1.7915		− 91	.32
		488		−1165			1272		
.33	.6832		.5166		− 2	1.9187		− 98	.33
		502		−1167			1174		
.34	.6330		.6333		+ 2	2.0361		−106	.34
		518		−1165			1068		
0.35	−0.5812		−0.7498		6	2.1429		−116	0.35
		532		−1159			952		
.36	.5280		.8657		9	2.2381		−124	.36
		548		−1150			828		
.37	.4732		−0.9807		12	2.3209		−132	.37
		562		−1138			696		
.38	.4170		−1.0945		15	2.3905		−139	.38
		578		−1123			557		
.39	.3592		1.2068		21	2.4462		−147	.39
		592		−1102			410		
0.40	−0.3000		−1.3170		22	2.4872		−154	0.40
		608		−1080			256		
.41	.2392		1.4250		27	2.5128		−159	.41
		622		−1053			+ 97		
.42	.1770		1.5303		31	2.5225		−165	.42
		638		−1022			− 68		
.43	.1132		1.6325		35	2.5157		−170	.43
		652		− 987			− 238		
.44	−0.0480		1.7312		39	2.4919		−173	.44
		668		− 948			− 411		
0.45	+0.0188		−1.8260		43	2.4508		−178	0.45
		682		− 905			− 589		
.46	.0870		1.9165		48	2.3919		−179	.46
		698		− 857			− 768		
.47	.1568		2.0022		51	2.3151		−182	.47
		712		− 806			− 950		
.48	.2280		2.0828		57	2.2201		−181	.48
		728		− 749			−1131		
.49	.3008		2.1577		60	2.1070		−183	.49
		742		− 689			−1314		
0.50	+0.3750		−2.2266		67	1.9756		−178	0.50

246 CALCULATION OF MUTUAL INDUCTANCE AND SELF-INDUCTANCE

AUXILIARY TABLE 4. DIFFERENTIAL COEFFICIENTS OF ZONAL HARMONICS (*Continued*)

ν	$P_3'(\nu)$	Δ'	$P_5'(\nu)$	Δ'	Δ''	$P_7'(\nu)$	Δ'	Δ''	ν
0.50	0.3750		−2.2266		67	1.9756		−178	0.50
		758		−622			−1492		
.51	.4508		2.2888		70	1.8264		−177	.51
		772		−552			−1669		
.52	.5280		2.3440		75	1.6595		−173	.52
		788		−477			−1842		
.53	.6068		2.3917		80	1.4753		−167	.53
		802		−397			−2009		
.54	.6870		2.4314		85	1.2744		−161	.54
		818		−312			−2170		
0.55	0.7688		−2.4626		91	1.0574		−153	0.55
		832		−221			−2323		
.56	.8520		2.4847		96	0.8251		−142	.56
		848		−125			−2465		
.57	0.9368		2.4972		101	0.5786		−133	.57
		862		− 24			−2598		
.58	1.0230		2.4996		106	0.3188		−117	.58
		878		+ 82			−2715		
.59	1.1108		2.4914		112	+0.0473		−105	.59
		892		194			−2820		
0.60	1.2000		−2.4720		118	−0.2347		− 88	0.60
		908		312			−2908		
.61	1.2908		2.4408		123	0.5255		− 69	.61
		922		435			−2977		
.62	1.3830		2.3973		129	0.8232		− 51	.62
		938		564			−3028		
.63	1.4768		2.3409		136	1.1260		− 21	.63
		952		700			−3049		
.64	1.5720		2.2709		140	1.4309		− 4	.64
		968		840			−3053		
0.65	1.6688		−2.1869		147	−1.7362		+ 26	0.65
		982		987			−3027		
.66	1.7670		2.0882		154	2.0389		54	.66
		998		1141			−2973		
.67	1.8668		1.9741		160	2.3362		87	.67
		1012		1301			−2886		
.68	1.9680		1.8440		165	2.6248		120	.68
		1028		1466			−2766		
.69	2.0608		1.6974		172	2.9014		156	.69
		1042		1638			−2610		
0.70	2.1750		−1.5336		180	−3.1624		200	0.70
		1058		1818			−2410		
.71	2.2808		1.3518		186	−3.4034		237	.71
		1072		2004			−2173		
.72	2.3880		1.1514		193	3.6207		285	.72
		1088		2197			−1888		
.73	2.4968		0.9317		198	3.8095		333	.73
		1102		2395			−1555		
.74	2.6070		0.6922		206	3.9650		383	.74
		1118		2601			−1172		
0.75	2.7188		−0.4321		214	−4.0822		439	0.75

AUXILIARY TABLE 4. DIFFERENTIAL COEFFICIENTS OF ZONAL HARMONICS (*Concluded*)

ν	$P_3'(\nu)$	Δ'	$P_5'(\nu)$	Δ'	Δ''	$P_7'(\nu)$	Δ'	Δ''	Δ'''	ν
0.75	2.7188		−0.4321		214	−4.0822		439		0.75
		1132		2815			−733		57	
.76	2.8320		−0.1506		220	4.1555		496		.76
		1148		3035			−237		60	
.77	2.9468		+0.1529		228	4.1792		556		.77
		1162		3263			+319		66	
.78	3.0630		0.4792		234	4.1473		622		.78
		1178		3497			941		66	
.79	3.1808		0.8289		244	4.0532		688		.79
		1192		3741			1629		76	
0.80	3.3000		1.2030		249	−3.8903		764		0.80
		1208		3990			2393		71	
.81	3.4208		1.6020		258	3.6510		835		.81
		1222		4248			3228		83	
.82	3.5430		2.0268		265	3.3282		918		.82
		1238		4513			4146		89	
.83	3.6668		2.4781		273	2.9136		1002		.83
		1252		4786			5148		82	
.84	3.7920		2.9567		281	2.3988		1084		.84
		1268		5067			6232		99	
0.85	3.9188		3.4634		289	−1.7756		1183		0.85
		1282		5356			7415		93	
.86	4.0470		3.9990		297	1.0341		1276		.86
		1298		5653			8691		101	
.87	4.1768		4.5643		304	−0.1650		1377		.87
		1312		5957			10068		110	
.88	4.3080		5.1600		315	+0.8418		1487		.88
		1328		6272			11555		108	
.89	4.4408		5.7872		320	1.9963		1595		.89
		1342		6592			13150			
0.90	4.5750		6.4464		332	3.3113					0.90
		1358		6924						
.91	4.7108		7.1388		338	4.7965					.91
		1372		7262						
.92	4.8480		7.8650		347	6.4644					.92
		1388		7609						
.93	4.9868		8.6259		357	8.3276					.93
		1402		7966						
.94	5.1270		9.4225		364	10.3991					.94
		1418		8330						
0.95	5.2688		10.2555		375	12.6925					0.95
		1432		8705						
.96	5.4120		11.1260		383	15.2218					.96
		1448		9088						
.97	5.5568		12.0348		392	18.0016					.97
		1462		9480						
.98	5.7030		12.9828		401	21.0474					.98
		1478		9881						
0.99	5.8508		13.9709		410	24.3746					0.99
		1492		10291						
1.00	6.0000		15.0000			28.0000					1.00

Chapter 23

FORMULAS FOR THE CALCULATION OF THE MAGNETIC FORCE BETWEEN COILS

The calculation of the magnetic attraction between two coils, carrying current, is a subject closely related to the calculation of their mutual inductance. Since their mutual energy is equal to the product of their mutual inductance by the currents in the coils, the component of the magnetic force (attraction or repulsion) in any direction is equal to the differential coefficient of the mutual inductance, taken with respect to that coordinate, and multiplied by the product of the currents. If the mutual inductance is given in abhenrys and the currents in abamperes, the mutual force will be in dynes. Evidently the force may be calculated by simple differentiation in any case where a general formula for the mutual inductance is available, expressed as a function of the coordinate along which the force is required. Only a few of the more important cases will here be considered.

Force between Two Coaxial Circular Filaments. This case has been treated by a number of authors. Maxwell [94] gave a formula in terms of elliptic integrals and a table was prepared by Lord Rayleigh.[95] This was recalculated at the Bureau of Standards and included in the article on the absolute determination of the international ampere by Rosa,[96] Dorsey, and Miller. This covers only a portion of the range of possible circles. Nagaoka [97] has expressed the force in q series and has given tables to aid in calculations. Further very complete tables have been prepared by Nagaoka and Sakurai.[98]

Let the two circular filaments have radii a and A and let d be the distance between their planes. Currents i_1 and i_2 abamperes flow in the filaments.

The mutual force between the filaments exerted along their common axis is

$$F = i_1 i_2 F_0 \qquad (193)$$

in which F_0 is the force when unit current flows in each. F_0 is a function of the ratios of the radii and distance of the planes. With

$$\alpha = \frac{a}{A} \quad \text{and} \quad \delta = \frac{d}{A},$$

FORMULAS FOR MAGNETIC FORCE BETWEEN COILS

calculate the value of

$$k^2 = \frac{4aA}{(A+a)^2 + d^2} = \frac{4\alpha}{(1+\alpha)^2 + \delta^2}. \tag{194}$$

Then,

$$F_0 = \frac{\delta}{\sqrt{\alpha}} \cdot P, \tag{195}$$

where P is to be taken from Table 49 as a function of k^2. Table 49 is an abridgment of the table of Nagaoka and Sakurai [98] (page 161 of the reference cited).

Interpolation from Table 49 is satisfactory except for values of k^2 greater than 0.9 and for values of k^2 less than 0.2. More accurate values for these cases may be calculated from Nagaoka's formulas in q series, which are very convergent.

For values of k^2 greater than 0.9

$$P = \frac{\pi}{16} \frac{1}{q_1} \left[(1 + 12q_1 - 192q_1^2 + 1232q_1^3 + \cdots) \right.$$
$$\left. - 12q_1 (1 - 10q_1 + 60q_1^2 - 300q_1^3 + \cdots) \log_e \frac{1}{q_1} \right]. \tag{196}$$

The quantity q_1 and $\log \left(\log_e \frac{1}{q_1} \right)$ may be obtained from Table 50, which is an abridgment of one by Nagaoka and Sakurai [99] for values of $k'^2 = 1 - k^2$.

For values of k^2 less than about 0.2, values of P are readily calculated by Nagaoka's formula.[97]

$$P = 192\pi^2 q^{\frac{1}{2}} [1 + 20q^2 + 225q^4 + \cdots]. \tag{197}$$

The variable q is the same function of k^2 as q_1 is of k'^2 and may be taken from Table 50.

Maximum Value of the Force. The force between equal circular filaments is, of course, a maximum with the filaments indefinitely close together: for unequal circles the force is less for the coplanar position than for some greater axial separation. A knowledge of the spacing for maximum value of the force is of great importance in the theory of the current balance. The distance between planes for maximum value of the force, and the value of the maximum force for unit currents in the filaments is a function of the ratio of the radii alone. Considerable attention has been paid to the calculation of the constants for this position of the coils and very accurate values may be computed.[100, 101, 102] Table 51 gives values of the ratio [101] $\delta_m = \frac{d}{A}$ and F_{0m} for maximum force for certain values of $\alpha = \frac{a}{A}$.

250 CALCULATION OF MUTUAL INDUCTANCE AND SELF-INDUCTANCE

TABLE 49. VALUES OF P. FORCE BETWEEN COAXIAL CIRCULAR FILAMENTS

k²	P	Δ'	Δ''	k²	P	Δ'	Δ''	k²	P	Δ'	Δ''	k²	P	Δ'	Δ''
0	0			0.25	0.0826			0.50	0.7637			0.75	4.740		
		19				100				584				394	
0.01	0.000019		69	.26	.0926		9	.51	.8221		40	.76	5.134		41
		88				109				624				435	
.02	.000107		105	.27	.1035		9	.52	.8845		43	.77	5.509		47
		193				118				667				482	
.03	300		130	.28	.1153		9	.53	0.9512		47	.78	6.051		53
		323				127				714				535	
.04	623		157	.29	.1280		11	.54	1.0226		49	.79	6.586		63
		480				138				763				598	
0.05	0.001103		180	0.30	0.1418		10	0.55	1.0989		53	0.80	7.184		71
		660				148				818				669	
.06	1763		203	.31	.1566		12	.56	1.1807		57	.81	7.853		85
		863				160				875				754	
.07	2626		228	.32	.1726		12	.57	1.2682		63	.82	8.607		100
		1091				172				938				854	
.08	3717		250	.33	.1898		13	.58	1.3620		68	.83	9.461		119
		1341				185				1006				973	
.09	.005058		274	.34	.2083		13	.59	1.4626		73	.84	10.434		146
		1615				198				1079				1119	
0.10	0.00667		298	0.35	0.2281		15	0.60	1.5705		79	0.85	11.551		173
		191				213				116				1292	
.11	.00859		33	.36	.2494		15	.61	1.686		9	.86	12.843		218
		224				228				125				1510	
.12	.01082		35	.37	.2723		17	.62	1.811		9	.87	14.353		272
		259				245				134				1782	
.13	1341		37	.38	.2968		18	.63	1.945		10	.88	16.135		349
		296				263				144				2131	
.14	1637		41	.39	.3230		18	.64	2.089		11	.89	18.266		455
		337				281				155				2586	
0.15	0.01974		43	0.40	0.3510		19	0.65	2.244		13	0.90	20.85		609
		380				300				168				320	
.16	2354		46	.41	.3811		21	.66	2.412		13	.91	24.05		83
		426				321				181				403	
.17	2780		49	.42	.4132		23	.67	2.593		15	.92	28.08		122
		475				343				196				525	
.18	3255		53	.43	.4475		24	.68	2.789		17	.93	33.33		181
		528				367				213				706	
.19	3783		56	.44	.4842		25	.69	3.002		18	.94	40.39		293
		584				392				231				999	
0.20	0.04367		59	0.45	0.5234		27	0.70	3.233		20	0.95	50.38		
		643				419				250					
.21	5010		64	.46	.5653		29	.71	3.483		23	.96	67.50		
		707				448				274					
.22	5717		68	.47	.6101		30	.72	3.757		25	.97	90.94		
		775				478				298					
.23	6492		71	.48	.6579		33	.73	4.055		29	.98	142.28		
		846				511				327					
.24	7337		76	.49	.7091		35	.74	4.382		31	0.99	297.66		
		922				546				358					
0.25	0.08259		80	0.50	0.7637		38	0.75	4.740		36	1.00	∞		

FORMULAS FOR MAGNETIC FORCE BETWEEN COILS

TABLE 50. VALUES OF q_1 (OR q) FOR VALUES OF k'^2 (OR k^2)

k'^2	q_1	Δ'	Δ''	$\log q_1$	Δ'	Δ''	$\log\left(\log_e \dfrac{1}{q_1}\right)$	Δ'	Δ''
0	0								
		3133							
0.005	0.0003133		15	$\bar{4}.49594$			0.90679		
		3148						−3916	
.010	6281		17	.79806			.86763		
		3165						−2472	
.015	.0009446		16	$\bar{4}.97525$			.84291		
		3181						−1849	
.020	.0012627		15	$\bar{3}.10129$			.82442		
		3196						−1494	
0.025	0.0015823		18	$\bar{3}.19930$			0.80948		231
		3214			8030			−1263	
.030	19037		16	.27960		−1224	.79685		163
		3230			6806			−1100	
.035	22267		16	.34766		− 895	.78585		121
		3246			5911			− 979	
.040	25513		18	.40677		− 683	.77606		98
		3264			5228			− 881	
.045	28777		17	.45905		− 540	.76721		72
		3281			4688			− 809	
0.050	0.0032058		17	$\bar{3}.50593$		− 435	0.75912		62
		3298			4253			− 747	
.055	35356		17	.54846		− 360	.75165		52
		3315			3893			− 695	
.060	38671		18	.58739		− 302	.74470		44
		3333			3591			− 651	
.065	42004		18	.62330		− 258	.73819		38
		3351			3333			− 613	
.070	45355		18	.65663		− 221	.73206		32
		3369			3112			− 581	
0.075	0.0048724		19	$\bar{3}.68775$		− 194	0.72625		29
		3388			2918			− 552	
.080	52112		17	.71693		− 168	.72073		26
		3405			2750			− 526	
.085	55517		19	.74443		− 151	.71547		23
		3424			2599			− 503	
.090	58941		19	.77042		− 133	.71044		20
		3443			2466			− 483	
.095	62384		19	.79508		− 121	.70561		18
		3462			2345			− 465	
0.100	0.0065846			$\bar{3}.81853$			0.70096		

252 CALCULATION OF MUTUAL INDUCTANCE AND SELF-INDUCTANCE

TABLE 51. SPACING RATIO AND FORCE FOR MAXIMUM POSITION COAXIAL CIRCULAR FILAMENTS

α	δ_m	Δ'	Δ''	δ_m by series	F_{om}	$\log F_{om}$
0	0.50000			0.50000	0	
		− 113				
0.05	.49887		−225	.49886	0.042458	$\bar{2}$.62796
		− 338				
.10	.49549		−227	.49549	.17086	$\bar{1}$.23264
		− 565				
.15	.48984		−229	.48984	.38369	$\bar{1}$.58924
		− 794				
.20	.48190		−233	.48190	0.70052	$\bar{1}$.84542
		−1027				
0.25	0.47163		−237	0.47163	1.11563	0.04752
		−1264				
.30	.45899		−242	.45899	1.64555	.21631
		−1506				
.35	.44393		−249	.44393	2.30667	.36298
		−1755				
.40	.42638		−255	.42638	3.12162	.49438
		−2010				
.45	.40628		−265	.40631	4.12190	.61510
		−2275				
0.50	0.38353		−271	0.38360	5.35201	0.72852
		−2546				
.55	.35807		−283	.35815	6.87654	.83737
		−2829				
.60	.32978		−290	.3299	8.79239	0.94411
		−3119				
.65	.29859		−298	.2987	11.2516	1.05121
		−3417				
.70	.26442		−304	.2645	14.5068	1.16157
		−3721				
0.75	0.22721		−304	0.2271	19.0115	1.27902
		−4025				
.80	.18696		−294	.1864	25.670	1.40943
		−4319				
.85	.14377		−272	.1413	36.587	1.56332
		−4591				
.90	.09786		−228	0.0945	58.062	1.76389
		−4819				
0.95	0.04967		−148		121.550	2.08475
		−4967				
1.00	0					

FORMULAS FOR MAGNETIC FORCE BETWEEN COILS 253

For values of α not greater than about 0.75 the spacing ratio δ_m for maximum force is given by the formula

$$\delta_m = \tfrac{1}{2}(1 - \tfrac{9}{10}\alpha^2 - \tfrac{1}{8}\alpha^4). \tag{198}$$

The accuracy of this expression suffices in the range for which it converges well for all except the most precise work. Curtis [100] has shown that an uncertainty of not more than a part in a thousand in δ_m gives rise to an error not exceeding a part in a million in the maximum force. This is explained by the very small change in the force for small displacements from the critical position.

Example 86: For a pair of circular filaments whose radii are in the ratio of $\alpha = 0.5$, formula (198) gives for the spacing ratio for maximum force $\delta_m = 0.3836$. The more exact value in Table 51 is 0.38353.

Calculating k^2 with $\alpha = 0.5$ and $\delta_m = 0.3836$, the value found is $k^2 = 0.83432$. To interpolate for this value in Table 49 requires the consideration of second and third differences. The value of P thus found is 9.865, so that formula (195) yields $F_{0m} = 5.352$. The more accurate value given in Table 51 is 5.35201.

Example 87: To illustrate the use of formula (196), let us calculate the value of P for $k^2 = 0.90$, that is, for $k'^2 = 1 - k^2 = 0.10$. Table 50 gives for this value

$$q_1 = 0.006585, \quad \log\left(\log_e \frac{1}{q_1}\right) = 0.70096,$$

$$1 + 12q_1 + 192q_1^2 = 1.0707,$$

$$1 - 10q_1 + 60q_1^2 = 0.9368,$$

$$12q_1(0.9368)\log_e \frac{1}{q_1} = 0.3719.$$

Therefore, by formula (196)

$$P = \frac{\pi}{16}\frac{1}{q_1}(1.0707 - 0.3719) = 20.83.$$

The value given in Table 49 is 20.852.

Example 88: To calculate P for distant circular filaments for which $k^2 = 0.08$. Corresponding to this value, Table 50 gives, using k^2 in place of k'^2 and q in place of q_1, the values

$$q = 0.005211, \quad \log q = \bar{3}.71693,$$

$$1 + 20q^2 = 1.00054, \quad \log q^{\frac{1}{2}} = \bar{6}.29232,$$

and $P = 0.003717$, which agrees exactly with the value given in Table 49. Thus, in cases where k^2 is so small that interpolation from Table 49 is not satisfactory, formula (197) furnishes an easy solution.

Force between Two Coaxial Coils of Rectangular Cross Section.

For this case the formulas for coaxial circular filaments are basic. If the cross sectional dimensions of each coil are small compared with its mean radii, the force between them is given to a good approximation by the formula

$$F = N_1 N_2 i_1 i_2 F_0 \tag{199}$$

in which i_1 and i_2 are the currents in abamperes and N_1 and N_2 are the numbers of turns on the two coils.

If the currents are I_1 and I_2 in amperes, this formula may be written in the forms

$$F = N_1 N_2 \frac{I_1 I_2}{100} F_0 \text{ dynes,}$$

$$= \frac{N_1 N_2 I_1 I_2 F_0}{98{,}000} \text{ grams weight,}$$

$$= 2.25(10^{-8}) I_1 I_2 N_1 N_2 F_0 \text{ lb. weight.} \tag{200}$$

The calculation of F_0 is made by the methods of the preceding section.

If the cross sectional dimensions are not small compared with the mean radii of the coils and the distance between their planes, or if a high degree of precision is desired, each coil may be divided into sections and each section may be replaced by a circular filament using the method of Lyle (page 12) to calculate the radii and spacings of the equivalent filaments.

Example 89: For two of the coils of the Bureau of Standards current balance the turns numbers are $N_1 = 647$ and $N_2 = 72$, and these are wound in square channels of, respectively, 2 cm. on a side and 1 cm. on a side. The approximate values of the mean radii are, respectively, 25 cm. and 12.5 cm.

Thus the ratio of the radii is $\alpha = 0.5$ and the spacing ratio for the maximum force $\delta_m = 0.38353$. The coils, joined in series, carry 0.7676 amperes.

For this case. $k^2 = \dfrac{2}{2.3971} = 0.83434$. Interpolating in Table 49, $P = 9.8677$, and from formula (195), $F_0 = 5.352$. Substituting in formula (200) the force is found to be equal to the weight of 1.499 grams.

To estimate the correction to take into account the finite cross sections of the coils, each coil was imagined to be divided into four sections. Each section was replaced by its equivalent circular filament and the force between the coils per unit current calculated by averaging the values of F_0, calculated for each pair of filaments, one in each coil. The corrected value of F_0 for these coils was 50 parts in 1,000,000 less than the value for the central filaments. Therefore, in general, for the spacing for maximum force, it will suffice merely to replace the coils by their central filaments.

Direction of the Force. If the currents in the two filaments are in the same direction around the common axis, the force between the two filaments is one of attraction. Starting with very small values when the filaments are widely separated, the force builds up in value as they approach, reaches a maximum at a certain distance, depending upon the ratio of the radii, as has been discussed above, and becomes zero when the filaments are in the same plane. This is in agreement with Maxwell's rule of maximum flux linkage.

The equilibrium in this position is stable, since for any displacement a force appears tending to resist the displacement.

If the currents in the two filaments are in opposite directions around the axis, the force between them is one of repulsion. The force is zero, it is true, when the circular filaments are coplanar, but this is a position of unstable equilibrium.

Force between a Solenoid and a Coaxial Circular Filament. The solenoid is assumed to have a winding density n_1 and an axial length x. The radii are a and A, A being the larger. It is, however, immaterial whether the filament has a radius smaller or larger than that of the solenoid.

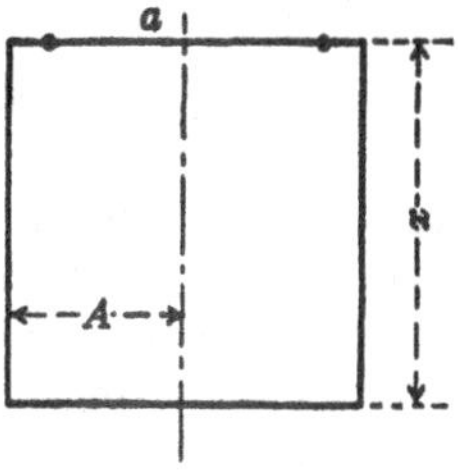

Case 1. Center of Circle at Center of End Face of Coil. The filament is in the end plane of the solenoid.

The force F_0 in dynes when one abampere flows in solenoid and coaxial filament is

$$F_0 = n_1[m(0) - m(x)] \quad (201)$$

in which $m(x)$ is the mutual inductance in abhenrys of two coaxial circles of radii a and A with a distance x between their planes, and $m(0)$ is the corresponding value when the circular filaments are coplanar. These values of mutual inductance are to be calculated by formula (77) and Table 13 on page 79. If currents i_1 and i_2 in abamperes flow in the solenoid and filament, the force between them is

$$F = i_1 i_2 F_0 \text{ dynes.}$$

Fig. 63

If the currents are in the same direction around the common axis, the force is one of attraction: the filament is attracted toward the median plane of the solenoid. Evidently the force as given by formula (201) is greater, the longer the solenoid. Furthermore, the force on the filament is a maximum when it is placed in the end plane of the solenoid.

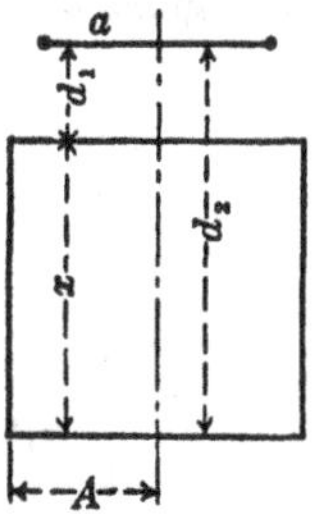

Case 2. Filament Outside the Solenoid. The plane of the filament is separated from the end plane of the solenoid by a distance d_1. Placing $d_2 = x + d_1$, the formula for the force is now

$$F = i_1 i_2 n_1 [m(d_1) - m(d_2)] \text{ dynes,} \quad (202)$$

Fig. 64

in which $m(d_1)$ and $m(d_2)$ are the mutual inductances of circular filaments of radii a and A with distances between their planes of d_1 and d_2, respectively. If the currents are in the same direction around the common axis, the force is directed toward the center of the solenoid. The force approaches zero as d_1 is increased, the values of $m(d_1)$ and $m(d_2)$ becoming smaller and smaller and approaching equality.

Case 3. Filament Inside the Solenoid. In this case the force on the filament may be regarded as the resultant of two opposing forces, one due to a solenoid of length d_1 and the other and greater that due to a solenoid of length d_2 greater than d_1 with the filament in the end face of each. By formula (201) the resultant is seen to be

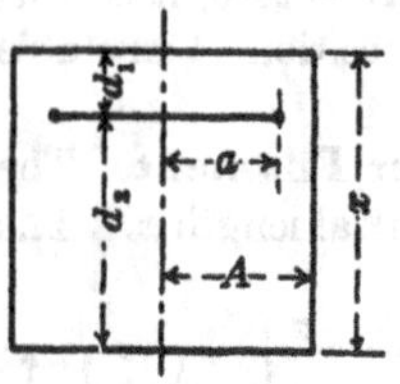

$$F = i_1 i_2 F_0 = i_1 i_2 n_1 [m(d_1) - m(d_2)], \qquad (203)$$

with the same nomenclature as before and the relation

$$d_1 + d_2 = x,$$

Fig. 65

and the understanding that d_1 is the smaller distance.

If the currents have the same direction about the common axis, the force is directed toward the median plane. For the median plane $d_1 = d_2$, and the force becomes zero.

If the circular filament has a radius larger than that of the solenoid, the formula is unchanged. The force, with currents in the same direction around the axis, tends to move the filament toward the median plane, that is, coplanar with the center of the solenoid.

Force between a Single-layer Coil and a Coaxial Coil of Rectangular Cross Section. A good approximation is obtained by the use of the formulas for the preceding cases if the equivalent length of the single-layer coil of N_1 turns and a winding density n_1 per cm. be taken for x and the mean radius of the winding is put for the radius of the solenoid.

If N_2 is the number of turns of the coil of rectangular cross section, these are assumed to act at the center of the cross section, so that for a in the formulas of the preceding section is put the radius of the central filament of the coil.

Having calculated F_0 by formula (201), (202), or (203) for these assumed dimensions, the force between single-layer coil and coil of rectangular cross section is

$$F = \frac{N_2 I_1 I_2 F_0}{100} \text{ dynes,}$$

$$= \frac{N_2 I_1 I_2 F_0}{98,000} \text{ grams weight,}$$

$$= 2.250(10^{-8}) N_2 I_1 I_2 F_0 \text{ lb. weight,} \qquad (204)$$

in which I_1 and I_2 are in amperes.

Formula (204) will be sufficiently accurate for most purposes. If, however, the dimensions of the rectangular cross section are relatively large, a more accurate value will be found by means of the method of Lyle (page 12), which replaces the coil by two or more filaments not at the center. For the important case of a square cross section of side c, the equivalent filament has

FORMULAS FOR MAGNETIC FORCE BETWEEN COILS 257

a radius $r = a\left(1 + \dfrac{1}{24}\dfrac{c^2}{a^2}\right)$, which should be used in place of a in formulas, (201), (202), or (203).

Example 90: Assume a single-layer coil wound with 80 turns with a winding density of 10 turns per centimeter to give a single-layer coil of mean diameter $A = 10$ cm.

A circular coil of 144 turns wound in a square channel, 1.2 cm. on a side, has a mean radius of $a = 5$. It is placed coaxial with the single-layer coil so that its mean plane is distant 2 cm. from the nearer end of the single-layer coil. Calculate the force between the coils when 5 amperes flow in the single-layer coil and 2 amperes in the coil with square cross section.

From the given data we have, for a first approximation,

$$n_1 = 10, \quad a = 5, \quad A = 10, \quad x = \tfrac{80}{10} = 8,$$

$$d_1 = 2, \quad \text{and} \quad d_2 = 2 + 8 = 10.$$

Formula (77) (page 77) is to be used to find the mutual inductances of two pairs of coaxial circles, namely,

$$a = 5, \quad A = 10, \quad d = 2,$$

$$a = 5, \quad A = 10, \quad d = 10.$$

The corresponding values of k'^2 in formula (78) are 0.12663 and 0.38462, and from Table 13 (page 79) $f = 0.007104$ and $f = 0.002288$, respectively. Thus, from (77),

$$m(d_1) - m(d_2) = \sqrt{50}[7.104 - 2.288]$$

$$= 34.05 \text{ abhenrys.}$$

Therefore, from (202), $F_0 = 340.5$ dynes with 1 abampere in each coil.

Assuming that the 144 turns are concentrated in the position of the central filament, the force is, by (204),

$$F = \frac{144(5)(2)(340.5)}{98{,}000},$$

$$= 5.003 \text{ grams weight.}$$

To take into account more accurately the finite cross section, we calculate the equivalent radius of the coil with square cross section by the Lyle formula (page 12). Here $c = 1.2$ and $a = 5$, so that $r = 5\left[1 + \dfrac{1}{24}\left(\dfrac{1.2}{5}\right)^2\right] = 5(1.0024)$.

Thus, for the second approximation, we have to repeat the calculation with the new value of $\alpha = \dfrac{a}{A} = 0.5012$ and the same values of d_1 and d_2 as before.

These result, for d_1 and d_2, in

$$k'^2 = 0.12592 \quad \text{and} \quad k'^2 = 0.38383,$$

which give in Table 13,

$$f = 0.0071315 \quad \text{and} \quad f = 0.0022957,$$

respectively. Multiplying these by $\sqrt{10(5.012)}$ and by 1000 to reduce to abhenrys, formula (202) yields

$$F_0 = 10\sqrt{50.12}(7.1315 - 2.2957)$$

$$= 342.3 \text{ dynes,}$$

compared with 340.5 for the central filament. That is, the correction to take into account the finite cross section is about $\frac{1}{2}$ per cent.

Force between Two Coaxial Single-layer Coils. Let the radii of the solenoids be a and A, their winding densities n_1 and n_2, their equivalent axial lengths b and B, respectively, and the distance between centers s.

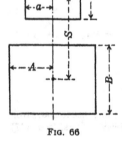

Fig. 66

If the total number of turns on the coils are N_1 and N_2, the equivalent lengths are $b = \dfrac{N_1}{n_1}$ and $B = \dfrac{N_2}{n_2}$.

Calculate the distances

$$x_1 = s + \frac{b}{2} + \frac{B}{2}, \quad x_3 = s + \frac{b}{2} - \frac{B}{2},$$

$$x_2 = s - \frac{b}{2} + \frac{B}{2}, \quad x_4 = s - \frac{b}{2} - \frac{B}{2}.$$

Then the force between the coils when each carries one abampere is

$$F_0 = n_1 n_2 [Q(x_2) - Q(x_1) - Q(x_4) + Q(x_3)], \tag{205}$$

in which $Q(x)$ is the function that enters into the calculation of the mutual inductance of a solenoid of length x and a coaxial circular filament in its end plane.

Making use of formula (103) and Table 27 (page 115), the force F in dynes between the solenoids when they carry currents of i_1 and i_2 abamperes may be written

$$F = 2\pi^2 a \alpha n_1 n_2 i_1 i_2 [x_2 \rho_2 Q_0(x_2) - x_1 \rho_1 Q_0(x_1)$$

$$- x_4 \rho_4 Q_0(x_4) + x_3 \rho_3 Q_0(x_3)], \tag{206}$$

in which $\alpha = \dfrac{a}{A}$, $\rho_n{}^2 = \dfrac{A^2}{A^2 + x_n{}^2}$, and $Q_0(x_n)$ is taken from Table 27 for the values of α and $\rho_n{}^2$ as parameters.

When one coil is partly or wholly inside the other, one or more of the distances x_n may be negative, and is so to be regarded in formula (206).

When the currents have the same directions about the common axis the force is attractive and tends to move the coils so as to cause their centers to

FORMULAS FOR MAGNETIC FORCE BETWEEN COILS 259

coincide ($s = 0$). Evidently in that position $x_4 = -x_1$ and $x_3 = -x_2$, so that in formula (206) the terms cancel in pairs, making $F = 0$ (as should be the case) since the value of $Q_0(x)$ is an even function of x.

When the two solenoids are relatively far apart, formula (206) suffers under the disadvantage that the four terms are individually much larger than their combination. In such cases a more accurate value of the force may be obtained by replacing the shorter solenoid by a number of filaments and calculating the force between each filament and the solenoid, assuming that the number of turns of the second solenoid are equally shared between the filaments. The total force is the sum of these individual forces.

Example 91: Two solenoids are assumed with the following data:

$$n_1 = 15, \quad A = 10, \quad N_1 = 225, \quad B = 15,$$
$$n_2 = 20, \quad a = 9, \quad N_2 = 300, \quad b = 15,$$

with a distance $s = 5$ between their centers.

For this case

$$\alpha = 0.9, \quad 2\pi^2 a\alpha = 16.2\pi^2,$$
$$x_1 = 20, \quad x_2 = x_3 = 5, \quad x_4 = -10,$$
$$\rho_1^2 = 0.2, \quad \rho_2^2 = \rho_3^2 = 0.8, \quad \rho_4^2 = 0.5,$$

and from Table 27 these are interpolated:

$$Q_0(x_1) = 1.01115, \quad Q_0(x_2) = Q_0(x_3) = 1.2320, \quad Q_0(x_4) = 1.07105,$$
$$\rho_1 x_1 Q_0(x_1) = 9.0440, \quad \rho_2 x_2 Q_0(x_2) = 5.5090, \quad \rho_4 x_4 Q_0(x_4) = -7.5733,$$

so that

$$F_0 = 15(20)(16.2\pi^2)[2(5.5090) - 9.0440 + 7.5733]$$
$$= 4860\pi^2(9.5473)$$
$$= 457{,}980 \text{ dynes.}$$

If each solenoid carries one ampere the force of attraction is 457,980 divided by 98,000 = 4.67 grams weight.

This result was checked by computing the forces on five equally spaced filaments replacing the solenoid of radius a, assuming 60 turns to be concentrated in each filament. The result found was $F_0 = 468{,}700$, but the force changes rapidly with the position of filaments near the plane of the upper face of the large solenoid, so that this last value is not very accurate.

Example 92: As an example of the relations that exist when the solenoids are farther apart, assume the case

$$n_1 = 10, \quad a = 20, \quad b = 4, \quad \alpha = 0.8,$$
$$n_2 = 20, \quad A = 25, \quad B = 6, \quad s = 30.$$

260 CALCULATION OF MUTUAL INDUCTANCE AND SELF-INDUCTANCE

Details of the calculation follow

$x_1 = 35$ $\qquad$ $x_2 = 31$ $\qquad$ $x_3 = 29$ $\qquad$ $x_4 = 25$

$\rho_1^2 = 0.33784$ $\qquad$ $\rho_2^2 = 0.39408$ $\qquad$ $\rho_3^2 = 0.42633$ $\qquad$ $\rho_4^2 = 0.5$

$Q_0(x_1) = 1.0253$ $\qquad$ $Q_0(x_2) = 1.0341$ $\qquad$ $Q_0(x_3) = 1.04075$ $\qquad$ $Q_0(x_4) = 1.0571$

and

$$F_0 = 32\pi^2(10)(20)[20.124 - 20.858 - 18.686 + 19.706]$$
$$= 6400\pi^2(0.286) = 18{,}030 \text{ dynes}.$$

To obtain a three-figure accuracy, each of the four terms ought to be more accurately calculated than the tabular values of the Q_0 allow.

However, better results are obtained by replacing the shorter solenoid by four equally spaced filaments with 10 turns associated with each and calculating the forces by formula (202) for solenoid and coaxial circular filament. The result is

$$F_0 = (10)(20)\sqrt{20(25)}[1.1567 + 1.0780 + 1.0053 + 0.9385]$$
$$= 18{,}690 \text{ dynes}.$$

If the shorter solenoid be assumed to be replaced by two Lyle equivalent filaments with 20 turns associated with each, the result found is 18,758. Finally, an accuracy sufficient for most purposes would in this example be found if the whole 40 turns of the shorter solenoid were assumed to be concentrated in a circular filament in its median plane. The result in that case is 18,620. It appears evident that the accurate value is not far from 18,700 dynes.

Chapter 24

HIGH FREQUENCY FORMULAS

General Considerations. The inductance formulas given in the preceding sections are based on the assumption that the current is uniformly distributed over the cross section of the wire, that is, they apply strictly only to direct current or alternating currents of low frequency.

With higher frequencies, the current density ceases to be uniform. There is a tendency for the current to forsake the interior of the cross section and to crowd into the portions nearer the surface of the wire. This is the so-called "skin effect." It may be explained by the fact that the inductance of the filaments near the center of the cross section is greater than for the filaments near the surface. Consequently, the reactance of the inner filaments is greater than that of the outer, and not only is this true, but the ratio of the reactance to the resistance is greater for the inner and increases progressively toward the center.

Electromagnetic energy enters the surface of the wire and is more and more attenuated and retarded in phase as the center is approached. At very high frequencies, the attenuation is so great that the current amplitude becomes inappreciable after the wave has penetrated into the wire only a small fraction of a millimeter. At indefinitely high frequency, the distance of penetration approaches zero: that is, there are no internal flux linkages. For this reason, the last term in the direct current formula for the inductance of a straight wire [formula (8), page 35] is lacking in the formula for the wire at indefinitely high frequencies. The maximum change in the inductance of a copper wire in going from zero frequency to very high frequencies amounts to a decrease from $2l \left[\log_e \frac{2l}{\rho} - \frac{3}{4} \right]$ to $2l \left[\log_e \frac{2l}{\rho} - 1 \right]$ abhenrys. This decrease is usually a relatively small change.

Skin effect may, however, cause a large change in the resistance of the conductor, since it effectively reduces the area of cross section of the current path. If the frequency is increased without limit, the resistance increases without limit also. At any given frequency, a cylindrical conductor may be regarded as replaced by a tube whose outer diameter is that of the wire and whose thickness is directly related to the effective distance of penetration of the current into the conductor. For the calculation of the high frequency resistance, formulas are usually given for the *resistance ratio*, that is, the ratio of the high frequency resistance to the resistance with direct current. Knowing this ratio and the resistance with direct current, the high frequency resistance is the product of the two.

The distribution of current density in an isolated wire depends upon the distance of a point in the cross section from the axis but is independent of its orientation around the axis. Accordingly, the magnetic field due to the wire at points outside the wire depends upon the distance of the point from the axis, just as in the direct current case. This is practically true also in the case of two wires both of which carry high frequency currents, provided they are sufficiently distant to exert no appreciable effect on the distribution of currents in the cross sections. In such cases the mutual inductance is essentially the same as the value with direct current. Only in the case where the distance between the wires is not very different from the cross sectional dimensions is the mutual inductance a function of the frequency. With wires close together, however, the distribution of current in one conductor is affected by the distribution of current in the other. At very high frequencies the current is not only confined to the surface of the wires but is distributed around the axis in conformity with the law of distribution of the charges in the corresponding electrostatic problem.[103] That is, if current flows in opposite directions in two parallel wires, spaced not very far apart, the current density in each wire is a maximum at the nearest points of the cross sections of the wires. This has the effect of a reduction of the effective spacing of the wires and gives rise to a reduction of the mutual inductance below the direct current value.

The limiting inductance of such a system, considered as a go-and-return circuit, may be calculated by means of the reciprocal relation between inductance and capacitance, which may be stated as follows. If the frequency is so high that the current may be considered as essentially confined to the surface of the conductors, then an electromagnetic wave is propagated along the wires with the velocity of light. Under this condition, the inductance of the go-and-return circuit, expressed in abhenrys per unit length of the circuit, is equal to the reciprocal of the capacitance of per unit length of the conductors, expressed in electrostatic units. The usefulness of this relation lies in the fact that accurate capacitance formulas are known for certain simple cases.

For example, in the important case of a coaxial cable, the current, with increasing frequency, tends to be confined to the outer surface of the inner conductor and to the inner surface of the outer conductor. If the outer tube and the inner conductor are accurately coaxial, the distribution of current is symmetrical around the common axis. If the axis of the inner conductor is somewhat displaced from the axis of the outer tube, the current density is a maximum on the nearer portions of the two conductors and the inductance of the cable is decreased. The limiting value of the inductance with any desired axial displacement may be evaluated by the reciprocal theorem. It must be remembered that with direct currents, the inductance of the cable is not affected by an axial displacement. The change with higher frequencies results from the mutual distorting effect of one conductor on another in giving rise to a variation of current density around the axes.

Except for the simple cases thus far considered, the evaluation of the changes of inductance and resistance with change of frequency are problems which offer great mathematical difficulties. Even in such a simple circuit element as a single-layer coil is this true. The mere coiling of the wire introduces effects between one turn and another so that the high frequency resistance of the coil may easily be several times the resistance of the same wire stretched out straight and measured at the same frequency. Several authors [104] have treated this case mathematically but certain simplifying assumptions have to be made to overcome mathematical difficulties, and the formulas by different authors do not agree closely with one another and with the results of measurement.

In the single-layer coil the simple skin effect of an isolated round wire is modified by the mutual effects of the turns one on the other, so that the current in the cross section is crowded toward the axis of the coil. This is equivalent to a reduction of the mean radius of the winding, which of course means a reduction of the inductance. The action of the current in one turn on the current distribution on the adjacent turns is, except for the end turns, a balanced effect, so that the effective pitch of the winding is unaltered. The ratio of the effective cross sectional diameter to the pitch is, however, reduced and this gives rise to an increase of inductance. Except for very open windings this will be a second order effect. The effective reduction in the mean radius of the winding is of major importance relative to this, but evidently no change of effective mean radius is possible greater than the radius of cross section of the wire. Except in the case of very thick wire the change in the inductance of a single-layer coil due to skin effect is bound to be small.

A further difficulty for the evaluation of the effect of frequency on the inductance and resistance of coils lies in the effect of the self-capacitance of the coil. Although it is actually a distributed capacitance, it influences the behavior of the coil, considered as a circuit unit, much as a lumped capacitance in parallel with a series arrangement of inductance and resistance. As is well known, such a parallel resonant combination offers a resistance which increases rapidly as the frequency is increased for values of frequency below the resonant frequency. In the neighborhood of the resonant frequency, the inductive reactance changes rapidly from a maximum to zero and then passes through a maximum of capacitive reactance. For still higher frequencies, the coil acts as a condenser whose capacitive reactance falls as the frequency rises still higher. Naturally, coils are not useful except for frequencies considerably below their natural (resonance) frequency, but it is evident that the resistance is increased in addition to the skin effect. A further increase results from energy losses in the dielectric between the turns. The inductance of the coil is also increased by the capacitance of the coil for frequencies below the resonant frequency, and this condition may overcome the opposite result of skin effect. Capacity effects are especially prominent with multilayer coils.

At low frequencies the resistance and inductance of a coil are altered by amounts that may be estimated by the formulas

$$R \cong R_0(1 + 2\omega^2 L_0 C),$$
$$L \cong L_0(1 + \omega^2 L_0 C),$$

in which direct current values are designated by the zero subscript, $\omega = 2\pi$ times the frequency, and C is the coil capacitance assumed as a lumped parallel capacitance. It should be emphasized that these formulas are to be used only for frequencies that make these correction factors little different from unity. The assumption of a lumped capacitance becomes increasingly inaccurate as the frequency is increased.

All in all, the high frequency resistance and inductance of coils cannot be accurately calculated and should be measured at the desired frequencies. For this purpose simple standards of forms for which the resistance and inductance may be accurately treated mathematically are necessary. Resistance measurements may be based on standard resistances consisting of fine straight wires in which skin effect is small or negligible.

In the sections that follow are included high frequency formulas for the inductance and resistance ratio for the most useful forms of standard.

Straight Cylindrical Conductor.[105]

Let l = length of the conductor in cm.,
ρ = radius of the conductor in cm.,
σ = resistivity of the material,
μ = permeability of the material ($\mu = 1$ for nonmagnetic materials),
f = frequency in cycles per second.

R' and L' are the resistance and inductance, respectively, at frequency f, while R and L are the values with direct current.

The skin effect depends upon the parameter

$$x = 2\pi\rho\sqrt{\frac{2\mu f}{\sigma}} \tag{207}$$

HIGH FREQUENCY FORMULAS 265

in which σ is expressed in absolute c.g.s. electromagnetic units. The high frequency resistance is readily obtained from the measured direct current value and a knowledge of the resistance ratio $\dfrac{R'}{R}$. The latter may be interpolated from Table 52 for the given value of x. This table is taken from the *Bureau of Standards Circular* **74**, "Radio Instruments and Measurements," page 309 (1918), and is an abridgment of Table XXII of the *Bureau of Standards Scientific Paper* 169.

It is evident that the skin effect is greater the greater the value of the permeability of the material, but this will not, in general, be accurately enough known to permit of very accurate values of the resistance ratio. Fortunately, in most practical cases the permeability of the material is closely equal to unity.

For copper at 20° C., $\sigma = 1721$, and the parameter x may be written as

$$x_c = 0.2142\rho\sqrt{f} = m\rho. \tag{208}$$

To facilitate the calculation of x_c, Table 53 [106] will be found useful. It gives the value x_0 for copper wire 1 mm. in diameter at different frequencies. For a copper wire of diameter δ mm. the value of x_c is obtained by multiplying x_0 by δ. (Attention should be paid to the fact that here the diameter rather than the radius is used, and the millimeter is used rather than the centimeter. The relation between δ and ρ in formula (208) is $\delta = 20\rho$.) The range of Table 53 may be extended by remembering that x_c is proportional to the square root of the frequency. For example, the value of x_c at 10 megacycles per second is 10 times as great as the value at 100 kilocycles, etc.

For very high frequencies

$$\frac{R'}{R} \cong \frac{x}{2\sqrt{2}} \tag{209}$$

which for copper becomes [107]

$$R' = 4.15\,\frac{1}{\rho}\,\sqrt{f}\cdot 10^{-8} \text{ ohms per cm.} \tag{210}$$

For high frequency resistance measurements, resistance standards constructed of a short length of resistance wire of small diameter are useful. For these the resistance ratio may be calculated by formula (207) and Table 52, using the proper resistivity value of the resistance alloy employed for the standard. Since the parameter x varies inversely as the square root of the resistivity, the ratio $\dfrac{R'}{R}$ may be kept little different from unity, if only the wire thickness is chosen small. Table 54 [108] gives the maximum diameter of wires of different materials, used at different frequencies, if the ratio is to be no more than 1 per cent greater than unity.

266 CALCULATION OF MUTUAL INDUCTANCE AND SELF-INDUCTANCE

TABLE 52. HIGH FREQUENCY RESISTANCE AND INDUCTANCE OF STRAIGHT WIRES

z	$\frac{R'}{R}$	Δ	T	Δ	z	$\frac{R'}{R}$	Δ	T	Δ	z	$\frac{R'}{R}$	Δ	T	Δ
0	1.0000		1.0000		5.2	2.114		0.5351		14.0	5.209		0.2016	
		3		−2			+.070		−194			0.177		−169
0.5	1.0003		0.9998		5.4	2.184		.5157		14.5	5.386		.1947	
		4		1			70		181			.176		−165
.6	1.0007		.9997		5.6	2.254		.4976		15.0	5.562		.1882	
		5		3			70		167			0.353		−117
.7	1.0012		.9994		5.8	2.324		.4809		16.0	5.915		.1765	
		9		5			70		157			.353		104
.8	1.0021		.9989		6.0	2.394		0.4652		17.0	6.268		.1661	
		13		6			69		−146			.353		92
0.9	1.0034		.9983		6.2	2.463		.4506		18.0	6.621		.1569	
		18		9			70		138			.353		82
1.0	1.005		0.9974		6.4	2.533		.4368		19.0	6.974		.1487	
	0.003			−12			+.070		129			.354		74
1.1	1.008		.9962		6.6	2.603		.4239		20.0	7.328		0.1413	
		3		16			70		122			0.353		−67
1.2	1.011		.9946		6.8	2.673		.4117		21.0	7.681		.1346	
		4		19			70		115			.353		61
1.3	1.015		.9927		7.0	2.743		0.4002		22.0	8.034		.1285	
		5		25			70		−109			.353		56
1.4	1.020		.9902		7.2	2.813		.3893		23.0	8.387		.1229	
		6		31			71		103			.354		51
1.5	1.026		0.9871		7.4	2.884		.3790		24.0	8.741		.1178	
	0.007			−37			.070		98			.353		47
1.6	1.033		.9834		7.6	2.954		.3692		25.0	9.094		0.1131	
		9		44			70		93			0.353		−44
1.7	1.042		.9790		7.8	3.024		.3599		26.0	9.447		0.1087	
		10		51			70		88			0.70		−77
1.8	1.052		.9739		8.0	3.094		0.3511		28.0	10.15		.1010	
		12		59			71		−85			.71		68
1.9	1.064		.9680		8.2	3.165		.3426		30.0	10.86		.0942	
		14		−69			70		80			.71		−58
2.0	1.078		.9611		8.4	3.235		.3346		32.0	11.57		.0884	
	0.033			−163			.071		77			.70		52
2.2	1.111		.9448		8.6	3.306		.3269		34.0	12.27		.0832	
		41		200			70		73			.71		47
2.4	1.152		.9248		8.8	3.376		.3196		36.0	12.98		.0785	
		49		235			70		70			.71		41
2.6	1.201		.9013		9.0	3.446		0.3126		38.0	13.69		.0744	
		55		268			71		−68			.71		37
2.8	1.256		.8745		9.2	3.517		.3058		40.0	14.40		0.0707	
		62		293			.070		64			0.70		−34
3.0	1.318		0.8452		9.4	3.587		.2994		42.0	15.10		.0673	
	0.067			−312			71		62			.71		30
3.2	1.385		.8140		9.6	3.658		.2932		44.0	15.81		.0643	
		71		322			70		59			.71		28
3.4	1.456		.7818		9.8	3.728		.2873		46.0	16.52		.0615	
		73		325			.071		−57			.70		26
3.6	1.529		.7493		10.0	3.799		0.2816		48.0	17.22		.0589	
		74		320			.176		−134			0.71		−23
3.8	1.603		.7173		10.5	3.975		.2682		50	17.93		0.0566	
		75		310			.176		120			3.54		−95
4.0	1.678		0.6863		11.0	4.151		.2562		60	21.47		.0471	
	0.074			−295			.176		110			3.53		67
4.2	1.752		.6568		11.5	4.327		.2452		70	25.00		.0404	
		74		279			.177		102			3.54		50
4.4	1.826		.6289		12.0	4.504		0.2350		80	28.54		.0354	
		73		261			.176		93			3.53		40
4.6	1.899		.6028		12.5	4.680		.2257		90	32.07		.0314	
		72		243			.176		87			3.54		−31
4.8	1.971		.5785		13.0	4.856		.2170		100	35.61		0.0283	
	0.072			−225			.177		−80						
5.0	2.043		0.5500		13.5	5.033		0.2090		∞	∞		0	

HIGH FREQUENCY FORMULAS

TABLE 53. VALUES OF z_0 FOR COPPER WIRE 1 MM. DIAMETER, FREQUENCIES 1 TO 100 KC.

f_{kc}	z_0	Δ	f_{kc}	z_0	Δ	f_{kc}	z_0	Δ	f_{kc}	z_0	Δ	f_{kc}	z_0	Δ
1.0	0.3387		5.0	0.7573		25	1.6934		50	2.3948		75	2.9331	
		165			370			335			239			195
1.1	.3552		5.5	.7943		26	1.7269		51	2.4187		76	2.9526	
		158			353			329			236			193
1.2	.3710		6.0	.8296		27	1.7598		52	2.4423		77	2.9719	
		152			339			323			233			193
1.3	.3862		6.5	.8635		28	1.7921		53	2.4656		78	2.9912	
		145			326			317			232			191
1.4	.4007		7.0	.8961		29	1.8238		54	2.4888		79	3.0103	
		141			314			312			229			189
1.5	0.4148		7.5	0.9275		30	1.8550		55	2.5117		80	3.0292	
		136			304			307			227			189
1.6	.4284		8.0	.9579		31	1.8857		56	2.5344		81	3.0481	
		132			295			302			226			188
1.7	.4416		8.5	0.9874		32	1.9159		57	2.5570		82	3.0669	
		128			286			297			223			186
1.8	.4544		9.0	1.0160		33	1.9456		58	2.5793		83	3.0855	
		124			279			292			222			186
1.9	.4668		9.5	1.0439		34	1.9748		59	2.6015		84	3.1041	
		122			271			289			219			184
2.0	0.4790		10.0	1.0710		35	2.0037		60	2.6234		85	3.1225	
		233			523			284			218			183
2.2	.5023		11	1.1233		36	2.0321		61	2.6452		86	3.1408	
		224			499			280			216			182
2.4	.5247		12	1.1732		37	2.0601		62	2.6668		87	3.1590	
		214			479			277			214			181
2.6	.5461		13	1.2211		38	2.0878		63	2.6882		88	3.1771	
		206			461			273			212			180
2.8	.5667		14	1.2672		39	2.1151		64	2.7094		89	3.1951	
		199			445			269			209			179
3.0	0.5866		15	1.3117		40	2.1420		65	2.7305		90	3.2130	
		192			430			266			210			178
3.2	.6058		16	1.3547		41	2.1686		66	2.7515		91	3.2308	
		187			417			263			207			177
3.4	.6245		17	1.3964		42	2.1949		67	2.7722		92	3.2485	
		181			405			260			206			176
3.6	.6426		18	1.4369		43	2.2209		68	2.7928		93	3.2661	
		176			394			257			205			175
3.8	.6602		19	1.4763		44	2.2466		69	2.8133		94	3.2836	
		172			383			253			203			174
4.0	0.6774		20	1.5146		45	2.2719		70	2.8336		95	3.3010	
		167			374			251			202			174
4.2	.6941		21	1.5520		46	2.2970		71	2.8538		96	3.3184	
		163			366			249			200			172
4.4	.7104		22	1.5886		47	2.3219		72	2.8738		97	3.3356	
		160			357			246			199			172
4.6	.7264		23	1.6243		48	2.3465		73	2.8937		98	3.3528	
		156			349			243			197			170
4.8	.7420		24	1.6592		49	2.3708		74	2.9134		99	3.3698	
		153			342			240			197			170
5.0	0.7573		25	1.6934		50	2.3948		75	2.9331		100	3.3868	

268 CALCULATION OF MUTUAL INDUCTANCE AND SELF-INDUCTANCE

TABLE 54. MAXIMUM DIAMETER OF CONDUCTORS IN CM. FOR RESISTANCE RATIO 1.01

Frequency ÷ 10^6	0.1	0.2	0.4	0.6	0.8	1.0	1.2	1.4	1.6	1.8	2.0	3.0
Copper	0.0356	0.0251	0.0177	0.0145	0.0125	0.0112	0.0102	0.0095	0.0089	0.0084	0.0079	0.0065
Silver	.0345	.0244	.0172	.0141	.0122	.0109	.0099	.0092	.0086	.0082	.0077	.0063
Gold	.0420	.0297	.0210	.0172	.0149	.0133	.0121	.0112	.0105	.0099	.0094	.0077
Platinum	.1120	.0793	.0560	.0457	.0396	.0354	.0323	.0300	.0280	.0264	.0250	.0205
Mercury	.264	.187	.132	.1080	.0936	.0836	.0763	.0706	.0661	.0623	.0591	.0483
Manganin	.1784	.1261	.0892	.0729	.0631	.0564	.0515	.0477	.0446	.0420	.0399	.0325
Constantan	.1892	.1337	.0946	.0772	.0664	.0598	.0546	.0506	.0473	.0446	.0423	.0345
German Silver	.1942	.1372	.0970	.0792	.0692	.0614	.0560	.0518	.0485	.0458	.0434	.0354
Graphite	0.765	0.541	0.383	0.312	.271	.242	.221	.204	.191	.180	.171	0.140
Carbon	1.60	1.13	0.801	.654	.566	.506	.462	.428	.400	.377	.358	0.292
Iron (μ = 1000)	0.00263	.00186	.00131	.00108	.00094	.00083	.00076	.00070	.00066	.00062	.00059	.00048
Iron (μ = 500)	.00373	.00264	.00187	.00152	.00132	.00118	.00108	.00100	.00093	.00088	.00084	.00068
Iron (μ = 100)	0.00838	.00590	.00418	.00340	.00295	.00264	.00241	.00223	.00209	.00197	.00186	0.00152

The high frequency inductance of the straight isolated wire is given by

$$L' = 2l \left[\log_e \frac{2l}{\rho} - 1 + \frac{\mu}{4} T \right] \text{ abhenrys,} \qquad (211)$$

in which the function T may be taken from Table 52 for the same parameter x as is used in finding the resistance. For very high frequencies, this function approaches zero, so the limiting inductance is

$$L'_\infty = 2l \left[\log_e \frac{2l}{\rho} - 1 \right] \text{ abhenrys.} \qquad (212)$$

Since the direct current inductance for a copper wire is

$$L = 2l \left[\log_e \frac{2l}{\rho} - 1 + \frac{\mu}{4} \right],$$

the maximum fractional decrease of the inductance, as the frequency is indefinitely increased, is for copper wire,

$$\left(\frac{\Delta L}{L} \right)_\infty = - \frac{1}{4 \log_e \frac{2l}{\rho} - 3}. \qquad (213)$$

Values [109] are given in Table 55 for different values of $\frac{2l}{\rho}$. In general, for lower and moderate frequencies, the fractional decrease of inductance is given by

$$\frac{\Delta L}{L} = - \left(\frac{\Delta L}{L} \right)_\infty (1 - T). \qquad (214)$$

Example 93: The high frequency constants will be found for a copper wire 200 cm. long and of diameter 0.25 cm. at a frequency of 500 kc. per second.

From Table 53 is found for a copper wire 0.1 cm. in diameter at 5 kc. per second the value $x_0 = 0.7573$, so that at 500 kc. $x_0 = 7.573$. For the wire in question, therefore, we find $x_5 = 2.5 (7.573) = 18.93$. Using this value for x in Table 52, there is found for the resistance ratio $\frac{R'}{R} = 6.949$. The high frequency resistance is readily found by multiplying the easily calculated direct current resistance of the wire by this number.

For this wire the maximum fractional decrease of inductance $\left(\frac{\Delta L}{L} \right)_\infty$ due to skin effect, as the frequency is indefinitely increased, is obtained from Table 55, using the value $\frac{2l}{\rho} = 3200$. The value is -0.03415. For the parameter $x = 18.93$, calculated above, Table 52 yields $T = 0.1493$, so that from formula (214)

$$\frac{\Delta L}{L} = - \left(\frac{\Delta L}{L} \right)_\infty (1 - 0.1493) = -0.02905.$$

270 CALCULATION OF MUTUAL INDUCTANCE AND SELF-INDUCTANCE

TABLE 55. LIMITING FRACTIONAL CHANGE OF INDUCTANCE WITH FREQUENCY

Single Wire			Parallel Wires			Circular Rings		
$2l/\rho$	$(\Delta L/L)_\infty$	Δ	d/ρ	$(\Delta L/L)_\infty$	Δ	$8a/\rho$	$(\Delta L/L)_\infty$	Δ
50	0.0791	−143	5	0.1344	−120			
100	.0648	−98	6	.1224	−86	100	0.0876	−171
200	.0550	−45	7	.1138	−65	200	.0705	−73
300	.0505	−28	8	.1073	−51	300	.0632	−43
400	.0477	−19	9	.1022	−44	400	.0589	−29
500	0.0458	−15	10	0.0978	−207	500	0.0560	−22
600	.0443	−12	20	.0771	−86	600	.0538	−17
700	.0431	−10	30	.0685	−50	700	.0521	−14
800	.0421	−8	40	.0635	−34	800	.0507	−12
900	.0413	−7	50	0.0601	−25	900	.0495	−10
1000	0.0406	−41	60	.0576	−20	1000	0.0485	−58
2000	.0365	−21	70	.0556	−16	2000	.0427	−27
3000	.0344	−12	80	.0540	−14	3000	.0400	−18
4000	.0332	−10	90	.0526	−11	4000	.0382	−13
5000	0.0322	−8	100	0.0515	−64	5000	0.0369	−9
6000	.0314	−6	200	.0451	−31	6000	.0360	−8
7000	.0308	−5	300	.0420	−19	7000	.0352	−7
8000	.0303	−4	400	.0401	−14	8000	.0345	−5
9000	.0299	−3	500	0.0387	−11	9000	.0340	−5
10000	0.0296	−23	600	.0376	−8	10000	0.0335	−28
20000	.0273	−12	700	.0368	−7	20000	.0307	−15
30000	.0261	−7	800	.0361	−7	30000	.0292	−9
40000	.0254	−6	900	.0354	−5	40000	.0283	−7
50000	0.0248	−4	1000	0.0349	−31	50000	0.0276	−6
60000	.0244	−4	2000	.0318	−15	60000	.0270	−4
70000	.0240	−3	3000	.0303	−10	70000	.0266	−4
80000	.0237	−2	4000	.0293	−8	80000	.0262	−3
90000	.0235	−3	5000	0.0285	−6	90000	.0259	−3
100000	0.0232		6000	.0279	−4	100000	0.0256	
1000000	0.0191		7000	.0275	−4	1000000	0.0207	
			8000	.0271	−4			
			9000	.0267	−3			
			10000	0.0264				

HIGH FREQUENCY FORMULAS

Isolated Tubular Conductor. In general, the use of a tubular conductor in place of a solid wire is to be recommended in cases where a marked skin effect is to be expected, since the central portions of a solid conductor will, in that case, conduct little of the current. If the thickness of the wall of the tube is small, compared with the outer radius of the tube, the resistance ratio will, at a given frequency, be smaller than that of a solid conductor of the same radius of cross section as the outer radius of the tube.

The exact formula [110] for the resistance ratio of the tube in Bessel functions is well known, and numerical calculations, although not simple, may be made by it, using suitable tables of these functions such as those of Dwight.[111] A formula that suffices for a moderate accuracy, such as will be suitable for engineering purposes, has been given by Dwight.[112] This may be put in the following convenient form.

If q = the inner radius of the tube, in cm.,
r = the outer radius,
and
t = the thickness of the tube = $r - q$,
then

$$\frac{R'}{R} = \tau\left(1 - \frac{t}{2r}\right)\left[1 + \frac{1}{2\tau}\frac{t}{r} + \frac{3}{16}\frac{1}{\tau^2}\frac{t^2}{r^2} + \cdots\right.$$

$$+ 2\epsilon^{-2\tau}\cos 2\tau \left\{1 - \frac{3}{4}\frac{1}{\tau}\frac{t^2}{qr} + \frac{3}{64}\frac{1}{\tau^2}\frac{t^2}{r^2}\left(7 - 6\frac{r}{q} + 3\frac{r^2}{q^2}\right) + \cdots\right\}$$

$$+ 2\epsilon^{-2\tau}\sin 2\tau \left\{1 + \frac{3}{64}\frac{1}{\tau^2}\frac{t^2}{r^2}\left(7 - 6\frac{r}{q} + 3\frac{r^2}{q^2}\right) + \cdots\right\}$$

$$\left. + \text{terms in } \epsilon^{-4\tau} + \cdots \right] \qquad (215)$$

in which the parameter is $\tau = \frac{mt}{\sqrt{2}}$ and $m = 2\pi\sqrt{\frac{2\mu f}{\sigma}}$, which for copper has the value $0.2142\sqrt{f}$. Table 53 can be used to calculate τ by the relation $\tau = 20x_0$ times $t/\sqrt{2}$.

Formula (215) is of course more convergent, the greater the parameter τ. Furthermore, terms in powers of $\frac{t}{r}$ and terms with higher exponents of the base have been neglected. This formula is Dwight's formula 26 on page 174 of the reference quoted.[112] For values of τ greater than 3, the exponential terms may be neglected and the first line of (215) will suffice. In the region of τ between about 0.5 and unity, the formula is only approximate and it is best to obtain the resistance ratio graphically from a curve calculated for higher values extrapolated to the value unity at about $\tau = 0.5$. For greater accuracy the Bessel functions formula should be used.

Since, in addition to the frequency, the ratio of the thickness to the outer radius of the tube enters as a parameter, a table of values for the resistance ratio would not be simple. However, for an accuracy in the resistance ratio to a few per cent in the important case of a thin tube $\left(\dfrac{t}{r} \text{ small}\right)$, the data of Table 56 may be used. These have been obtained from Dwight's formula for the limiting case $\dfrac{t}{r} = 0$. The value F_0 interpolated from this table, multiplied by the factor $\left(1 - \dfrac{t}{2r}\right)$, gives the resistance ratio for the tube.

Table 56 is essentially in agreement with the ordinates of a curve calculated by Whinnery [113] and published also in a paper by Race and Larrick.[107]

TABLE 56. VALUES OF FACTOR F_0 FOR APPROXIMATE CALCULATIONS BASED ON (215) AND (227)

τ	F_0	τ	F_0	τ	F_0
0.5	1.00	1.5	1.37	2.5	2.48
0.6	1.005	1.6	1.46	2.6	2.59
0.7	1.02	1.7	1.56	2.7	2.70
0.8	1.04	1.8	1.67	2.8	2.80
0.9	1.07	1.9	1.78	2.9	2.91
1.0	1.10	2.0	1.90	3.0	3.01
1.1	1.14	2.1	2.01		
1.2	1.19	2.2	2.13		
1.3	1.24	2.3	2.25		
1.4	1.30	2.4	2.36		
1.5	1.37	2.5	2.48		

For large skin effect, values of τ greater than about 3, the first two terms of the first line of (215) will suffice, and if the direct current resistance per unit length of the wire $R = \dfrac{\sigma}{\pi(r^2 - q^2)} = \dfrac{\sigma}{2\pi t r \left(1 - \dfrac{1}{2}\dfrac{t}{r}\right)}$ is substituted, the high frequency resistance per unit length of the conductor may be written

$$R' = \frac{\sqrt{f\sigma}}{r}\left(1 + \frac{1}{2}\frac{1}{\tau}\frac{t}{r}\right) \text{ ohms per cm.} \qquad (216)$$

HIGH FREQUENCY FORMULAS

For copper this becomes

$$R' = \frac{4.15}{r} 10^{-8} \left(1 + \frac{1}{2}\frac{1}{\tau}\frac{t}{r}\right) \sqrt{f} \text{ ohms per cm.,} \qquad (217)$$

and for still larger values of τ, the formula [107]

$$R' = \frac{4.15}{r} \sqrt{f} \cdot 10^{-8} \text{ ohms per cm.} \qquad (218)$$

suffices.

The change in inductance of a tubular conductor caused by skin effect is relatively smaller than with a solid conductor. At very high frequencies the same limiting formula as for the solid conductor may be used if the radius of the tube be placed for the radius of the solid wire in (212). That is,

$$L_\infty = 2l \left[\log_e \frac{2l}{r} - 1 \right] \text{ abhenrys.} \qquad (219)$$

If the tube is very thin, the internal linkages of the flux with the material of the conductor may be safely neglected. For thicker tubes, a correction for internal linkages may be made by imagining the tube to be replaced by one having a thickness equal to the equivalent penetration of the high frequency current, $\delta_1 = \frac{\sqrt{2}}{m}$. The inductance of this equivalent tube is calculated by the direct current formula (11) (page 36) for a tube of radii $\rho_1 = r$ and $\rho_2 = r\left(1 - \frac{\sqrt{2}}{mr}\right)$. In (11), then, $\log_e \zeta$ is taken from Table 4, using the value $\left(1 - \frac{\sqrt{2}}{mr}\right)$ for $\frac{\rho_2}{\rho_1}$.

Example 94: Assume the case of a tubular conductor, 200 cm. long, with an outer radius of 1 cm. and a thickness of 0.1 cm. Thus $r = 1.0$, $t = 0.1$, and $q = 0.9$.

For a frequency of 25,000 cycles per second, Table 53 gives $x_0 = 1.6934$. Multiply this by 20, and $m = 33.87$ is found. Thus $\tau = \frac{33.87(0.1)}{\sqrt{2}} = 2.395$. Interpolating in Table 56 for this value of τ, $F_0 = 2.358$, so that the resistance ratio of the conductor is $\frac{R'}{R} = 2.358 (1 - \frac{1}{2}0.1) = 2.28$. The direct current resistance is 2.88×10^{-6} ohms per cm., so that, at 25,000 cycles per second, the resistance is 6.57×10^{-6} ohms per cm.

If the frequency is raised to one megacycle per second, then $m = 0.2142\sqrt{10^6} = 214.2$ and $\tau = 15.15$, $\frac{1}{\tau}\frac{t}{r} = 0.0066$, and by formula (215),

$$\frac{R'}{R} = 15.15 \, (1 - 0.05)(1.0066) = 14.48.$$

Accordingly, the resistance at one megacycle is 14.48 times 2.88×10^{-6} or 4.17×10^{-5} ohms per cm. The approximate formula (218) yields the value $R' = 4.15\sqrt{f}\cdot 10^{-8} = 4.15\cdot 10^{-5}$ ohms per cm. directly, and gives sufficient accuracy for this case.

The inductance of the conductor at the higher frequency may safely be calculated by the limiting formula (219).

$$L_\infty = 400\ (\log_e 400 - 1) = 400(4.992)\text{ abhenrys.}$$

The direct current value for the tube, expressed in abhenrys, may be derived by (11) and Table 4, using the values $\rho_1 = r$ and $\rho_2 = q$ so that $\dfrac{\rho_2}{\rho_1}$ in Table 4 is 0.9.

$$L = 400[4.992 + 0.0333] = 400[5.025]\text{ abhenrys,}$$

or 6.6 parts in 1000 greater than the limiting high frequency value.

At 25,000 cycles per second, the equivalent tube has a thickness of $\delta_1 = \dfrac{\sqrt{2}}{m} = \dfrac{\sqrt{2}}{33.87} = 0.0417$ cm. Using the direct current formula for this tube, $\dfrac{\rho_2}{\rho_1} = (1 - .0417) = 0.9583$ in Table 4, and $L = 400\ (4.992 + .0139) = 400\ (5.006)$, or 4.2 parts in 1000 less than the value at limiting frequencies.

Go-and-return Circuit of Round Wire. The wires are assumed to have equal cross sections and to be placed parallel with a spacing d between their centers. The current flows in opposite directions in the two wires.

The inductance of the go-and-return circuit is given by twice the inductance of one wire minus twice their mutual inductance. If the wires are far enough apart for negligible disturbing effect of the current in one wire on the current distribution of current in the other, then the inductance of the wires per cm. length of the circuit is

$$\frac{L'}{l} = 4\left[\log_e \frac{d}{\rho} + \frac{\mu}{4} T\right]\text{ abhenrys per cm.,} \qquad (220)$$

in which T is obtained from Table 52 for the parameter x calculated as in the previous sections.

Since the direct current value of the inductance per unit length is

$$\frac{L}{l} = 4\left[\log_e \frac{d}{\rho} + \frac{\mu}{4}\right]\text{ abhenrys per cm.,} \qquad (221)$$

we find for the fractional change of inductance for copper wires, referred to the direct current value,

$$\frac{\Delta L}{L} = -\left(\frac{\Delta L}{L}\right)_\infty \cdot (1 - T), \qquad (222)$$

in which the limiting fractional change $\left(\dfrac{\Delta L}{L}\right)_\infty$ with indefinitely high frequency may be obtained from column 6 of Table 55 for the given value of $\dfrac{d}{\rho}$.

The *resistance ratio*, in the absence of proximity effect, is calculated just as for an isolated round wire as described in the preceding section.

For wires where the spacing is little greater than the diameter of the wires, the distribution of the current in the wires is appreciably altered by the proximity of the wires. This situation has the effect of increasing the resistance ratio above the value for the isolated wires. Proximity effect brings about also a reduction in the inductance of the wires, in addition to that caused by skin effect, in the case of the wires with no proximity effect.

Formulas enabling the resistance ratio to be accurately calculated to take account of the skin effect have been derived.[114] They give the resistance as a series of terms involving Bessel functions and will not be given here. It will suffice to present numerical data for certain cases published by Dwight[115] in the form of curves. These are summarized in tabular form in Table 57 and give the factor that must be multiplied into the resistance ratio, calculated for isolated wires, to give the resistance ratio with the proximity effect taken into account. The argument of the table is the parameter x in the formula (208) calculated for copper wires, and the factor is given for a number of different values of the parameter:

$$\frac{d}{\rho} = \frac{\text{spacing between centers}}{\text{radius of the wire}}.$$

The change of *inductance* due to proximity effect is small and not simple to calculate. However, the maximum magnitude of the proximity effect may be evaluated for different spacing ratios, under the assumption of indefinitely high frequencies.

Assuming the current to flow entirely in the periphery of the cross section but to be symmetrically distributed about the axis of the wire (no proximity effect), the limiting value of the inductance per centimeter length of the circuit,

$$\frac{L'}{l} = 4 \log_e \frac{d}{\rho} \text{ abhenrys per cm.} \qquad (223)$$

Using the reciprocal relation of inductance and capacitance and the known formula [116] for the capacitance of two parallel wires, there results

$$\frac{L_\infty'}{l} = 4 \cosh^{-1} \frac{d}{2\rho} \qquad (224)$$

$$= 4 \left(\log_e \frac{d}{\rho} - \frac{\rho^2}{d^2} - \frac{3}{2} \frac{\rho^4}{d^4} - \cdots \right) \text{abhenrys per cm.} \qquad (225)$$

The limiting difference between the inductance calculated neglecting proximity effect, formula (223), and that taking proximity effect into account may be evaluated in a given case by use of these formulas employing tables

276 CALCULATION OF MUTUAL INDUCTANCE AND SELF-INDUCTANCE

of hyperbolic functions and logarithms. However, unless the spacing of the wires is very small, formula (225) may be employed together with (223) to show that the limiting fractional change of inductance given by Table 55 should merely be multiplied by the correction factor $\left[1 + 4\dfrac{\rho^2}{d^2} + 6\dfrac{\rho^4}{d^4}\right]$ to obtain the maximum fractional change of inductance occasioned by skin effect with the proximity effect taken into account.

Example 95: Consider a go-and-return circuit formed by two equal parallel wires of the same diameter 0.25 cm. as was assumed in Example 93. The wires are to be spaced 10 cm. between centers, and their constants are to be found at a frequency of 500 kc.

Here, $\dfrac{d}{\rho} = \dfrac{10}{0.125} = 80$. The parameter x_c is equal to 18.93 and the resistance ratio of each wire, isolated, is 6.949, just as shown in Example 93. It is evident from the data in Table 57 that for the values of the parameters $\dfrac{d}{\rho}$ and x_c which apply here the proximity correction is negligible and the resistance ratio is very closely that already calculated for each wire alone.

TABLE 57. CORRECTION FACTOR FOR PROXIMITY EFFECT PARALLEL ROUND WIRES

x_c	$d/\rho = 2.06$	$d/\rho = 2.12$	$d/\rho = 2.54$	$d/\rho = 3.36$	$d/\rho = 4$	$d/\rho = 6$	$d/\rho = 13$	$d/\rho = 36$
0	1.00	1.00	1.00	1.00	1.00	1.00	1.00	1.00
1	1.04	1.03	1.02	1.01	1.005	1.00	1.00	1.00
2	1.24	1.17	1.10	1.07	1.04	1.02	1.01	1.00
3	1.43	1.36	1.23		1.07	1.04	1.01	1.01
4	1.61	1.51	1.36		1.09	1.06	1.02	1.01
5	1.76				1.11		1.03	1.02
6	1.91				1.12		1.04	1.02
7	2.04				1.12		1.04	1.02
8	2.13				1.12		1.04	1.02
9	2.19				1.13		1.04	1.02
10					1.13		1.04	1.02

The limiting decrease in *inductance* of the circuit is given by Table 55, for the parameter $\dfrac{d}{\rho} = 80$, as 0.0540 of the direct current value. This value should be multiplied by the correction factor given above, which is here $[1 + 4(\tfrac{1}{80})^2] = 1 + \tfrac{1}{1600}$. This is a practically negligible proximity effect. At 500 kc. the value of T for the given parameter x_c is given by Table 52 as 0.1493, as in Example 93, so that the fractional decrease of the inductance at this frequency, referred to the direct current value, is

$$\frac{\Delta L}{L_0} = -0.0540(1 - 0.1493) = -0.0459.$$

At this frequency the proximity effect is still smaller than the limit already calculated, and is therefore negligible.

In the rather extreme case of a spacing of only 0.5 cm. between centers $\frac{d}{\rho} = 4$. From Table 57, it is evident that the proximity factor is little different from 1.13 for the given value of $x_c = 18.93$, so that the resistance ratio is $\frac{R'}{R} = 1.13(6.949)$ = 7.85. The proximity correction factor for the inductance is for this spacing $[1 + 4(\frac{1}{4})^2 + 6(\frac{1}{4})^4] = 1 + \frac{1}{4} + \frac{3}{128} \cong 1.27$ and this is for indefinitely high frequencies. For the frequency of the problem this factor multiplied by the value already found for the case of large spacing of the wires will give an upper limit of the inductance change for the smaller spacing.

Go-and-return Circuit of Parallel Tubular Conductors. The two conductors are supposed to consist of tubes of inner and outer radii of $(r - t)$ and r, respectively, and are to be spaced a distance d between their axes.

The *resistance ratio* of the circuit is to be calculated first, just as for a single tubular conductor. This value is to be multiplied by a correction factor to be taken from [117] Table 58 for the parameter $m\sqrt{r^2 - q^2}$ and spacing ratio of the case in question, to take proximity effect into account.

TABLE 58. CORRECTION FACTOR FOR PROXIMITY EFFECT IN PARALLEL TUBULAR CONDUCTORS

$m\sqrt{r^2 - q^2}$	$d/r = 2$	$d/r = 3$	$d/r = 4$
0.5	1.005	1.00	1.00
1.0	1.04	1.02	1.01
1.5	1.13	1.06	1.02
2.0	1.33	1.13	1.05
2.5	1.55		1.08
3.0	1.77		1.11
3.5	1.96		1.13
4.0	2.13		1.14
4.5	2.24		

The *high frequency inductance* of the circuit is equal to twice the inductance of one conductor minus twice the mutual inductance of the conductors.

The high frequency inductance of a single conductor is to be calculated by the formulas and methods already given for a tubular conductor. The mutual inductance of the tubular conductors, in the absence of proximity effect, is equal to the mutual inductance of the filaments along their axes, just as for solid conductors, that is, it is the same as for direct current. Having evaluated the inductance of the circuit ignoring proximity effect, the

278 CALCULATION OF MUTUAL INDUCTANCE AND SELF-INDUCTANCE

maximum proximity effect correction factor, which is the same as for parallel solid conductors, is to be applied.

Coaxial Cable. A coaxial cable consisting of an inner solid round wire and an outer tubular conductor will be considered.

Let a, b, and c be, respectively, the radius of cross section of the inner conductor, the inner radius of the tubular conductor and its outer radius, all expressed in centimeters, and let $t = (c - b)$ be the thickness of the tube. The current flows down one conductor and returns by the other.

Since usually the cross section of the inner conductor and the cross section of the tube will be of the same order of magnitude, the thickness of the tube will be small compared with the radii of the tube. Consequently, the skin effect in the tube will be smaller than in the inner conductor. In fact, at certain frequencies, the high frequency resistance of the tube may be little different from its direct current value while in the inner conductor a marked skin effect may exist.

High Frequency Resistance of Coaxial Cable.[118] The resistance ratio of the inner conductor is calculated just as for an isolated wire. That is, the parameter ma is to be computed by the relation

$$ma = 2\pi a \sqrt{\frac{2\mu f}{\sigma}},$$

$$= 0.2142 a \sqrt{f} \text{ for copper.} \quad (226)$$

This computation may be facilitated by the use of Table 53.

Using the value of ma in place of x in Table 52, the resistance ratio $\left(\dfrac{R'}{R}\right)$ for the inner conductor is obtained.

Skin effect has the effect in the outer conductor of causing the current density to increase toward the inner portions of the cross section, so that the current density is a maximum at the inner surface.

An exact formula for the resistance ratio of the outer conductor is known.[110] It involves Bessel functions of both the first and second kinds and will not be considered here. The formulas that follow are simpler and are sufficiently accurate for most practical purposes.

The parameter $\tau = mt/\sqrt{2}$ is to be calculated by (226) with $t/\sqrt{2}$ in place of a. If τ is not greater than about 0.5, the resistance of the outer conductor at the frequency in question is not appreciably different from its direct current value.

For larger values of τ, an asymptotic formula given by Dwight[112] [formula (27), page 175, of the reference cited] may be used. Expressed in terms of the parameter τ it reads

HIGH FREQUENCY FORMULAS

$$\left(\frac{R'}{R}\right)_{outer} = \tau\left(1 + \frac{t}{2b}\right)\left[1 - \frac{1}{2}\frac{1}{\tau}\frac{t}{b} + \frac{3}{16}\frac{1}{\tau^2}\frac{t^2}{b^2} + \cdots\right.$$

$$+ 2\epsilon^{-2\tau}\cos 2\tau\left\{1 - \frac{3}{4}\frac{1}{\tau}\frac{t^2}{bc} + \frac{3}{16}\frac{1}{\tau^2}\frac{t^2}{b^2}\left(1 + \frac{3}{4}\frac{t^2}{c^2}\right)\right\}$$

$$+ 2\epsilon^{-2\tau}\sin 2\tau\left\{1 - \frac{3}{16}\frac{1}{\tau^2}\frac{t^2}{b^2}\left(1 + \frac{3}{4}\frac{t^2}{c^2}\right)\right\}$$

$$\left. + \text{ terms in } \epsilon^{-4\tau} + \cdots \right]. \qquad (227)$$

For small values of t/b, often the case in practice, the resistance ratio may be found from the values given by this expression, for t/b equal to zero, multiplied by the factor $\left(1 + \frac{t}{2b}\right)$. For this calculation the value taken from Table 56 for the value of τ in question may be multiplied by $\left(1 + \frac{t}{2b}\right)$ to give a good approximation to the resistance ratio for the outer conductor.

Having found the resistance ratio for both conductors, the high frequency resistance per cm. of the cable length is given, to a good approximation, by the formula

$$R' = 0.548 \cdot 10^{-6}\left[\frac{1}{a^2}\left(\frac{R'}{R}\right)_{inner} + \frac{1}{2tb}\left(\frac{R'}{R}\right)_{outer}\right] \text{ ohms per cm.} \qquad (228)$$

In this the area of the cross section of the thin tubular conductor $\pi(c^2 - b^2)$ has been approximated by the value $2\pi bt$.

For still higher frequencies, where the value of ma is greater than about 20 and τ is greater than 3, the depth of penetration of the current is so small for both conductors that the formula

$$R'/l = \sqrt{\sigma\mu f}\left(\frac{1}{a} + \frac{1}{b}\right)10^{-9}$$

$$= 4.15\sqrt{f}\left(\frac{1}{a} + \frac{1}{b}\right)10^{-8} \text{ ohms per cm. for copper} \qquad (229)$$

suffices and is simple to use. Formulas (228) and (229) are given by Race and Larrick.[107]

Example 96: Calculations will be made for one of the coaxial cables for which Race and Larrick [107] give the dimensions

$$a = 0.325 \text{ cm.}, \quad b = 1.27 \text{ cm.}, \quad \text{and} \quad t = 0.025 \text{ cm.}$$

The resistance per cm. of length of the cable will be calculated for two frequencies.

280 CALCULATION OF MUTUAL INDUCTANCE AND SELF-INDUCTANCE

For $f = 25{,}000$ cycles per second, the value of ma, calculated by (226) is $ma = 11$, and by Table 52, $\left(\dfrac{R'}{R}\right)_{\text{inner}} = 4.15$. The parameter $\tau = mt/\sqrt{2}$ is equal to about 0.6, so that the resistance of the outer conductor is sensibly the same as its direct current value, that is, $\left(\dfrac{R'}{R}\right)_{\text{outer}} = 1$. Making use of formula (228), the resistance per unit length of the cable is

$$0.548 \times 10^{-6}\left\{\frac{4.15}{(0.325)^2} + \frac{1}{.05(1.27)}\right\} = 0.303 \times 10^{-4} \text{ ohms per cm.}$$

At a frequency of 315 megacycles $= 3.15 \times 10^8$, there are found $ma = 1230$ and $\tau = 67.2$. This case therefore comes well under the conditions of formula (229).

$$\frac{1}{a} + \frac{1}{b} = \frac{1}{0.325} + \frac{1}{1.27} = 3.864.$$

$$R'/l = 4.15(3.864)\sqrt{f}\cdot 10^{-8} = 28.46 \times 10^{-4} \text{ ohms per cm.}$$

This value is about 95 times the resistance at the lower frequency!

High Frequency Inductance of Coaxial Cable. For a coaxial cable in which the current flows in opposite directions in the two conductors, the inductance of the cable is equal to the self-inductance of the inner conductor plus the inductance of the outer conductor minus twice their mutual inductance. The mutual inductance of the two conductors is equal to the self-inductance L_2 of the outer conductor. If, therefore, L_1 denotes the inductance of the inner conductor, the inductance of the return circuit is $L = L_1 + L_2 - 2L_2 = L_1 - L_2$.

The inductance of the inner conductor is

$$L_1 = 2l\left[\log_e \frac{2l}{a} - 1 + \frac{\mu}{4}T\right] \text{ abhenrys.} \qquad (230)$$

Since the outer conductor is usually very thin, its internal linkages may be neglected without serious error, so that we may write

$$L_2 \cong 2l\left[\log_e \frac{2l}{b} - 1\right] \text{ abhenrys.} \qquad (231)$$

Accordingly, the inductance of the return circuit, per unit length of cable, is

$$L/l = 2\left[\log_e \frac{b}{a} + \frac{\mu}{4}T\right]$$

$$= 2\left[\log_e \frac{b}{a} + \frac{1}{4}T\right] \text{ for copper (abhenrys per cm.).} \qquad (232)$$

The function T is taken from Table 52 for the parameter x, calculated by (226) or by Table 53, using the radius a of the inner conductor.

HIGH FREQUENCY FORMULAS

For very high frequencies, where the current is confined to a thin layer at the outer surface of the inner conductor and the inner surface of the outer conductor, the limiting formula for the inductance of the return circuit is

$$L_\infty = 2l \left[\log_e \frac{b}{a} \right] \text{ abhenrys.} \tag{233}$$

With direct current, and for very low frequencies with negligible error, the inductance of a coaxial cable is not changed if the axis of one conductor is displaced from the other. In the high frequency case this is not true. The effect of a displacement d between the axes may readily be evaluated, at limiting high frequencies, by the use of the reciprocal relation between capacitance and inductance. From the known formula [119] for the capacitance per unit length of two cylinders, one within the other and with their axes displaced, we find the inductance per cm. of the return circuit formed by the two conductors:

$$L/l = 2 \cosh^{-1} \beta, \tag{234}$$

in which

$$\beta = \frac{b^2 + a^2 - d^2}{2ab}. \tag{235}$$

Formula (234) may be computed exactly by the use of tables of hyperbolic functions. For small values of the displacement as compared with the difference of the radii, the following approximate expression is satisfactory.

$$L/l \cong 2 \left[\log_e \frac{b}{a} - \log_e \left\{ 1 - \frac{d^2}{b^2 - a^2} \right\} \right] \tag{236}$$

By comparing the value of the inductance calculated by (234) or (236) with the value calculated by formula (233), for the coaxial case, the limiting effect of the displacement in reducing the inductance may be evaluated. This difference is to be regarded as the upper limit possible. At lower frequencies the effect is smaller.

Example 97: Consider the same coaxial cable treated in Example 96, that is, $a = 0.325$, $b = 1.27$, $t = 0.025$, all in cm.

At the frequency of 315 megacycles, the limiting formula (233) may safely be used. It gives $L/l = 2 \left[\log_e \frac{1.27}{0.325} \right] = 2[1.3629]$. At 25,000 cycles per second, however, the parameter $\tau = \frac{mt}{\sqrt{2}} = 0.6$, which shows that the current distribution in the outer conductor is sensibly the same as with direct current. Using the ratio $\frac{1.27}{1.295}$ for $\frac{\rho_2}{\rho_1}$ in Table 4 (page 23), $\zeta = 0.0065$, and this may be used in (11) to calculate the outer inductance as for a tube.

282 CALCULATION OF MUTUAL INDUCTANCE AND SELF-INDUCTANCE

For the inner conductor there was found $ma = 11$ in the preceding example, and for this value, used as x in Table 52, there is found $T = 0.2562$. Therefore, by formula (11),

$$L/l = 2\left[\log_e \frac{1.295}{0.325} + \frac{1}{4}(0.2562) - 0.0065\right]$$

$$= 2[1.440] \text{ abhenrys per cm.},$$

or more than 5 per cent greater than the value at the higher frequency. Formula (232) gives 2[1.427].

The direct current value of the inductance of the coaxial cable is found from formula (24) (page 42), using for ρ_1, ρ_2, and ρ_3 the values 1.295, 1.27, and 0.325, respectively. The value of $\frac{\rho_2}{\rho_1}$ is 0.9807, and for this Table 4 (page 23) gives $\zeta = 0.0065$.

$$L/l = 2\left[\log_e \frac{1.295}{0.325} + \frac{2(0.9807)^2}{1 - 0.9617}\log_e \frac{1.295}{1.27} - \frac{3}{4} + 0.0065\right]$$

$$= 2[1.3824 + 0.9794 - 0.75 + 0.0065] = 2[1.618] \text{ abhenrys per cm.}$$

Comparing this value with that for very high frequencies, the spread is seen to be more than 15 per cent of the direct current value.

If now the axes of the two conductors were displaced by 1 mm., $d = 0.1$ cm., and in (236), $1 - \frac{d^2}{b^2 - a^2} = 1 - \frac{.01}{1.507} = (1 - 0.00664)$, and $\log_e\left[1 - \frac{d^2}{b^2 - a^2}\right] = 0.00662$. Therefore,

$$L/l \text{ (displaced)} = 2[1.3629 - 0.0066] = 2[1.3563],$$

or a reduction in the inductance of only about $\frac{1}{2}$ per cent.

In the rather extreme case of $\frac{d}{b} = \frac{1}{2}$, or $d = 0.635$ cm.

$$\beta = \frac{(1.27)^2 + (0.325)^2 - (0.635)^2}{2(0.325)(1.27)} = 1.5955,$$

and by (234) $L/l = 2\cosh^{-1}\beta = 2(1.0435)$ abhenrys per cm. That is, at very high frequencies, this displacement reduces the inductance to about 0.765 of its value in the coaxial position.

REFERENCES

[1] Maxwell, *Elect. and Mag.* II, 701.
[2] Lorenz, *Wied. Ann.*, 7, 161, 1879; *Oeuvres Scient. de L. Lorenz*, Tome 2, 1, p. 196; *B. of S. Sci. Paper* 169, 118 (1912).
[3] J. Viriamu Jones, *Phil. Trans. Roy. Soc.* 182 (1891A); *Proc. Roy. Soc.* 63, 192; *Phil. Mag.*, Jan. 1889.
[4] Nagaoka, *Jour. Coll. Sci. Tokyo* 27, Art. 6 (1909); Terezawa, *Tokyo Math. Jour.* 10, Nos. 1, 2 (1916); Olshausen, *Phys. Rev.* 31, 617 (1910); *Phys. Rev.* 35, 150 (1912).
[5] *B. of S. Sci. Paper* 169, 6–32 (1912); Grover, *B. of S. Sci. Paper* 320, 538–543 (1918); Grover, *B. of S. Jour. of Res.* 1, 488–502 (1928).
[6] *B. of S. Sci. Paper* 320, 545–551 (1918).
[7] Dwight and Purssell, *Gen. Elect. Rev.* 33, 401 (1930).
[8] Weinstein, *Ann. der Phys.* 21, 329 (1884); Stefan, *Ann. der Phys.* 22, 113 (1884); Lyle, *Phil. Trans.* 213A, 421–435 (1914); and *B. of S. Sci. Paper* 320, 557 (1918).
[9] Dwight and Purssell, *Gen. Elect. Rev.* 33, 401 (1930).
[10] Rosa, *B. of S. Bull.* 2, 337 (1906); Rowland, *Collected Papers*, p. 162; and *Amer. Jour. Science* (3) XV (1878).
[11] Rayleigh, see Gray, *Absolute Meas.* II, part II, 322; *B. of S. Bull.* 2, 370 (1906); and *Sci. Paper* 169, 34.
[12] Lyle, *Phil. Mag.* 3, 310 (1902); *B. of S. Bull.* 2, 374–378 (1906).
[13] Grover, *B. of S. Sci. Paper* 455, 470 (1922); Rosa, *B. of S. Sci. Paper* 169, 41, 71 (1912).
[14] Rosa, *B. of S. Bull.* 2, 161 (1906); *B. of S. Sci. Paper* 169, 122 (1912).
[15] Maxwell, *Elect. and Mag.* II, §693; Rosa, *B. of S. Bull.* 3, 37 (1907); *B. of S. Sci. Paper* 169, 140 (1912).
[16] Maxwell, *Elect. and Mag.* II, 691–693. Maxwell gives the values of the g.m.d. for a number of different cases.
[17] Grover, *Proc. I.R.E.* 17, 2055 (1929).
[18] A. Gray gives formulas for a number of important cases, some previously given by Maxwell, and especially for parallel rectangles, in *Abs. Mea.* II, part II, 294–303. Rosa, *B. of S. Bull.* 3, 5 (1907), gives formulas for squares arranged in a row or obliquely situated. He points out in a footnote some misprints in Gray's article.
[19] A collection of geometric distance formulas is given also in *B. of S. Sci. Paper* 169, 166–170 (1912).
[20] E. B. Rosa, *B. of S. Bull.* 4, 301 (1907); *B. of S. Sci. Paper* 169, 151, formula (98) (1912).
[21] *B. of S. Sci. Paper* 169, 150, formula (94) (1912).
[22] *B. of S. Sci. Paper* 169, 159, formula (121) (1912).
[23] *B. of S. Sci. Paper* 169, 151, formula (100) (1912).
[24] Everitt, *Communication Engineering*, McGraw-Hill, 1st ed., pp. 113, 126.
[25] O. Schurig, *Trans. A.I.E.E.*, pp. 479–87 (1941).
[26] *B. of S. Sci. Paper* 169, 158 (1912).

[27] A. Russell, *Jour. I.E.E.* **62**, 1 (1923); and **69**, 270 (1931).
[28] Silsbee, *B. of S. Sci. Paper* **281** (1916); *B. of S. Jour. Res.* **4**, 73 (1930); *Res. Paper* **133**.
[29] Eccles, *Wireless Handbook* (1917); *B. of S. Circular* **74**, "Radio Instruments and Measurements," 271 (1918).
[30] *B. of S. Sci. Paper* **169**, 152, formula (102) (1912).
[31] G. A. Campbell, *Phys. Review*, June, 1915.
[32] F. F. Martens, *Ann. der Phys.* **29**, 963 (1909).
[33] G. A. Campbell, *Phys. Review*, June, 1915.
[34] *B. of S. Sci. Paper* **169**, 154 (1912).
[35] Grover, *B. of S. Sci. Paper* **468**, 751 (1923); Koga, *Jour. Inst. E.E. of Japan*, Oct. 1924.
[36] Bashenoff, *Proc. I.R.E.*, 1027, Dec. 1927; *Proc. I.R.E.* **16**, 1553 (1928).
[37] F. E. Neumann, *Allg. Gesetze der Inducirter Ströme*, Abhand. Berl. Akad (1845); Kirchhoff, *Gesamm. Abhand.*, 176; *Pogg. Ann.* **121** (1864); Niwa, *Res. Paper* **73**, Electrotech. Lab., Tokyo (1918).
[38] Grover, *loc. cit.*, pp. 746–751.
[39] Niwa, *Res. Paper* **141**, Electrotech. Lab., Tokyo (1924).
[40] Niwa, *Res. Paper* **141**, 39–43.
[41] Maxwell, *Elect. and Mag.* **II**, § 701.
[42] *B. of S. Sci. Paper* **169**, 6–19 (1912); *Sci. Paper* **320**, 538–544 (1918); Grover, *B. of S. Jour. of Res.* **1**, 487–502 (1928).
[43] Grover, *B. of S. Sci. Paper* **498** (1924).
[44] Curtis and Sparks, *B. of S. Sci. Paper* **492** (1924).
[45] Brooks, *B. of S. Jour. Res.* **7**, 293, 294 (1931); *Res. Paper* **342**.
[46] Rosa, *B. of S. Bull.* **4**, 348 (1907), also *loc. cit.*, p. 342; *B. of S. Sci. Paper* **169**, 39, 40 (1912).
[47] Maxwell, *Elect. and Mag.* **2**, Sec. 706; Perry, *Phil. Mag.* **30**, 223 (1890); Weinstein, *Ann. der Phys.* **21**, 329 (1884); Stefan, *Ann. der Phys.* **22**, 113 (1884); Butterworth, *Proc. Phys.* (Lond.) **27**, 371 (1915); *Phil. Mag.* **29**, 578 (1915); Dwight, *Trans. A.I.E.E.* **38**, part 2, 1675 (1919); *Elec. World* **71**, 300 (1918); Lyle, *Phil. Trans.* **213A**, 421 (1914); *B. of S. Sci. Paper* **320**, 557 (1918); Grover, *B. of S. Sci. Paper* **455** (1922).
[48] Stefan, *Ann. der Phys.* **22**, 113 (1884).
[49] Maxwell, *Elect. and Mag.*, 3rd Edit. **2**, 345, 346.
[50] Rosa, *B. of S. Bull.* **3**, 37 (1907); *B. of S. Sci. Paper* **169**, 140.
[51] Brooks, *B. of S. Jour. Res.* **7**, 301–305 (1931); *Res. Paper* **342**.
[52] Grover, *B. of S. Bull.* **18**, 457 (1922); *Sci. Paper* **455**.
[53] J. V. Jones, *Phil. Mag.* **27**, 61 (1889); *Trans. Roy. Soc.* **182A** (1891); *Proc. Roy. Soc.* **63**, 198 (1898).
[54] Rosa, *B. of S. Bull.* **3**, 209 (1907); *B. of S. Sci. Paper* **169**, 101.
[55] Lorenz, *Wied. Ann.* **25**, 1 (1885).
[56] Clem, *Jour. A.I.E.E.* **46**, 814 (1927).
[57] Dwight and Grover, *Trans. A.I.E.E. 1937*, p. 327, formula (21).
[58] A. Campbell, *Proc. Roy. Soc.* **79**, 428 (1907).
[59] Nagaoka, *Jour. Coll. Sci. Tokyo* **27**, art. 6 (1909); *B. of S. Sci. Paper* **169**, 64.
[60] Terezawa, *Tokyo Math. Jour.* **10**, 73 (1916).
[61] Olshausen, *Phys. Rev.* **31**, 617 (1910); *B. of S. Sci. Paper* **169**, 73.
[62] J. V. Jones, *Proc. Roy. Soc.* **63**, 192 (1898).
[63] *B. of S. Sci. Paper* **169**, 53–68.
[64] Clem, *Jour. A.I.E.E.* **46**, 814 (1927).
[64a] Grover, *Proc. I.R.E.* **21**, 1039 (1933).

REFERENCES

[65] Searle and Airey, *Lond. Elect.* **56**, 318 (1905); Rosa, *B. of S. Bull.* **3**, 224 (1907); Dwight and Grover, *Elect. Eng.* **56**, 347 (1937).

[66] Havelock, *Phil. Mag.* **15**, 343 (1908); *B. of S. Sci. Paper* **169**, 68; Dwight and Grover, *loc. cit.*, p. 349.

[67] *B. of S. Sci. Paper* **169**, 87–90, incl.

[68] Dwight and Grover, *loc. cit.*, p. **350**, formula (19).

[69] *B. of S. Sci. Paper* **169**, 71, 72.

[70] Dwight and Grover, *loc. cit.*, equats. (11) and (6), pp. 347–8.

[71] Lorenz, *Wied. Ann.* **7**, 161 (1879); *B. of S. Sci. Paper* **169**, 117.

[72] *B. of S. Sci. Paper* **169**, 116–122; *Sci. Paper* **320**, 551; Grover, *B. of S. Jour. Res.* **1**, 502–509 (1928).

[73] Nagaoka, *Jour. Coll. Sci. Tokyo* **27**, 18–33, art. 6 (1909); *B. of S. Sci. Paper* **169**, 119–121.

[74] Nagaoka, *loc. cit.*; *B. of S. Sci. Paper* **169**, 224, Table XXI.

[75] Rayleigh and Niven, *Proc. Roy. Soc.* **32**, 104 (1881); Coffin, *B. of S. Bull.* **2**, 113 (1906).

[76] Rosa, *B. of S. Bull.* **2**, 161 (1906); *B. of S. Sci. Paper* **169**, 122.

[77] Coffin, *loc. cit.*; *B. of S. Sci. Paper* **169**, 129.

[78] Grover, *Proc. I.R.E.* **12**, 193–208 (1924).

[79] Grover, *Proc. I.R.E.* **17**, 2053 (1929); *B. of S. Res. Paper* **90** (1929).

[80] *B. of S. Circular* **74**, "Radio Inst. and Meas.," 251 (1917).

[81] *B. of S. Sci. Paper* **169**, 124 (1912).

[82] Niwa, *Electrotech. Lab. Tokyo, Res. Paper* **73** (1918), and **141** (1924).

[83] Grover, *B. of S. Sci. Paper* **468**, 743–751 (1923); Koga, *Jour. Inst. E.E. Japan*, Oct. 1924.

[84] Grover, *loc. cit.*, pp. 746, 748, 751.

[85] Koga, *loc. cit.*; Grover, *B. of S. Jour. Res. Paper* **90**, 172–174.

[86] Butterworth, *Phil. Mag.* **31**, 443 (1916); *B. of S. Sci. Paper* **320**, 546–548.

[87] Snow, *B. of S, Jour. Res.* **3**, 255 (1929).

[88] *B. of S. Sci. Paper* **320**, 546, formula (10A).

[89] Sohon, *Engineering Mathematics*, D. Van Nostrand, 1944, p. 151.

[90] Grover, *Proc. I.R.E.* **32**, 620–629 (1944).

[91] Snow, *Bur. St. Jour. Res.* **1**, 694 (1928).

[92] Dwight and Purssell, *G.E. Rev.* **33**, 401 (1930).

[93] Clem, Two formulas communicated to the writer. These have never been published before.

[94] Maxwell, *Elect. & Mag.* II, sec. 701.

[95] Rayleigh, *Phil. Trans.* **175**, 411–460 (1884); Rayleigh's *Sci. Papers* **2**, 327.

[96] Rosa, Dorsey and Miller, *B. of S. Bull.* **8**, No. 2, 392, Table XXX (1911).

[97] Nagaoka, *Phil. Mag.* **6**, 19 (1903); and *Tokyo Math. Phys. Soc.* **6**, 154 (1911).

[98] Nagaoka and Sakurai, *Sci. Papers of the Tokyo Institute of Physical and Chemical Research*, Table 2, Sept. 1927.

[99] Nagaoka and Sakuari, *Sci. Papers of Tokyo Inst.*, vol. II, p. 1, Table 1, p. 75.

[100] H. L. and R. W. Curtis, *B. of S. Jour. of Research* **12**, 665 (1934); *Research Paper* **685**.

[101] Nagaoka, *Tokyo Math. Phys. Soc.*, 2 ser. **2X**, No. 4 (1917).

[102] Grover, *B. of S. Bull.* **12**, 317 (1915); *Sci. Paper* **255**.

[103] Russell, *Jour. I.E.E.* **69**, 270 (1931), section on the Heaviside Effect.

[104] Maxwell, *Elect. and Mag.*, II, sec. 690; Heaviside, *Elect. Papers* II, 64; Rayleigh, *Phil. Mag.* **21**, 381 (1886); Kelvin, *Math. and Phys. Papers* III, 491 (1889); *B. of S. Bull.* **8**, 173 (1912); *Sci. Paper* **169**, 173.

[106] Hickman, *B. of S. Bull.* **19,** 73 (1923); *Sci. Paper* 472; Sommerfeld, *Ann. der Phys.* **15,** 673 (1904); and **24,** 609 (1907); Butterworth, *Phil. Trans.* **222A,** 57 (1921); Howe, *Jour. I.E.E.* **54,** 473 (1916); *Proc. Roy. Soc.* **93A,** 468 (1917); Palermo and Grover, *Proc. I.R.E.* **19,** 1278 (1931); *B. of S. Circular* **74,** 299–308 (1918).
[106] Table 53 is adapted from Table XXIII of *B. of S. Sci. Paper* **169.**
[107] Race and Larrick, A.I.E.E. Meeting, Northeastern Section, May 1942.
[108] This is Table 18, p. 310, *B. of S. Circular* **74,** "Radio Instruments and Measurements," 1918.
[109] Table 55 is an adaptation of Table XXIII, *B. of S. Bull.* **8,** 229 (1912); *Sci. Paper* **169.**
[110] Heaviside, *Elect. Papers,* **II,** 69, equation (50b); Dwight, *Jour. A.I.E.E.,* Aug. 1923, p. 827, equations (4) and (10); Russell, *Theory of Alt. Currents,* Camb. Univ. Press, London, p. 207 (1914).
[111] Dwight, "Bessel Functions for A.C. Problems," *Trans. A.I.E.E.,* July 1929 p. 812; *Tables of Integrals and Other Math. Data,* Macmillan, p. 211.
[112] Dwight, *Electrical Coils and Conductors,* McGraw-Hill, 1945.
[113] Whinnery, *Electronics,* Feb. 1942.
[114] Manneback, *Jour. Math. Phys.,* April 1922; Dwight, *Trans. A.I.E.E.,* 850 (1923); Nicholson, *Phil. Mag.* **18,** 417 (1909); Curtis, *B. of S. Bull.* **16,** 93 (1920); *Sci. Paper* **374.**
[115] Dwight, p. 225, reference 112, above.
[116] Russell, *Alt. Currents* **1,** 102, formula (8) (1904).
[117] Dwight, p. 226, reference 112, above.
[118] Russell, *Phil. Mag.* **17,** 524 (1909).
[119] Russell, *Alt. Currents* **1,** 104, formulas (3) and (4) (1904).